CRITICAL FACTORS IN THE APPLICATION OF DIESEL ENGINES

The Institution of Mechanical Engineers

Proceedings 1969–70 · Volume 184 · Part 3P

CRITICAL FACTORS IN THE APPLICATION OF DIESEL ENGINES

A Symposium arranged by the
Combustion Engines Group and the
Automobile Division of the Institution of Mechanical Engineers
and the Diesel Engineers and Users Association
22nd–24th September 1970

1 BIRDCAGE WALK · WESTMINSTER · LONDON · S.W.1

© *The Institution of Mechanical Engineers 1971*

ISBN 0 85298 040 X

CONTENTS

		PAGE
Introduction		vii

Paper 1 Critical factors in the application of dual-fuel engines, by P. W. A. Eke, C.Eng, M.I.Mech.E., J. H. Walker, B.A., C.Eng., F.I.Mech.E., and M. A. Williams, C.Eng. 1

Paper 2 Developing a diesel for an industrial market sector, by N. R. Senior, C.Eng., M.I.Mech.E., and D. J. Clarke, C.Eng., M.I.Mech.E. . 10

Paper 3 The effect of ambient and environmental atmospheric conditions, by W. Lowe, B.Sc., C.Eng., F.I.Mech.E. 18

Paper 4 Critical factors in the application of diesel engines to rail traction, by W. Petrook, C.Eng., F.I.Mech.E., and W. A. Stewart, C.Eng., F.I.Mech.E 25

Paper 5 Application of diesel engines in the Royal Navy, by W. H. Sampson, C.Eng., M.I.Mech.E. 32

Paper 6 Alignment investigation following a medium-speed marine engine crankshaft failure, by D. Castle, C.Eng., F.I.Mech.E. . . . 44

Paper 7 The turbo-charged diesel as a road transport power unit, by E. Holmér, M.Sc.(Eng.), and B. Häggh, M.Eng. 55

Paper 8 The diesel engine as a source of commercial vehicle noise, by P. E. Waters, B.Sc.(Eng.), N. Lalor, B.Sc.(Eng.), M.Sc.(Eng.), and T. Priede, Inž.Mech., Ph.D., D.Sc.(Eng.), M.I.Mech.E. . . 63

Paper 9 Fuel-injection system requirements for different engine applications, by P. Howes, C.Eng, F.I.Mech.E. 73

Paper 10 Fuel limitations on diesel engine development and application, by H. E. Howells, B.Sc., and S. T. Walker, C.Eng., M.I.Mech.E. . 81

Paper 11 Diesel generation for Ascension Island, by A. K. Mackenzie, B.Sc.(Eng.) 90

Paper 12 Improvements to conventional diesel engines to reduce noise, by M. F. Russell, B.Sc., M.Sc. 98

Paper 13 Some aspects of diesel exhaust emissions, especially in confined spaces, by C. Lunnon, B.Sc., C.Eng., M.I.Mech.E. . . . 106

Paper 14 Diesels and the generation of electricity, by F. Ratcliffe, C.Eng. . 112

CONTENTS

		PAGE
Paper 15	Diesel engine piston ring design factors and application, by D. A. Law, C.Eng., F.I.Mech.E.	119
Paper 16	Critical factors in the application of diesel engines to fighting vehicles, by D. H. Millar, B.Sc., C.Eng., M.I.Mech.E.	139
Paper 17	Further considerations in injector design for high specific output diesel engines, by J. P. S. Curran, B.Sc.	148

Summing up, by W. P. Mansfield, B.Sc., Ph.D, C.Eng., F.I.Mech.E. . 155

Discussion 159

Authors' replies 175

List of delegates 184

Index to authors and participants 186

Subject index 187

Critical Factors in the Application of Diesel Engines

A SYMPOSIUM was held in the Physics Lecture Theatre of The University, Southampton, from the 22nd to the 24th September 1970. It was sponsored by the Combustion Engines Group and the Automobile Division of the Institution and by the Diesel Engineers and Users Association. The symposium was opened by S. H. Henshall, B.Sc.(Eng.), C.Eng., M.I.Mech.E., and 160 delegates registered to attend.

The papers were divided into four Sessions for presentation and discussion.

On Wednesday, 23rd September:

Session 1; Chairman: Mr S. H. Henshall
Papers 7, 2, 14, 11 and 1.

Session 2; Chairman: Mr V. H. F. Hopkins
Papers 4, 16, 5, 6 and 3.

On Thursday, 24th September

Session 3; Chairman: Mr M. A. Williams
Papers 8, 12, 10, 13 and 15.

Session 4; Chairman: Mr C. C. J. French
Papers 9 and 17; Summing up.

Technical visits were arranged on Tuesday, 22nd September, to M.V.E.E., Christchurch; British Transport Docks, Southampton; Southern Gas Board, Hythe Works; C.E.G.B., Marchwood power station; and Wellworthy Ltd, Lymington. Delegates were also shown round the Institute of Sound and Vibration Research. A dinner was held at the Polygon Hotel.

ORGANIZING COMMITTEE

S. H. Henshall (*Chairman*)

V. H. F. Hopkins	M. A. Williams
T. Priede	W. L. Clifton
I. W. Goodlet	A. E. W. Austen

Paper 1

CRITICAL FACTORS IN THE APPLICATION OF DUAL-FUEL ENGINES

P. W. A. Eke*   J. H. Walker†   M. A. Williams†

A dual-fuel engine may be defined as a compression-ignition engine using mainly gaseous fuel but with a small quantity of fuel oil injected as an ignition source; the engine can be changed over instantaneously and under load to operate on liquid fuel alone. The recent availability of natural gas in this country once again attracts the attention of engineers towards gas as a fuel for internal-combustion engines.

This paper traces the development of dual-fuel engines, originally using sewage gas and more recently using natural gas, and considers their advantages, both technical and economic, compared with spark-ignited and diesel engines. The dual-fuel engines within the authors' experience are described. The critical factors in handling natural gas in its liquid form are considered, and the extended scope of dual-fuel engines and alternative fuel engines in mobile applications is briefly reviewed. Finally, the paper examines the future for dual-fuel engines and suggests directions in which further development is required.

INTRODUCTION

SINCE THE INCEPTION of the internal-combustion engine, flammable gases have been closely associated with its development; and in the early days of the oil engine it was an inevitable experiment that gas should be mixed with the induction air. Before 1900 Dr Rudolph Diesel, in his British patent, covered this feature but stated that 'the amount of illuminating gas which can be used with the air is limited to small proportions due to loss of combustion control' (1)‡. Diesel engines propelled the ill-fated airship *R.101*, and arrangements were made for mixing gas with combustion air. This was a novel way of consuming the 40 000 ft³ of hydrogen which had to be jettisoned with every ton of fuel consumed in order to maintain the trim of the ship. It was estimated that a 20 per cent saving in fuel consumption would be achieved by this means; but the airship met with disaster on her maiden voyage in 1930, so the scheme never progressed beyond the laboratory stage.

The exploitation of natural gas in the U.S.A. led the Nordberg Company, in 1928, to experiment with a gas-burning diesel engine in an endeavour to develop a gas-burning engine which would have the efficiency of a diesel;

it was found necessary to inject a small quantity of oil with the gas to ensure ignition. Several two-stroke air-blast injection engines of this kind were sold between 1935 and 1939; but they could not work on oil fuel only without modification, and the idea was not pursued.

In the U.K., experiments by Dr C. M. Walter (2) and Dr J. S. Clarke of the City of Birmingham Gas Department led to an engine operating on town gas with oil ignition, based on research in 1937 by Dr J. Rifkin of Birmingham University. Soon after, in February 1939, the National Gas and Oil Engine Company (3) tested a 400-hp engine successfully on dual fuel, and later the same year the company obtained permission to convert a 120-hp gas engine to dual fuel at the Coleshill Works of the Birmingham, Tame and Rea District Drainage Board. This, the first example of a dual-fuel engine operating on sewage gas, was so successful that the conversion of further engines was planned; however, these plans were held up by the war. Similar work was also going on at the Mogden works of the West Middlesex Drainage Undertaking and by 1941 two of the ten 650-hp Harland & Wolff gas engines there had been converted by the makers to dual fuel. The war also hindered further development work at Mogden (4).

Up to this time the development of dual-fuel engines in the U.K. was progressing, but using flammable gas other than natural gas. In America, however, the abundant availability of natural gas favoured the development of the

---

The MS. of this paper was received at the Institution on 18th February 1970 and accepted for publication on 17th March 1970. 24
* North Thames Gas Board, Romford Works, Romford, Essex.
† North Thames Gas Board, Monck St, London, S.W.1.
‡ *References are given in Appendix 1.1.*

spark-ignited gas engine, which is now available over a wide power range. However, it should be noted that, in 1944, the Worthington Corporation demonstrated the first four-stroke 'gas–diesel' engine in the U.S.A. operating with pilot mechanical injection, based on the system developed in this country. Dual-fuel engines of two- and four-stroke design are now common in the larger sizes.

In the sewage industry, where the quantity of gas produced by the sludge-digestion process is variable, the advantage of an engine which can be changed from dual-fuel operation to full diesel and back again while under load is obvious. In this country, interest was stimulated in 1949 when the National Gas and Oil Engine Company installed the first power station using dual-fuel engines at the Maple Lodge Works of West Hertfordshire Drainage Authority (5). Sewage plants serving a population of more than 50 000 can profitably operate a sludge-digestion plant, and the majority of such plants in this country are now equipped with dual-fuel engines. It was also during this period that the English Electric Company was developing its range of naturally aspirated engines, many of which were used on natural gas, which is so freely available at oil fields and refineries throughout the world (6).

The development of the liquid methane terminal at Canvey Island in Essex, which commenced some 10 years ago, involved the use of prime movers for various purposes. It was therefore natural for the engineers concerned with this project to consider using methane as an engine fuel. This installation has an interesting range of engines—alternative fuel engines, dual-fuel engines, spark-ignition engines, and straight diesel engines. Following the North Sea finds, natural gas is now rapidly becoming available throughout the U.K. as the Gas Council's network of high-pressure pipelines continues to spread. Concurrently, the Area Boards are developing their own domestic networks and converting their medium-pressure grid and low-pressure district systems to the new fuel. Thus, many power users are now considering the possibility of using natural gas as a fuel for their engines.

FUEL GASES

Many gases are flammable, but their suitability as an engine fuel varies enormously; this is so even when clean gases only are considered. The efficiency of any internal-combustion engine in which a gaseous mixture is compressed can be improved by raising the compression ratio; but the most important limitation to this improvement is the incidence of detonation. When the mixture is ignited a flame spreads outwards throughout the combustion chamber, compressing ahead of it the unburnt charge or 'end gas'. If the temperature of this end gas is raised by compression and radiation from the flame sufficiently to ignite spontaneously, a sudden rise of pressure will occur causing knock or detonation. The incidence of detonation is therefore most likely to occur when the flame speed is highest and the compression of the end gas most rapid. Of all fuel gases, methane has the lowest flame speed and therefore the best anti-knock properties. Natural gas, which contains 90–95 per cent methane, is therefore a most attractive engine fuel; similarly, sewage gas, which consists of 60–65 per cent methane, the remainder being inert gases which still further reduce the flame speed. The 'loss of combustion control' referred to by Dr Diesel in his patent was almost certainly due to detonation from the presence of hydrogen—a gas with a high flame speed very prone to knock—in the 'illuminating gas' used in his experiments.

The natural gas now being distributed in this country contains about 92 per cent methane, and this renders it an attractive fuel for engines. Table 1.1 gives typical analyses of gas from various fields in the North Sea and also, for comparison, of imported Algerian liquid natural gas and of Dutch natural gas. Since these figures were compiled, further finds in the North Sea have been announced and known reserves are estimated to last for at least 30 years at three times the present national rate of gas consumption. Over 1000 miles of high-pressure distribution mains, most of large diameter—between 18 in and 36 in—have already been laid to distribute the gas (Fig. 1.1). The final network will extend to some 2500 miles,

Table 1.1. Natural gas, some typical analyses*

	North Sea gas					Lockton Home Oil	Algerian LNG vaporized	Dutch Slochteren	Pure methane
	West Sole, B.P.	Hewett, Phillips	Hewett, ARPET	Indefatigable, AMOCO	Leman, AMOCO				
Nitrogen	1·2	2·3	8·9	3·7	1·3	2·6	0·4	14·0	—
Helium	0·03	0·06	0·1	0·05	0·03	0·04	Nil	Nil	—
Carbon dioxide	0·5	0·04	0·1	0·65	0·04	0·26	Nil	0·9	—
Methane	94·1	91·9	81·8	91·8	94·8	93·7	87·7	81·7	100·0
Ethane	3·2	3·9	6·0	2·9	3·0	2·6	8·6	2·7	—
Propane	0·6	0·9	2·5	0·5	0·5	0·4	2·3	0·4	—
Butane etc.	0·4	0·9	0·6	0·4	0·3	0·4	1·0	0·3	—
Gross C.V., Btu/ft^3	1041	1060	1020	1012	1041	1024	1132	869	1012·6
S.G. (air = 1)	0·593	0·609	0·653	0·604	0·587	0·593	0·634	0·643	0·555
Wobbe No.	1352	1358	1262	1302	1359	1330	1422	1120	1360
Air req. (vol./vol.)	9·8	9·9	9·6	9·6	9·8	9·6	10·8	8·4	9·57
Hydrogen sulphide, p.p.m.	Nil	Nil	415	Nil	Nil	410	Nil	Nil	—

* Gas Council Research Comms 150 and 155, *Trans. Instn Gas Engrs* 1969 **9** (Nos 6 and 7).

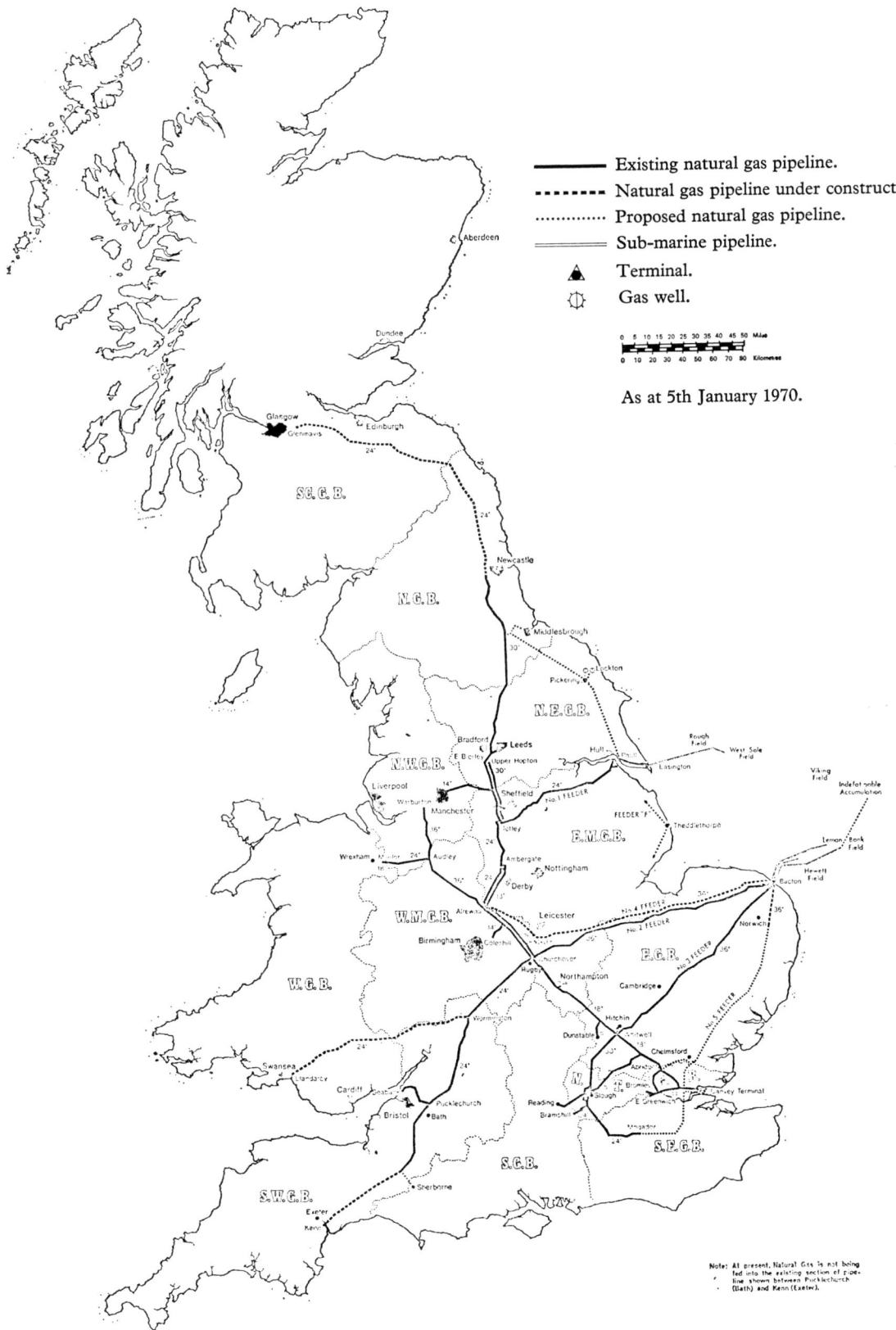

Fig. 1.1. The Gas Council natural gas transmission system

providing a natural gas supply to the greater part of the U.K.—a supply which is drawn from several independent sources. Thus a secure supply of fuel of high anti-knock rating, of high calorific value, and free from corrosion-forming impurities is available to industry, and often at high pressure.

SPARK IGNITION, DUAL FUEL, OR DIESEL

The cost of a gas is necessarily a critical factor in its application as fuel. Dual-fuel engines may be self-selecting in applications where fuel gas, although cheap, is not constant in quality or availability, or both. In the sewage-treatment industry the quantity available may vary, and so may the calorific value. The gas, however, is available as a by-product, the use of which will improve the overall efficiency of the enterprise—the choice of a dual-fuel engine which can, if required, burn up to 100 per cent liquid fuel is therefore an obvious one. Furthermore, natural gas may be a cheap form of fuel to the large industrial user who can accept a supply on the basis of an interruptible tariff.

Apart from these major considerations, and although the spark-ignition gas engine has been much improved recently, the following factors still favour dual fuel:

(*a*) The compression ignition of the pilot oil provides hundreds of times as much energy as an electric spark, even when generated by the most modern of solid-state electronic systems. Ignition is thus more certain and variations in peak cylinder pressure are reduced with consequently less cyclic variation in speed (**7**).

(*b*) An improvement in fuel consumption of between $\frac{1}{2}$ and 1 per cent is obtained. This is chiefly because ignition is initiated at a large number of liquid-fuel droplets in the combustion chamber. Thus, since the flame path is shorter, end-gas detonation is less likely to occur and a higher compression ratio can be used (**8**). Other factors leading to better fuel consumption are (i) less critical gas/air ratio because of higher ignition energy, (ii) more rapid combustion allowing later ignition and more time for cylinder scavenging and mixing of the charge, (iii) because of the more rapid combustion, a closer approach to the constant-volume combustion of the ideal Otto cycle.

(*c*) Maintenance of fuel-injection equipment is cheaper than that of electric ignition (**7**). Although injection nozzles are more complicated than sparking plugs, they require less frequent attention. In the authors' experience, until recently sparking plugs had a life of about 2000 h; an improved type of plug is now in use which requires cleaning and adjustment every 1500 h, and which has an expected life of 6000 h. Injection nozzles, however, if properly maintained will last almost indefinitely.

The spark-ignition engine does, however, have important applications. There are parts of the world where natural gas is so cheap, compared with diesel oil, that it is not economic to use dual fuel. The slightly higher efficiency of the dual-fuel engine does not compensate for the extra cost of the pilot fuel, though the latter may account for only 5 per cent of the total fuel requirements of the engine. In addition, at some sites the delivery or storage of liquid fuel may present difficulties. A popular application of the spark-ignition engine in many parts of the world is in the gas industry for pumping natural gas; it is usually economic to bleed off as engine fuel some of the gas being pumped rather than to make provision for liquid fuel. The integral gas-engine compressor, often a two-stroke unit, has been developed for this duty.

How do natural gas burning dual-fuel and spark-ignition engines compare with the straight diesel engine? The following advantages are claimed for the gas-burning engines:

(*a*) The gas is sulphur free; hence, the products of combustion are not corrosive and cylinder wear is reduced.

(*b*) A less highly detergent lubricating oil is required, and because it does not become contaminated it has a longer life in the engine.

(*c*) They are noticeably less noisy than the pure diesel —particularly advantageous in residential areas or anywhere where noise is distracting.

(*d*) The exhaust is virtually free from smoke and smell and has a high standard of acceptability from the standpoint of atmospheric pollution.

(*e*) Carbon deposits in cylinders, turbo-chargers, and other components are minimal—thus reducing maintenance costs.

The diesel engine on the other hand is more efficient thermally, as is shown by Fig. 1.2, which relates to a particular turbo-charged unit, operating as a dual-fuel engine and as a straight diesel engine. At full load the difference in specific fuel consumption is about 7 per cent higher for dual fuel, but at half load the difference is nearer 25 per cent. The authors believe that this excessive fuel consumption at part load must be improved in order to render gas dual-fuel engines more attractive for variable load duties. In the U.S.A., where natural gas is very cheap compared with diesel oil, fuel consumption is of secondary importance, and the majority of industrial engines made in that country are either dual fuel or spark-ignited gas.

Gas-burning engines are also popular in those remoter parts of the world where gas is produced but where, because there are few uses for it, its local value is low. In this country, engines have recently come on the market with more sophisticated control systems, and the assessment of their performance is awaited with interest. The authors consider that there is room for more research into the automatic control of that vital balance, the gas/air ratio.

ECONOMIC APPRAISAL

Technical considerations may pre-dispose the potential user to select one form of engine or another, but in most

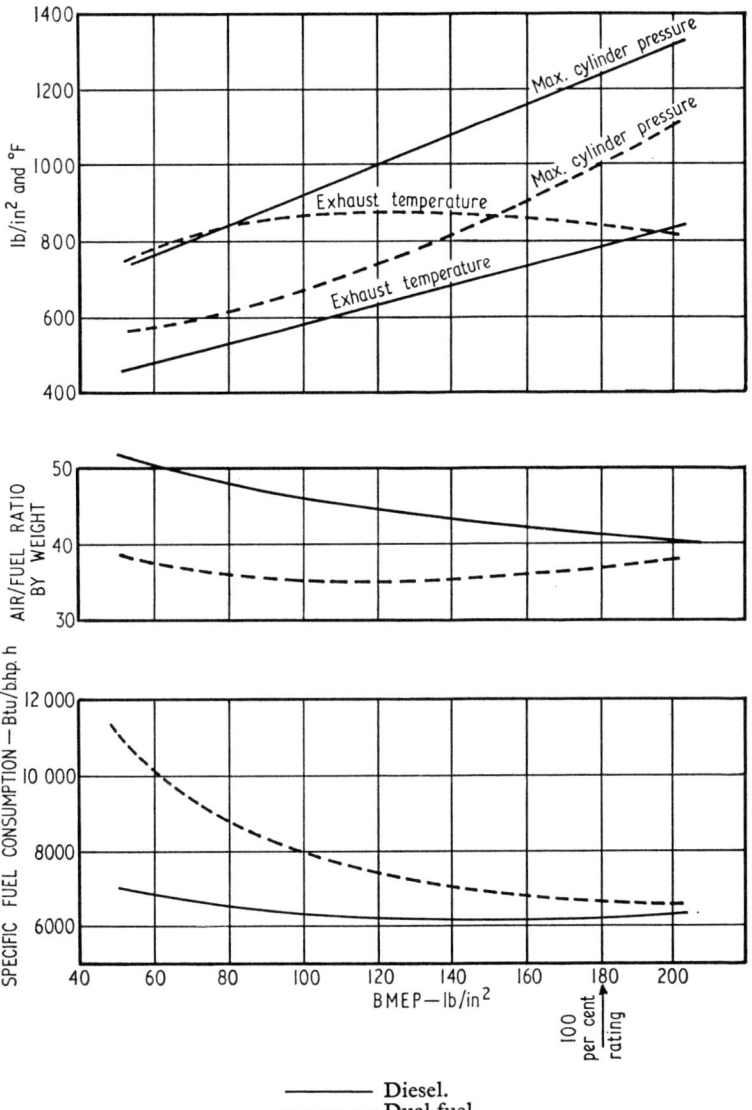

Fig. 1.2. Mirrlees KP major

cases economics will form the ultimate basis of his decision. In making comparisons it must be remembered that, in this country at least, the price of imported liquid fuel is very susceptible to international incidents, and that history shows oil products to have been a target for successive Chancellors of the Exchequer. Natural gas is not only an ideal engine fuel technically; it is also an established fuel in the industrial energy market of the U.K. Large quantities of gas have already been sold to industry, but much more will have to be sold during the next few years if the Gas Council's sales target is to be met. For this to be achieved, natural gas must compete with all existing industrial fuels over a wide range of applications in the premium and non-premium markets.

Reference has already been made to interruptible supply tariffs. Where these are admissible, natural gas can compete against any grade of oil for any class of business. Contracts for interruptible supplies exceeding one million therms per annum are now an important feature of the gas industry's marketing strategy and enable temperature-sensitive demand to be balanced with supply. Generally in these cases, the supplies of gas may be interrupted at previously agreed notice, the price depending on the length of notice and the period of interruptibility in any one year. The economic attractiveness of natural gas as an engine fuel will depend partly on the nature of the engine load and, in many cases, upon the magnitude of a potential customer's other needs for natural gas.

OPERATING EXPERIENCE WITH DUAL-FUEL ENGINES

As mentioned earlier, a wide variety of engines has been installed at the Canvey Island Methane Terminal, as listed in Table 1.2. This variety was determined in most

Table 1.2. Engines installed at Canvey Island Methane Terminal

Item no.	No. of engines	Maker	Maker's type no.	Type of fuel	Horse-power	Speed, rev/min	Duty
a	1	National	M4AG3	Diesel	100	1200	Gas compressor
b	1	National	RVG4	Gas or diesel	124	600	Gas compressor
c	2	Paxman	12YLC	Diesel	1500	750	Alternators
d	3	Meadows	6D970	Diesel	160	1100	Water pumps
e	2	Mirrlees National	R4AP	D/F	416	600	Gas compressors
f	2	Paxman	6RPH	Diesel	215	1065	Air compressors
g	2	Dorman	6QTCA	Diesel	330	1500	Alternators
h	2	Rolls-Royce	844600A	Diesel	290	1650	Fire-fighting pumps
i	3	English Electric	16CSVD	D/F	1455	530	Gas compressors
j	2	Thomassen	LGC35-6	Gas	1150	360	Gas compressors (integral)
k	6	Allen	GS37D	D/F	1210	428	Gas compressors
l	2	Ruston	6APC	Diesel	620	750	Gas compressors

cases by delivery date, but it has provided an opportunity of evaluating the different types of machines.

From this schedule the three dual-fuel installations, items (e), (i), and (k), are selected to give further relevant details:

(e) Mirrlees National dual-fuel engines, type R4AP, Engine Nos. 5923-1/2.
Normally aspirated, eight-cylinder, 416 hp at 600 rev/min.
Bore 9 in (229 mm), stroke 12 in (305 mm), displacement 763 in^3/cylinder (12·5 litres).
Compression ratio 13·8:1, mean piston speed 1200 ft/min (6·1 m/s), b.m.e.p. 90 lbf/in^2 (6·3 kg/cm^2).

These machines were intended to operate on natural gas with a variable fuel mix between 100 per cent diesel oil and 10 per cent diesel oil/90 per cent gas, the 10 per cent being the minimum recommended for ignition purposes. The engines were equipped with two fuel pumps and one injector per cylinder, one pump being of infinitely variable output whilst the auxiliary pump operated at a constant rate of 10 per cent of full-load requirements. The engines were commissioned on diesel fuel only, in October 1964, and subsequent attempts to run on gas were unsatisfactory because of the difficulty in maintaining a balanced load on the cylinders with the original gas-control linkage. During the first 18 months of service much trouble was experienced with broken fuel lines and cracked auxiliary-pump bodies. Eventually all these pumps, originally of cast aluminium, were changed to a cast steel design able to withstand higher pressures, and a permanent bleed line was installed in the fuel header. At the second major overhaul of each machine in 1967 the crank deflections showed misalignment due to movement of the compressors. The machines were re-aligned and at the same time the fuel gas control gear was modified. When running on gas the governor operates a two-part mechanical linkage to a control shaft and then individually adjustable linkages to butterfly valves controlling the gas to each cylinder. Gas at controlled pressure passes through the butterfly valves before being mixed with the normally aspirated air entering the cylinder; the auxiliary pump supplies ignition fuel at a constant rate. The frequency of top overhauls has been about 3000 h and of major overhauls 6000 h, dictated by the life of the non-metallic piston rings in the gas compressors.

(i) English Electric dual-fuel engines, type 16CSVD, engine nos. 1H 6916, 7290, and 7291.
Turbo-charged with charge-air cooling, 16-cylinder vee-form, 1455 b.h.p. at 530 rev/min.
Bore 10 in (254 mm), stroke 12 in (305 mm), displacement 941 in^3/cylinder (15·5 litres).
Compression ratio 11·7:1, mean piston speed 1060 ft/min (5·4 m/s), b.m.e.p. 144 lbf/in^2 (10·11 kg/cm^2).

These machines went to work in June 1968, running on diesel only. Troubles have mainly arisen from vibration due to gas pulsations at the compressors, and corrosion; the cylinder jackets are on an indirect cooling-water system but the charge-air coolers and lubricating-oil coolers use Thames estuary water, which is very corrosive. The flexible drive gear to all engine auxiliaries failed and has been modified. Similarly, the flexible drive to the gas-control camshaft has required changing to a stronger design.

(k) Allen dual-fuel engines, type GS37D.
Turbo-charged with charge-air cooling, nine cylinder 1210 b.h.p. at 428 rev/min.
Bore 12·8 in (325 mm), stroke 14·6 in (370 mm), displacement 1880 in^3/cylinder (30·8 litres).
Compression ratio 12·5:1, mean piston speed 1250 ft/min (6·35 m/s), b.m.e.p. 128 lbf/in^2 (9·00 kg/cm^2).

These engines have only recently been installed and no experience in their operation has been obtained.

With the extended distribution of North Sea gas, new engines installed at works of the North Thames Gas Board, other than Canvey, will probably be of the dual-fuel type. Three such engines have recently been installed at Beckton:

Mirrlees National dual fuel, type KP Major.
Turbo-charged with charge-air cooling, seven cylinders, 2163 hp at 428 rev/min.

Bore 15 in (381 mm), stroke 18 in (457 mm), displacement 3180 in³ (52 litres).
Compression ratio 11·35:1, mean piston speed 1284 ft/min (6·7 m/s), b.m.e.p. 180 lbf/in² (12·65 kg/cm²).

These engines have a sophisticated device for controlling the gas/air ratio, and the graphs shown in Fig. 1.2 were obtained from an engine of this type.

LIQUEFIED NATURAL GAS (LNG) AS AN ENGINE FUEL

The space occupied by natural gas can conveniently be reduced by liquefaction, and the transportation and storage of LNG are now well established so that its availability is rapidly increasing throughout the world.

Natural gas cannot be liquefied by pressure alone as its critical temperature is −116°F (−82°C), but if its temperature is lowered to −258°F (−161°C) it changes to the liquid state at atmospheric pressure. Storage tanks and pipelines have then only to be designed for this low temperature, and not for a combination of low temperature and high pressure.

LNG is a clear, colourless liquid occupying 1/600 of its gaseous volume at standard temperature and pressure (s.t.p.). The following is a comparison of its physical properties with those of diesel oil:

	LNG	Diesel oil
Specific gravity	0·46	0·82
Gross calorific value	23 000 Btu/lb	19 500 Btu/lb
or	1·06 therm/gal	1·60 therm/gal

Low temperature is the critical factor in the handling and storage of this fuel, and the following are some of the more important considerations:

(a) The materials in contact with the LNG must be capable of withstanding the low temperature without risk of brittle fracture. Common cryogenic materials are aluminium alloys, copper alloys, 18/8 stainless steel and 9 per cent nickel steel; other materials include epoxy resins, bonded laminates, foamed glass, foamed polyurethane, and concrete.

(b) to exclude heat, the storage container must be well insulated, and it is usually of double-wall construction. In large static tanks the interspace may be filled with expanded Perlite; in sea-going LNG tankers, where relative movement between the tank walls may occur, balsa wood has been used for insulation; in portable containers and road and rail tankers the air is usually exhausted from the interspace, which may be filled with an insulating material.

(c) LNG in the container boils and so maintains its liquid temperature by auto-refrigeration. The amount of boil-off gas depends on the efficiency of the tank insulation, and in a large well-insulated tank is of the order of 0·2 per cent of the contents of the tank per day. Means must therefore be provided for dealing with this boil-off, in the case of large installations either by liquefaction plant or compressors for pumping the gas away; with mobile plant the tank is usually designed to withstand the pressure built up by the boil-off occurring during a limited period, e.g. 24 h.

(d) Because of the wide temperature differential between an empty tank at ambient temperature and a full one at −258°F, allowance must be made for tank and pipeline contraction. This amounts to some 5 in/100-ft length in the case of aluminium alloys.

(e) A cryogenic liquid such as LNG must never be trapped in a confined space without the provision of adequate venting. Ingress of heat will inevitably cause evaporation, with consequent disruptive rise in pressure.

When vaporized, LNG is, of course, the gas already discussed, but it has additional properties which can be attractive in certain circumstances. Its high calorific value on a weight basis makes it of special interest to aircraft and hovercraft designers, though its low specific gravity may be a disadvantage in certain other applications. However, its greatest potential probably lies in the 'heat sink' value resulting from its low temperature.

Most of the early work on LNG as an engine fuel was done on alternative fuel engines or single-fuel engines, but it is important in the context of this paper because it is applicable to dual-fuel engines, which are already running successfully on natural gas. The problem with LNG is to harness it successfully as a source of natural gas for mobile engines.

It appears that the first use of liquid methane as a fuel in the U.K. was in 1938 when Professor Sir Alfred Egerton (9) initiated a programme to evaluate its use as a motor fuel in view of its wartime potential. The liquid methane was prepared from sewage gas, and experiments included laboratory work and a 25 000-mile road test of a Leyland motor bus. The trials proved that it was practicable to run mobile plant on LNG with a high degree of reliability, though of course at that time the cost of the fuel made the project uneconomic.

There are records of similar experiments after the war in other parts of Europe, notably in Russia (10), but it was not until the mid-1960s that interest really blossomed with the greater availability and cheapness of LNG in the U.S.A. Several companies, particularly on the West Coast (11) (12), have converted cars and trucks to run on LNG, obtained from the liquefaction plants which the gas supply companies have established for peak-load shaving. The purpose of these trials has been to assess the economics of LNG as an alternative to petrol, since it is found that the use of LNG virtually eliminates atmospheric pollution. In Vienna, the experimental conversion of diesel-engined buses to dual fuel (diesel/LPG), primarily to avoid air pollution, has been reported (13).

The cost of LNG, before tax, in the U.S.A. is about the same as that of petrol on an energy basis, and considerably less per gallon; and road tests show that more miles per gallon can be realized because a higher compression ratio can be used without detonation. In this country the cost of imported LNG is of the order of 7d. per therm (or per gallon) landed. At this cost it can be seen that, despite high capital costs for storage and engine conversion, it is an attractive fuel for specialized uses.

The user likely to benefit most from converting to LNG appears to be a company operating a large fleet of vehicles used daily from a fixed centre (**14**), e.g. lorries, buses, taxis, off-the-road transport, or even river craft.

Another exciting prospect for LNG as a fuel is in the aircraft industry (**15**), particularly for supersonic transport. When applied to turbojet propulsion, advantage can be taken of its high energy-to-weight ratio and of its heat sink and cooling properties, which are of value to the highly heat-stressed sections of the turbine. Using the Boeing Company Airport Activity Analysis for 1976 to 1980 as a base, it has been estimated that by the year 1990 Kennedy Airport is likely to have a daily LNG consumption of 25 000 tons, which is equivalent to two tanker loads per day of the size of the *Methane Progress* or *Methane Princess*.

In this country imported Algerian LNG has been regularly used for research at Farnborough for some five years now; while in U.S.A., Pratt and Whitney Aircraft Division have operated gas-turbine engines on LNG for several years, accumulating over 300 000 h of engine-operating time by early in 1968.

THE FUTURE OF DUAL-FUEL ENGINES

For land-based stationary installations the future appears secure wherever there is readily available an economic supply of methane. This may be by-product methane, as in sewage works, natural gas near overseas gas and oil fields, or natural gas purchased in industrial countries on a bulk tariff, which may be interruptible. The overall economics may be further enhanced by reduced maintenance costs.

The wider adoption of the dual-fuel engine would be encouraged if more development work were done. Improved mixture control on reduced outputs is required. An increase in brake mean effective pressure (b.m.e.p.) would be an advantage, and in this country there is room for more work on a two-cycle machine.

When total energy schemes are considered, the possible use of natural gas is now often investigated. It can, of course, be used in gas turbines or spark-ignited engines, but the dual-fuel engine must be a serious contender.

Marine propulsion is an area now being considered for dual-fuel plant. As long ago as 1904 it was forecast by Sir Dugald Clerk, in the Cantor Lectures, that in the future one could expect to find the ocean covered with gas-driven ships. He was thinking of a compact producer gas plant supplying gas engines, but today, with LNG available, ship designers are again considering gas as a possible fuel. Early applications may be specialist in character, but a forerunner of this development is the large two-cycle marine dual-fuel engine now available from Sulzer to run on natural gas (**16**). It is interesting to note that the development work was carried out using LNG which had been transported more than 1000 km overland.

Over the last 10 years it has been demonstrated that bulk LNG can be safely carried in a variety of cryogenic tank designs; and if a ship at sea is regarded as a closed community, the prospect of a total energy installation based on LNG is very attractive. If the cold potential is not required for refrigeration purposes it can possibly be used for power generation, using a propane cycle. The propane is evaporated by heat exchange with the sea water, and condensed by evaporating the LNG; the temperature difference between sea water and LNG is enough to provide a useful source of power.

Many marine terminals for LNG already exist around the world, and the number is increasing yearly. The numerous peak-shaving liquefaction plants springing up, particularly in the United States, where there are now at least 16, will make LNG more readily available for all transport applications. A 300-gal/day fully portable liquefaction plant which only requires to be coupled to a natural gas main had been marketed by the North American Phillips Company (**17**). Such facilities, together with the already well-proven techniques of transporting LNG by road or rail in cryogenic tanks, will make natural gas widely available.

CONCLUSIONS

Natural gas is the most useful gaseous fuel for internal-combustion engines, and its increasing availability, in one form or another, in many parts of the world and at market related prices, has caused an upsurge in thinking. Development work is in progress to improve the dual-fuel engine, making it even more attractive, and there is plenty of scope for ingenuity in its application.

ACKNOWLEDGEMENTS

The authors wish to thank the Chairman and the Director of Engineering of the North Thames Gas Board for permission to publish this paper and various colleagues for assistance in its preparation. Their thanks are also due to the Gas Council and various engine manufacturers for information used in the paper.

APPENDIX 1.1
REFERENCES

(**1**) FELT, A. E. and STEELE, W. A. 'Combustion control in dual-fuel engines', *S.A.E. Trans.* 1962 **70**, 644.

(**2**) WALTER, C. M. 'The application of town gas as a fuel for internal-combustion engines', *Proc. Instn mech. Engrs* 1939 **141**, 386.

(**3**) JONES, J. 'The position and development of the gas engine', *Proc. Instn mech. Engrs* 1944 **151**, 32.

(**4**) KEEP, G. A. and CUMMINS, K. M. 'Dual-fuel engines', *D.E.U.A. Publication S.231* 1954.

(**5**) ROUGHTON, J. H. 'Dual-fuel engines at Maple Lodge', *D.E.U.A. Publication 327* 1969.

(**6**) HOPKINS, V. H. F., TALBUTT, R. J. and WHITEHOUSE, N. D. 'The development of dual-fuel engines for powers above 1500 b.h.p.', *Int. Congr. Combust. Engng A.9 (CIMAC)* 1962.

(**7**) ULREY, L. 'Engine ignition considerations', *Proc. Am. Gas Ass.* 1964.

(**8**) LAND, M. L. 'Performance and operation of spark-ignited gas engines', *Proc. Instn mech. Engrs* 1966–67 **181** (Pt 1), 900.

(9) PEARCE, M. 'Liquid methane as a motor fuel', *Proc. Auto. Div. Instn mech. Engrs* 1949–50, 155.

(10) KELLY, —. 'Liquid methane behind the Iron Curtain', *Petrol. Times* 1962 (23rd February), **66**, 139.

(11) ENGLER, M. R. 'LNG utilization', Paper 43, Session 47, *First Int. Conf. on LNG, Chicago* 1967.

(12) ANON. 'Natural gas dual-fuel vehicles operating', *Gas* 1968 (December).

(13) ANON. 'Diesel buses converted to dual fuel', *Engineer* 1969 (January) **227** (No. 5894), 41.

(14) KARIM, G. A. 'Some aspects of the utilization of LNG for the production of power in i.c. engines', Paper 12, Session 6, *Cryogenics in Fuel and Power Technol. Symp., London* 1969.

(15) JOSLIN, C. L. 'LNG usage in aircraft engines', Paper 44, Session 7, *First Int. Conf. on LNG, Chicago* 1967.

(16) STEIGER, H. A. 'Large crosshead dual-fuel engines', *Gas Oil Pwr* 1969 (November/December).

(17) MOYLAN, J. E. 'LNG, a future fuel for the transportation industry', Paper 46, Session 7, *First Int. Conf. on LNG, Chicago* 1967.

Paper 2

DEVELOPING A DIESEL FOR AN INDUSTRIAL MARKET SECTOR

N. R. Senior[*]   D. J. Clarke[*]

This paper discusses the controlling of the processes of design and installation. The need for continuous communication between the design and development teams, and with the customer and specialist suppliers, is emphasized. It is considered that involvement with the customer throughout the design and development stages, followed by field testing and approval, will ensure that the new product will meet all market requirements.

INTRODUCTION

IT WILL BE APPRECIATED that the industrial market calls for a wide variety of engine designs, outputs, and specifications for powering both mobile and stationary machinery. This paper is concerned with the development of conventional high-speed diesel engines with power outputs of between 20–200 b.h.p. approximately, for an important sector of that market.

Mobile industrial applications which use diesels range from specialized carrying vehicles, such as dump trucks, aircraft baggage handlers, and towing tractors, to locomotives, with fork lift trucks, tractor shovels, and crawler tractors as intermediate stages. These applications, broadly speaking, require engines similar to automotive vehicle types in respect of both ratings and installation finishing parts, although they more often employ sophisticated transmissions.

Stationary plant includes generating and welding sets, refrigerators, air compressors, and water-pump sets; in addition, the lifting engines for cranes and excavators can be generally considered to fall within this category. Stationary applications usually require engine ratings similar to those applying in the marine work-boat field, with many variations in finishing parts.

There are, of course, many other types of industrial machine powered by engines of the type under discussion, and the number grows yearly. In general, however, it is not economic to build from scratch a complete engine specification to suit each separate machine, with the possible exception of engines for large specialized industries. The requirement can only be met in volume production by the provision of a base engine unit having the necessary fundamental design features, with the separate addition of the various finishing parts required to complete each individual installation.

BASE ENGINE

It is not easy to define exactly what constitutes a base engine design, as it will vary according to circumstances. However, the steps taken to evolve the design will give some indication of the parameters.

A general market investigation will show a requirement for a certain power unit or engine range for a particular industrial market sector. Such a survey, which in practice would be supplemented by day-to-day commercial and engineering contacts, would indicate what the power unit should be capable of in terms of speed and torque. At the same time, the type of combustion system and the maximum extreme pressure at which it is intended to operate will have given an indication of the swept volume of the unit. The survey would also have defined the maximum package size and hence the configuration of the engine itself, e.g. whether a 'vee-form' or 'in-line' type of engine is required and the number of cylinders, the design envelope being determined by mechanical and economic factors. Obviously, the aim will be to meet the widest range of usage.

The survey should give a guide on the engine life required to meet the known market needs, considered both in terms of total engine life and anticipated overhaul periods. It will also provide ideally an indication of the

---

The MS. of this paper was received at the Institution on 23rd February 1970 and accepted for publication on 24th March 1970. 33
[*] *Perkins Engine Company, Frank Perkins Way, Eastfield, Peterborough.*

expected minimum production life of the engine. A detailed consideration of all the points involved will determine the go-ahead for the project, and if this is given, the initial layout of the engine can commence.

Regarding specialized base engine components for suiting various applications, these tend to fall into two categories: those which must be considered an integral part of the base engine design, and those which are little more than variations on a theme completing the finished envelope of the complete engine.

It may well be necessary to increase the range of usefulness of the base engine by including in it components which have characteristics essential for one or two types of application only, but which cannot successfully be added in as finishing parts, e.g. crankshaft with end power take-off facilities. The justification for the inclusion of this type of component can be shown to be commercially viable if purchase, storage, and service problems are eased. This can still be true, even though the piece–part price of the component involved is raised above the level required for many applications.

The design of the base unit must allow the expected required finishing parts to be fitted with the required maximum interchangeability of interrelated components, so that while the base design is being formulated it is necessary to have an indication of what type of general finishing parts will be needed and their design, i.e. the base unit is not designed completely independently as a separate first stage.

Components which are not easy to change on an application-by-application basis will now be discussed. Such components must be considered as part of the base design, unless a very specialized need arises involving a sufficient quantity of power units, permitting the design and manufacture of more specialized components.

Considering first the cylinder block. It will be determined at a very early stage whether this unit is likely to become an integral part of a load-carrying structure, or whether it will be a component subject only to the working and operating forces of the engine itself, and to bending or other forces imposed by the immediate auxiliaries fitted to it. Obviously, if the first situation applies, then the block structure must be of a more robust design to meet the requirements for the type of machine into which the engine is to be fitted. If the second case is accepted, then the block structure will be designed to meet adequately the decreased loadings, with consequent weight decrease.

At this stage, too, the essential basic auxiliary equipment required and the position of the mounting faces must be determined. These faces must be located so that the maximum variety of optional equipment and fittings which can be carried and the number of alternatives of essential auxiliary positions are obtained.

Now the crankshaft will be considered. According to the type of machinery into which the engine will be fitted, it will be known (a) whether or not the shaft will have to be designed to permit engine power to be taken from either end, and (b) if it is to be subject to forces such as bending or severe end loading due to unusual methods of power transmission.

The piston and ring pack configuration will be also subject to considerable development. The speeds at which the engine is to run may permit changes in ring pack specification to give optimum results in certain fields, an instance being the need for a relatively early 'bed-in' of piston rings against a shorter total component life, e.g. for combine-harvester type applications.

The camshaft and valve train can be subject to considerable design work, and a modified form of camshaft may be introduced to give more acceptable results for certain types of operation. The engine low-temperature starting may be improved or, alternatively, the cam form may be modified to give acceptable operation of the valve train at high over-run speeds, which can occur with both modern types of synchromesh gearboxes and hydrostatic transmissions. It may be necessary to introduce modified valve springs or oil seal arrangements to cover prolonged operation at low speeds.

The fuel-injection system may require to be capable of technical adjustment to permit the greatest flexibility of the base engine rating characteristics, sometimes with modifications to the combustion system. By modifying fuel delivery and timing, in addition to the changes to the valve train and camshaft, it is possible to provide widely differing power and torque curves on the same base engine structure (Fig. 2.1). The fuel pumps must be capable of

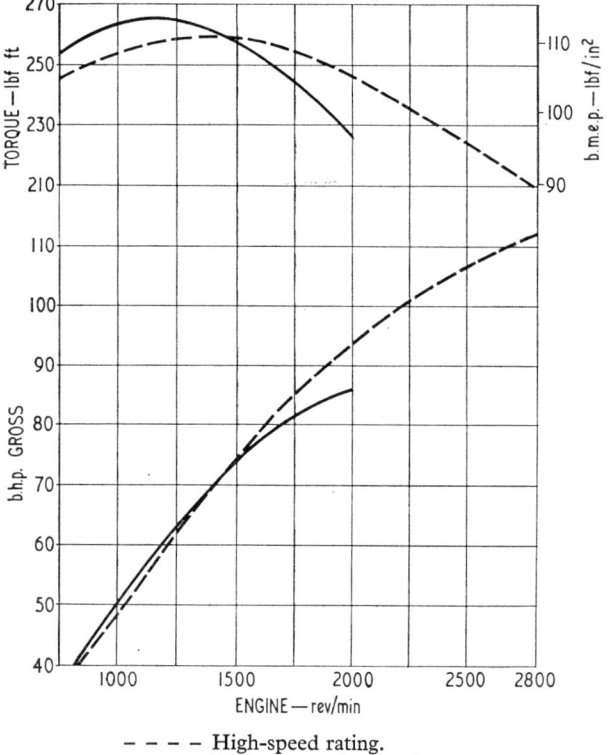

- - - - High-speed rating.
———— High-torque 'back-up' rating.

Fig. 2.1. Power curve showing two alternative rating forms from one base engine

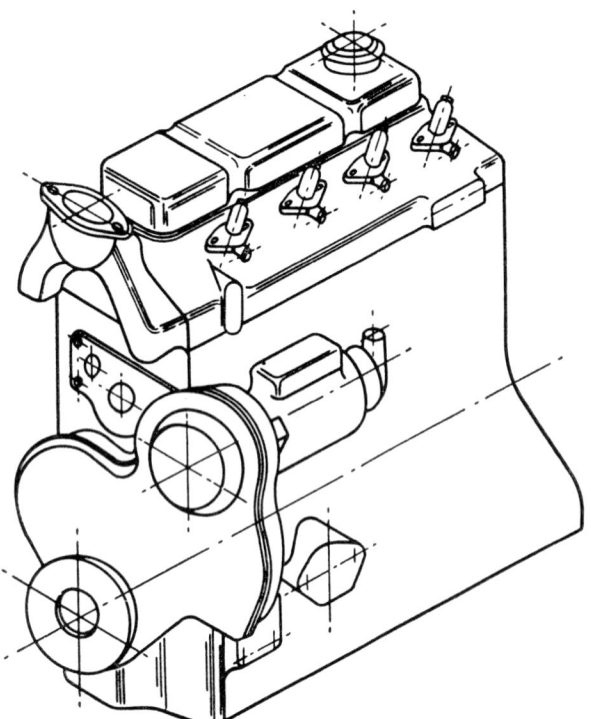

Fig. 2.2. Illustration of a typical base engine

which may make the need for specialized equipment, such as oil coolers, harmonic balancers, etc., essential. It must be decided whether or not these items are to be included as essential features of the base unit design, or if not, under what conditions they are to be added.

Should it be necessary to meet a specific market sector for unusual applications, such as flameproofed machinery, it may be essential for the design of the base unit to incorporate features to meet specific regulations; in this case, items such as minimum limits on certain joint-face widths, and certain piston and valve clearance requirements.

Fig. 2.2 is an illustration of a typical base engine configuration.

FINISHING PART COMPONENTS

Proceeding from a base unit outlined above, we shall now consider what additional features must be investigated to meet the particular industrial market requirements by way of finishing parts to complete the installation. Fig. 2.3 illustrates typical finishing parts.

It will never be possible to have available 'off-the-shelf' finishing parts to cater for the complete coverage with a range of preferred items for both original equipment manufacturers and distributor loose engine sales.

Considering first the flywheel. In the case of mobile applications this has to be of adequate inertia for starting purposes under the limiting temperature conditions envisaged, but not of a sufficiently high value to adversely affect the operation of the machine. However, for stationary plant, such as generating equipment, a flywheel

meeting certain market needs and often of being readily modified to provide different governing characteristics to permit the easy local modification of engines by distributors.

It is necessary to determine operational parameters

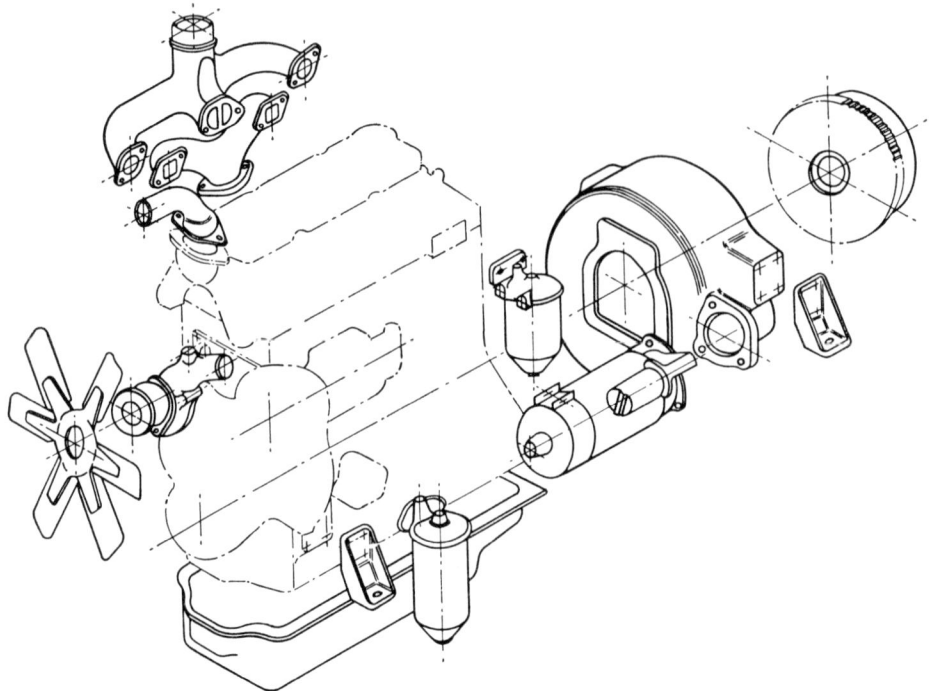

Fig. 2.3. Typical installation finishing components

Fig. 2.4. An engine test tilt rig

of large inertia is necessary to keep cyclic speed variations in the system within prescribed limits. Parameters affecting flywheel design include the maximum rev/min the flywheel is to achieve under both maximum governed speed and over-run conditions, and the bending imposed upon the crankshaft by large overhung wheels.

Associated closely with the flywheel is the flywheel housing, which is more adaptable if of the separately attached type. This must be a rigid structure without superfluous metal. It should be capable of maintaining the necessary alignment of the transmission or driven machinery, providing a suitable location for the engine starting equipment, and incorporating mounting faces of adequate size and strength. The fixings between it and the cylinder-block structure must be adequate under all conditions of operation.

It may be necessary to produce for certain applications, e.g. frameless machines or cantilevered units, a special rigid lubricating oil sump which, tied to the flywheel housing and timing case, forms a load-carrying structure, relieving the cylinder block of excessive loadings.

Next we will discuss the lubricating oil sump assembly. Apart from carrying adequate oil to meet the engine requirements, it must often be of a design which will permit satisfactory operation at large inclination angles. In order to determine angular limits at which sumps will be suitable, it is necessary to physically test samples in order to prove that design estimates of performance are correct. Fig. 2.4 shows an engine tilt rig.

For certain applications the sump capacity is increased above the maximum to provide longer oil-change periods, or to permit the engine to be run unattended for long periods of time, generally in association with automatic control gear.

Allied to the sump and its capacity is the lubricating oil filter. Many types are available, some consisting of a can fitted with a replacement element; others having a completely renewable element and can assembly. The type used will depend on market requirements, spares availability, etc. In addition, to meet certain conditions of operation, particularly where extended oil-change periods are required, the use is justified of a larger-than-standard primary filter and the introduction of a supplementary by-pass type.

Alternative induction and exhaust manifolds will be necessary. In addition to being of the required configuration, these must be designed in such a manner that the gas flow depressions or back pressures are at a level which can be tolerated by the base engine design; practical tests must be carried out to prove their capabilities. In the case of exhaust manifolds particularly, this practical running will give an indication of the strength of the component and its freedom from distortion under high-temperature conditions.

For military or emergency equipment, special designs are necessary. A typical example is a generating set required to be capable of complete inversion during transit without loss of engine coolant, fuel, or lubricant, the unit being ready for immediate use once righted. This necessitates the design of a special lubricating oil filler and dipstick, engine mountings and breather, as well as special fuel and cooling systems.

A further specialized application is the fork lift truck, considered to be a good example of a specially finished industrial diesel design engine. The general criteria applying to fork lift trucks are:

(1) High intermittent speeds of operation, coupled with the ability to idle over extended periods of time.

(2) The power unit should be capable of high rates of acceleration and be quickly responsive to changes in the throttle position, in order that the machine may be operated quickly and with safety.

(3) The power unit must be compact in order to fit into and operate satisfactorily in the smallest engine compartment.

(4) Provision must be made for several power take-off points on the engine itself, the total capacity of these being equal, or nearly equal, to the total engine output, for the smaller engine types.

Points (1) and (2) are dealt with within the base engine specification, a reasonably low-inertia flywheel being used.

To provide a compact power unit with the required power take-offs, the front end of the engine, which normally carries the fuel pump and camshaft drive, will have to be adapted so that gears of adequate size for the various hydraulic pump drives can be carried. There are several points which have to be taken into account.

The first step is to design a gear train of sufficient strength and at the same time to provide fixings on the cylinder block adequate to carry it; all to be contained within the confines of the base unit block design.

Considerable information is required, both from customers and proprietary equipment manufacturers, with regard to the type of hydraulic equipment it is intended to use. The gear case itself must be designed in such a manner that it leaves unchanged all the base unit components, such as fuel pump, camshaft drive, lube oil filter, and lube oil pump drive.

The power take-off positions must be laid out so that the pumps, together with their attendant fittings, can be quickly and easily fitted, both on the initial build of the truck at the customer's premises and as service replacements. The gears and the enclosures must be designed in such a manner that the widest possible range of proprietary equipment can be fitted without the need for additional specialized parts. Figs 2.5 and 2.6 show a specialized timing gear train, case, and adaptor for a fork lift truck.

Continuing with general applications, modified water pumps and crankshaft pulleys may have to be designed to permit the fitting of front-end drives of various types; or alternatively, to permit the use of fans of a much larger capacity than is usually used for engine cooling alone. In some compressor sets, for instance, the engine-driven fan cools not only the power unit itself but also the compressor lubricating oil. Other applications which use torque converters require additional fan capacity to cover the heat rejected from the converter under certain conditions of operation. Yet other components, such as radiators and power pack parts, may be required for the particular market.

Application of the engine

Following on the initial development stages of the base engine, as more customer application requirements are received, and more finishing parts designed, the testing

Fig. 2.5. Fork lift truck engine gear train

Fig. 2.6. Fork lift truck engine main hydraulic power take-off adaptor (below fuel pump)

will be extended to cover these new parts also. As the engine is brought to the notice of prospective manufacturers, their immediate and long-term needs become more clearly defined, and those requirements covering other items of supporting equipment become clearer.

MARKET INFORMATION

Since the development process covering all facets of operation is not one which is necessarily completed at the same time, it is primarily the responsibility of the sales department to inform manufacturers of the availability and relevant details of engines for certain types of applications, through the various stages of engine development.

To facilitate this, it is necessary that the engineering department provide the salesman with accurate technical data associated with the engine, usually found convenient in the form of an engine dossier. It enables prospective manufacturers to design or to obtain the necessary equipment, either from their own works or from specialist suppliers.

A good dossier should be clearly laid out and include, apart from general specification details of the power unit itself, the variations permitted in power and speed output, fuel consumption, the weight and centre of gravity of the unit, the type of power take-off points available and their capabilities, induction air requirements, cold starting performance, and the overall engine heat balance. In fact, all the data associated with the performance of the base engine unit and the variations permitted within it.

It is then necessary to expand this information for engines which are sold on a catalogue basis and show the range of specialized finishing parts available. The limiting features of operation or installation associated with any of the finishing parts must be clearly defined, so that when the power unit is finally installed, it will work satisfactorily under all conditions of operation expected of it.

Before this catalogue is prepared, a great deal of information will be collected from equipment manufacturers whose products will be used in conjunction with the proposed engine. Their own requirements, in terms of ratings, speeds, and specialized finishing parts, will be obtained, together with an indication of their future plans regarding the development of existing components and the design of new items. At the same time the sales department will be producing detailed specifications to match customers' immediate requirements, and these are all integrated to enable the design team to produce as few components as possible, consistent with the covering of the range of installations for which the engine is suitable. It cannot be too strongly emphasized that this type of approach, if carried to its logical conclusion, will produce components which are interchangeable one with another, and which do not rely entirely for their successful operation on being used in conjunction with any other differing type of finishing parts.

INSTALLATION PROCESS

The final stage of development is in the installation process, which is best described by following the progress of a typical example.

It is assumed that the sales department have received an enquiry from a prospective original equipment manufacturer customer for a certain quantity of engines to suit a specialized application. The first stage in the installation process is for the engineering department to review information and layout drawings provided by the customer, and to analyse the engine performance in the application, if needed. An estimate is made by the engineers of the time required to design and to test the specialized components. If at the same time the application requires modifications within the base unit itself, these too must be allowed for in the estimate of design and development time. From this appraisal the total costs for design and development can be estimated.

At this stage two separate evaluations of the project must be carried out before the design process commences. First, the engineering appraisal of the application must be favourable. This means that the customer's requirements can be met in terms of design, production, and component life, and service problems will be at a low level. Second, the commercial decision, which is to assess whether the cost of the engineering work associated with the proposed installation is justified in terms of the total business expected.

Using information from the customer as a basis, preliminary layout drawings are prepared which outline

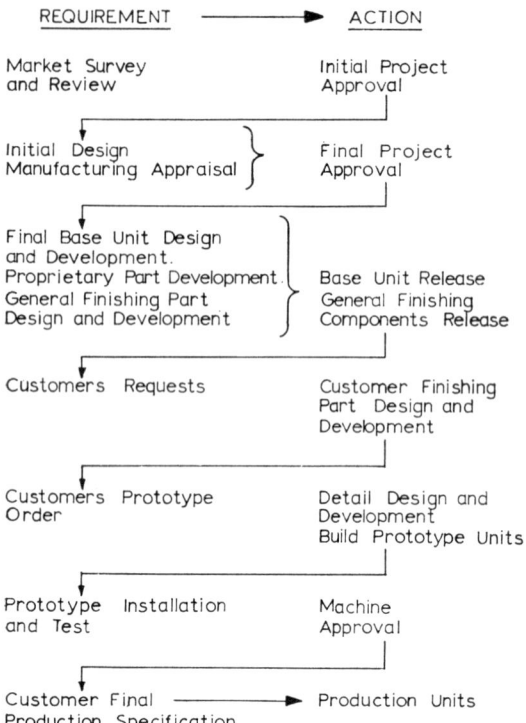

Fig. 2.7. Simplified chart illustrating the progress of an engine from conception to final production supply

the new mechanical components required to meet the customer's needs. Normally, before detail design work commences, the customer's approval of the preliminary layout work will be obtained. When this is received, the development work which may be required to produce new components within the base engine structure will be put in hand. Concurrent with the completion of the design work on the finishing parts, these items are ordered in the quantities necessary for prototype installation.

As this is a continuous process, close liaison with the customer must be maintained at all times. It is highly probable that the machine into which the engine is to be finally installed is, itself, within a similar design and development process, and the customer's needs may well change as his own development proceeds.

Next, engine development testing, simulating the maximum levels of load and stress expected within the installation, is carried out. From these tests alone, the need for further modifications to component or base unit design will become apparent. These modifications may, in turn, result in further changes to the application which have to be undertaken by the customer.

When the first engine is installed in the customer's prototype machine, the machine itself is then put on a series of field trials to prove, as far as is practical, the results of the simulated development tests. It will be tested under the most arduous load cycle expected and, where possible, at a level of loading above that anticipated in normal service.

During this part of operation, the capacity of the radiator and other heat exchange equipment will be checked and the limiting ambient temperatures under which the machine can operate will be determined. At the same time a thorough appraisal is carried out by the application engineers on the supporting equipment. The items checked out in detail would include the engine mountings, induction and exhaust systems, electrical system and cold start equipment, cooling system, fuel system, lube oil system, throttle linkage and stop control, main and auxiliary drives from engine, and any machine equipment affecting the functioning of the engine.

Experience gained from these appraisals can be useful in the development of the engine assembly, and the specialist advice of the diesel engineers is often of assistance to the customer.

SUMMARY

Fig. 2.7 shows the engine development process in abbreviated form. The completed installation on test is shown in Fig. 2.8.

Fig. 2.8. Completed installation on test

CONCLUSIONS

The purpose of this paper has been to show how the processes of design and installation are controlled, emphasizing the need for continuous communication between the design and development teams involved, with the customer and appropriate specialist suppliers. To be competitive, the aim throughout must be to produce the minimum number of component parts with the maximum degree of interchangeability. If this objective is achieved, then purchasing, production, storage, and field service problems will be reduced to the lowest levels, and good service thereby will be given to the end user.

The involvement with the customer throughout the design and development stages, followed by field testing and appraisal, will ensure that the end product will meet all market requirements.

Paper 3

THE EFFECT OF AMBIENT AND ENVIRONMENTAL ATMOSPHERIC CONDITIONS

W. Lowe*

The basis of existing national and international standards for derating naturally aspirated and turbo-charged diesel engines for temperature, altitude, and humidity is recapitulated. The effect of site ambient conditions on the permissible rating of modern turbo-charged and charge-air cooled engines, with fixed and rematched turbo-chargers, is discussed. The effect of turbo-charger and charge-air cooler limitations on the engine rating is examined, including the use of air-to-air charge cooling. Consideration is given to the installation details which affect the engine performance, e.g. air-intake temperature, cooling water, etc.

INTRODUCTION

STANDARDS LAYING DOWN RULES for the derating of engines have been in existence for many years, with periodical revisions in the light of improved knowledge and experience. The revisions have generally lagged behind the commercial practice of engine users and manufacturers because the recommendations are largely empirical and depend on service experience.

The greatly accelerated rate of development of diesel engines in recent years, as a result of improved technology and intensive research effort, has made it more difficult for the user to assess the validity of a manufacturer's rating of an engine which is to operate in an environment that may be greatly different from the one prevailing in the manufacturer's factory.

Fig. 3.1 shows how the catalogue ratings of medium-speed diesel engines of about 15 in bore have increased over the past years. The effects of turbo-charging in the 1930s and of charge-air cooling in the 1950s can be clearly seen, but they are overshadowed by the tremendous rate of increase of specific power in the 1960s. This latter increase is due entirely to improved technology in the design and construction of engine components, including turbo-chargers, bearings, fuel-injection equipment, etc., and in the manufacturer's better understanding of the functioning of the engine, its thermodynamics, aerodynamics, and mechanical and thermal stresses.

In recent years, efforts have been made by the International Organization for Standardization (I.S.O.) to bring together and bring up to date the national standards of member countries into one international standard for reciprocating internal combustion engines. This work has focused attention on the reasons for the derating methods used by different countries and by different manufacturers. Although progress towards an internationally agreed standard is understandably somewhat laborious, the measure of agreement which has been reached so far is most encouraging.

Notation

- B Brake power.
- I Indicated power.
- M Mass flow.
- N Rotational speed.
- P Ambient air pressure (absolute).
- P_s Saturated vapour pressure.
- T Ambient air temperature (absolute).
- T_w Charge-air coolant temperature (absolute).
- ϕ Relative humidity.

Suffix

- 0 Refers to standard reference conditions.

METHOD OF EXPRESSION OF DERATING FORMULAE

The method used in the British Standard specifications to express derating is a simple linear relationship between the rating and the ambient condition, e.g. B.S. 649 for naturally aspirated engines derates by $3\frac{1}{2}$ per cent/1000 ft

---

The MS. of this paper was received at the Institution on 28th January 1970 and accepted for publication on 4th March 1970. 23
* *Mirrlees-Blackstone Ltd, Stockport, Cheshire.*

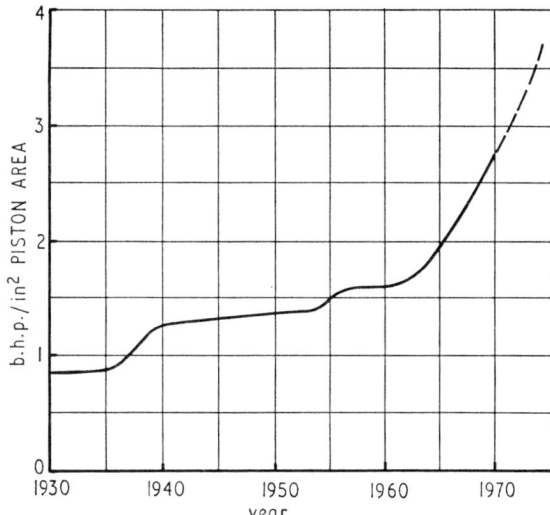

Fig. 3.1. Rated power of a medium-speed diesel engine

altitude above 500 ft above sea level, and 2 per cent/10 degF ambient temperature above 85°F.

The derating is a percentage of the brake power, and the amounts of derating for altitude and temperature are added together. The American D.E.M.A. procedure is similar, using a graphical presentation.

Continental practice is somewhat different, expressing the ratio of site power to standard power as a function of the ratio of site conditions to standard conditions, e.g.

DIN 6270

$$\frac{I}{I_0} \propto \left(\frac{T_0}{T}\right)^{0.75} \quad . \quad . \quad . \quad . \quad (3.1)$$

The derating is applied to indicated power, and a further calculation (see under 'Mechanical efficiency') is carried out to obtain brake power. The power ratios for altitude and temperature are multiplied together.

These two common methods of expression are not of the same form and hence cannot be exactly equated, even though they may have the same reference conditions of temperature and altitude. However, rating methods cannot be very exact, therefore the following approximate equivalents may be useful when comparing the two different methods of expressing derating:

$$\frac{B}{B_0} \propto \left(\frac{P}{P_0}\right)^a \quad . \quad . \quad . \quad (3.2)$$

is a derating of about $3.5a$ per cent/1000 ft.

$$\frac{B}{B_0} \propto \left(\frac{T_0}{T}\right)^b \quad . \quad . \quad . \quad (3.3)$$

is a derating of about $2b$ per cent/10 degF.

In the following sections the 'exponent' method of equations (3.1), (3.2), and (3.3) has been used to express derating, and the above approximate equivalents will enable rough comparisons with the 'percentage' method to be made.

MECHANICAL EFFICIENCY

The British Standard methods of derating assume that the mechanical efficiency of the engine does not change with changes in power output, whereas the continental methods often include an assumption for the way mechanical efficiency changes and give more complicated formulae of the form:

DIN 6270, CIMAC

$$B = \alpha B_0 \quad . \quad . \quad . \quad . \quad . \quad . \quad (3.4)$$

$$\alpha = k - 0.7(1-k)\left(\frac{1}{\eta}-1\right) \quad . \quad . \quad (3.5)$$

$$k = \frac{I}{I_0} = f\left[\left(\frac{T_0}{T}\right)\left(\frac{P}{P_0}\right)\right] \quad . \quad . \quad (3.6)$$

In equation (3.6) the derating is applied to the indicated power, according to some function of the ambient conditions.

In equation (3.5) it is assumed that the losses which reduce indicated power to brake power can be divided into two parts. At the reference conditions 70 per cent of the losses are assumed to be constant, and the other 30 per cent assumed to be proportional to the indicated power, i.e. at reference conditions,

$$\text{losses} = I_0 - B_0 \quad . \quad . \quad (3.7)$$

At the conditions being considered,

$$\text{losses} = I - B = (I_0 - B_0)\left(0.7 + 0.3\frac{I}{I_0}\right) \quad (3.8)$$

Putting

$$k = \frac{I}{I_0}, \quad \eta = \frac{B_0}{I_0} \quad \text{and} \quad \alpha = \frac{B}{B_0}$$

equation (3.8) reduces to equation (3.5).

For engines of relatively low mechanical efficiency, as might occur in some naturally aspirated engines, the difference between the indicated power derating (k) and the brake power derating (α) is appreciable; but for the high mechanical efficiencies that can be expected from pressure-charged engines, the difference is negligible.

For example, suppose the ambient conditions impose a reduction in indicated power of 10 per cent, i.e. $k = 0.9$.

If $\eta = 0.6$, then $\alpha = 0.85$, i.e. 15 per cent derating of brake power.

If $\eta = 0.9$, then $\alpha = 0.89$, i.e. 11 per cent derating of brake power.

Modern high-output turbo-charged and intercooled engines generally have mechanical efficiencies between 90 and 95 per cent; therefore the difference between indicated power and brake power deratings is negligible.

DERATING AND CORRECTION

When considering the effect of ambient conditions on an engine, it must be clearly understood whether alterations to the engine construction or settings have been included or excluded. It is well known, for example, that turbo-charged engines can have their turbo-chargers matched for the altitude condition in which the engine is to operate, and in many cases no derating will be necessary. It may then be essential to view the problem in reverse when considering an engine with an 'altitude' turbo-charger being tested at the manufacturer's factory at a lower altitude.

The expression 'derating' is considered by the author to allow changes in the engine design such as this to meet the general atmospheric pressure and temperature level which are to be expected in service.

The expression 'correction' is used when allowance is being made for day-to-day variations in environmental conditions and when the engine build and settings remain unchanged.

NON-TURBO-CHARGED ENGINES

Derating methods for naturally aspirated engines generally apply also to mechanically pressure-charged engines where the volumetric displacement is fixed by the engine speed, and hence the mass of air used by the engine depends directly on the ambient air pressure and temperature.

Smoke limit

When the engine output is limited by the onset of smoke, the weight of oxygen trapped in the cylinders determines the permissible output; therefore it is necessary to allow for the effect of humidity. A widely used formula (DIN 6270, CIMAC 'A') is:

$$\frac{I}{I_0} = \left(\frac{P - \phi P_s}{P_0 - \phi P_{s_0}}\right) \left(\frac{T_0}{T}\right)^{0.75} \quad . \quad . \quad (3.9)$$

The atmospheric-pressure adjustment is thus the ratio of the dry-air pressure to that at reference conditions. The temperature exponent appears to be a compromise between a simple assumption of constant volumetric efficiency, where the exponent would be unity, and that for a throttling process in which the air flow would be proportional to the square root of the absolute temperature, i.e. an exponent of 0·5. It seems reasonable to assume that there would be a change in volumetric efficiency with a change in air-intake temperature because of the difference in heating effect of the cylinder walls, and experimental evidence appears to support an exponent between 0·6 and 0·75.

Thermal limit

If the engine output is not limited by smoke but is restricted by thermal limitations, the methods of derating are generally based on maintaining constant air/fuel ratio. Derating is thus directly proportional to air density; humidity is not taken into account because it contributes towards the mass of air, and the reduction in oxygen mass is not limiting the output. The CIMAC 'T' formula is on this basis where:

$$\frac{I}{I_0} = \left(\frac{P}{P_0}\right)\left(\frac{T_0}{T}\right) \quad . \quad . \quad . \quad (3.10)$$

This basis includes the assumptions that the volumetric efficiency remains constant and that the indicated specific fuel consumption is unaltered.

B.S. 649 is rather more conservative than either of the above formulae (3.9) and (3.10) at high mechanical efficiencies, the equivalent exponent being about 1·1 for both pressure and temperature, with an allowance being made for humidity in the form of tables which over-compensate for the vapour pressure of water at moderate temperatures.

The American D.E.M.A. standard is also more conservative with equivalent exponents of about 1·2 for pressure and 1·4 for temperature, but with no allowance for humidity.

TURBO-CHARGED ENGINES WITHOUT CHARGE-AIR COOLING

Most national standards avoid specifying the amount of derating for turbo-charged engines, leaving it as a contractual arrangement between the manufacturer and customer. A notable exception is B.S. 649, which contains derating formulae for turbo-charged engines both with and without charge-air cooling. The values are based on operating experience and test data, using constant engine-exhaust temperature (or turbine-inlet temperature) as the criterion for the permissible rating of the engine.

If the B.S. formula is expressed in the exponent form as:

$$\frac{B}{B_0} = \left(\frac{P}{P_0}\right)^a \left(\frac{T_0}{T}\right)^b \quad . \quad . \quad . \quad (3.11)$$

the approximate values of the exponents are $a = 0.7$ and $b = 1.7$. Experimental evidence from a range of manufacturers tends to confirm that these values are reasonable for low- and medium-speed four-stroke engines, but there is no evidence that the numerical values would apply to two-stroke engines or high-speed four-stroke engines.

Rematching of the turbo-charger to suit site ambient temperature has a very small effect on exhaust temperature; therefore the method tends to derate rather more than necessary for such engines, but the effect is not marked. On the other hand, matching the turbo-charger to suit the site altitude has very marked effects on the engine and its exhaust temperature. It has therefore been general practice for some years for manufacturers to maintain their sea-level ratings up to a stated altitude and then to use the B.S. 649 derating from then on, the limiting altitude being determined by a limitation such as turbo-charger rotor speed. The actual altitude at which this limit is reached varies with the number of cylinders of the engine and the turbo-charger design. To simplify derating procedures it is usual for a particular manufacturer to standardize on one altitude for his range of engine sizes.

Without charge-air cooling, the turbine-inlet temperature is quite often, but not always, the limitation to engine output, so that although in some cases engines may be excessively or unnecessarily derated, this method provides a safeguard which will be effective in the majority of cases.

TURBO-CHARGED ENGINES WITH CHARGE-AIR COOLING

The B.S. 649 formula for derating turbo-charged engines with charge-air cooling includes a term for the temperature of the water to the charge-air cooler, in addition to those for ambient temperature and altitude. Again, constant exhaust temperature is used as the criterion for the permissible rating of the engine.

Extending equation (3.11) to include a term for charge-air coolant temperature will give:

$$\frac{B}{B_0} = \left(\frac{P}{P_0}\right)^a \left(\frac{T_0}{T}\right)^b \left(\frac{T_{w_0}}{T_w}\right)^c \quad . \quad . \quad (3.12)$$

The B.S. 649 values give equivalent exponents of $a = 0.7$, $b = 0.5$, and $c = 1.7$.

The very marked effect of charge-air cooling on exhaust temperature, and the development of turbo-chargers to permit higher pressure ratios, rotational speeds, and turbine-blade temperatures in recent years have meant that exhaust temperature is no longer a general criterion of the output limitation of modern high-output engines which invariably use charge-air cooling. In some high-speed engines, exhaust temperatures may still be a limitation; but in most medium-speed engines the limit is set either by component metal temperatures or by trapped air/fuel ratio. With improvements in component design, the limits of power due to component metal temperature are continually being lifted; examples of this can be seen in the growing popularity of ring-belt-cooled pistons and seat-cooled exhaust valves. Therefore the common factor to many modern engines has become the limiting trapped air/fuel ratio on which good combustion and low specific fuel consumption depend.

It is suggested, therefore, that a logical way in which to derate modern high-output engines is to maintain constant trapped air/fuel ratio until a point is reached, depending on the particular engine design, at which another limitation becomes critical.

To illustrate this point, Figs 3.2 and 3.3 show the effect of changes in ambient conditions on some of the operating characteristics of a particular medium-speed engine. A comparison of the two figures will show the great effect which is obtained by matching the turbo-charger build to suit the site conditions.

Fig. 3.2a shows the effect of ambient temperature. The turbo-charger has been matched for site conditions, and the maximum cylinder pressure has been maintained at its standard level. The engines could be derated on the constant trapped air/fuel ratio line, which has an exponent of $b = 0.2$, until a limiting parameter was reached. This might, for example, be a turbine-inlet temperature of 980°F, which would be reached at an ambient temperature of 105°F. The derating at ambient temperatures above 105°F would then be at a higher rate corresponding to the constant 980°F turbine-inlet temperature line, which has an exponent of $b = 1.1$. This second phase of derating would obviously depend on the particular engine design and its limiting criterion.

Fig. 3.2b shows the effect of altitude, similarly, and gives $a = 0.1$ up to 1700 ft above sea level and then $a = 0.75$.

Fig. 3.3 maintains the same turbo-charger build and the same fuel-injection timing. The same limiting parameters as before would give:

For ambient temperature

$b = 0.6$ up to 96°F, and then $b = 2.5$.

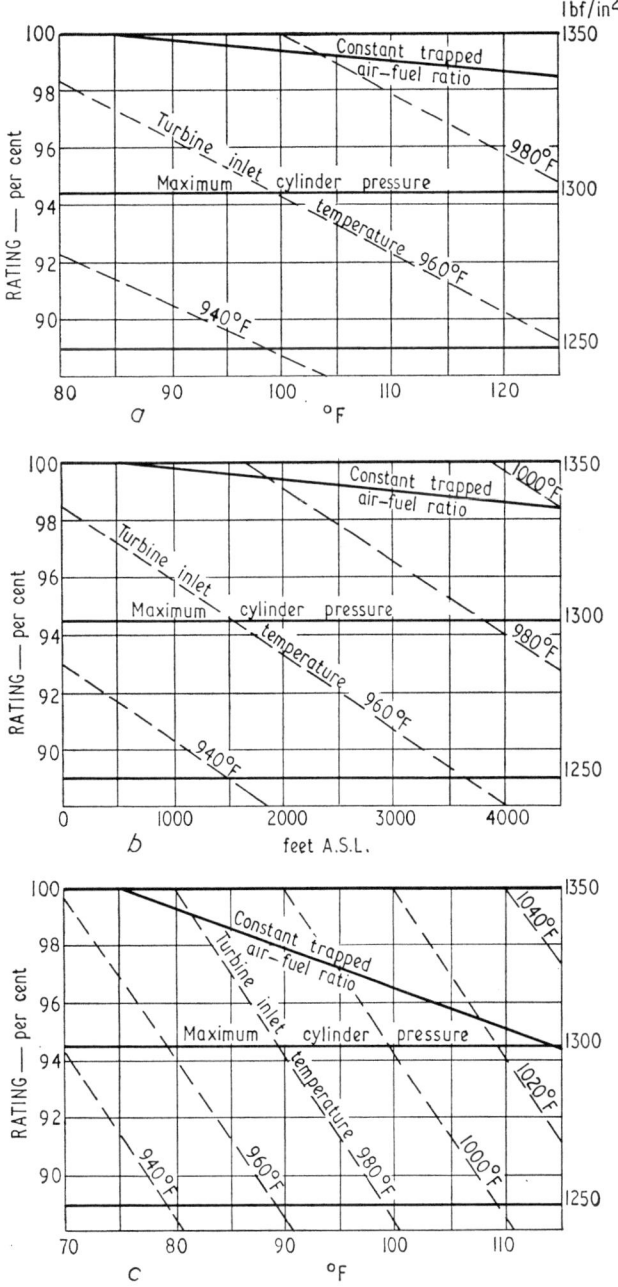

(a) Ambient air temperature.
(b) Altitude.
(c) Charge-air coolant temperature.

Fig. 3.2. Variable turbo-charger match

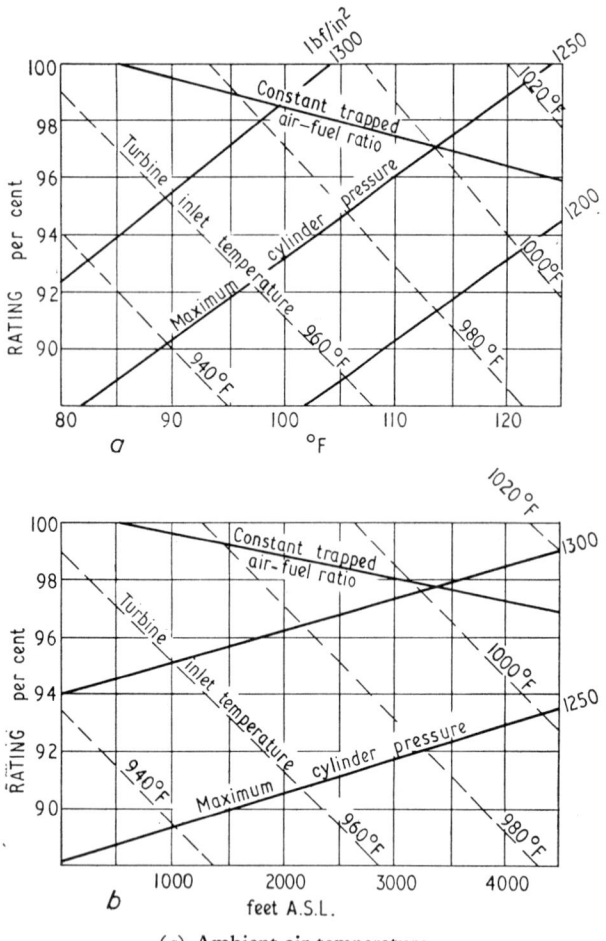

(a) Ambient air temperature.
(b) Altitude.

Fig. 3.3. Fixed turbo-charger match

For altitude

$a = 0.2$ up to 1500 ft above sea level and then $a = 1.1$.

The effect of increased charge-air coolant temperature has not been shown, because with a fixed turbo-charger build the onset of compressor surge imposes an immediate output limitation which is not comparable with the other parameters which have been illustrated.

The marked difference in slope between the lines of constant trapped air/fuel ratio and those of constant turbine-inlet temperature shows clearly that different engine designs demand vastly different amounts of derating for a given set of site ambient conditions. Component metal temperature changes tend generally to have the same degree of slope as those for turbine-inlet temperature, though of course at a different level of temperature.

In the case of the fixed turbo-charger (Fig. 3.3) it should be noted that not only are the slopes of constant trapped air/fuel ratio and constant turbine-inlet temperature much steeper than those for the variable match (Fig. 3.2) but also that a limiting parameter is reached much sooner.

The three graphs of Fig. 3.2 should be interpreted as a number of discrete points, each having a different turbo-charger build, rather than as continuous lines. Since Fig. 3.3 represents a constant build, it can be used progressively.

In practice, the standardization of nozzle ring steps, blade heights, etc., would mean a combination of Figs 3.2 and 3.3, so that a practical method of derating would have exponents between those given for the two conditions of fixed and rematched turbo-chargers.

CLIMATIC VARIATIONS

In the foregoing paragraphs, it has been assumed that an engine and turbo-charger combination can be matched to suit a specified set of site ambient conditions. Day-to-day and seasonal variations in temperature and barometric pressure are in many cases small enough to be neglected, since such variations will, of course, have occurred during the establishing of the standard rating of the engine. In some cases, wider variations may need to be considered in deciding the site rating, and one or two special considerations are worthy of mention.

Ambient air temperature

The paper has so far considered only the effect of an increase in ambient temperature on the rating of an engine. When a wide range of temperature is experienced it may be necessary to examine the low-temperature condition and its effect on maximum cylinder pressure and on the onset of turbo-charger surge.

Climatic changes must, of course, be the fixed turbo-charger match case, and reference to Fig. 3.3a shows that as the ambient air temperature decreases, the maximum cylinder pressure will increase. This is largely owing to an increase in charge-air pressure; and if maximum cylinder pressure becomes a limitation, it is often possible to avoid a reduction in load by blowing off some of the charge air, since the engine thermal conditions and air/fuel ratio are favourable. The same method can be used to avoid turbo-charger surge at low ambient temperatures, since the alternative of matching the turbo-charger for the low ambient temperatute would mean unnecessary derating when the temperature was high.

Fig. 3.4 shows an engine operating line at constant power during wide ambient temperature variations, superimposed on the turbo-charger compressor characteristic. Although the actual air mass flow is continually increasing as ambient temperature decreases, the non-dimensional mass flow begins to decrease at low temperatures, with the turbo-charger operating at substantially constant speed. The combined effect of reduced non-dimensional mass flow and increased pressure ratio causes the surge line to be approached as the ambient temperature decreases.

Charge-air coolant temperature

As already mentioned under 'Turbo-charged engines with charge-air cooling', an increase in charge-air coolant temperature will quickly reach a condition of compressor

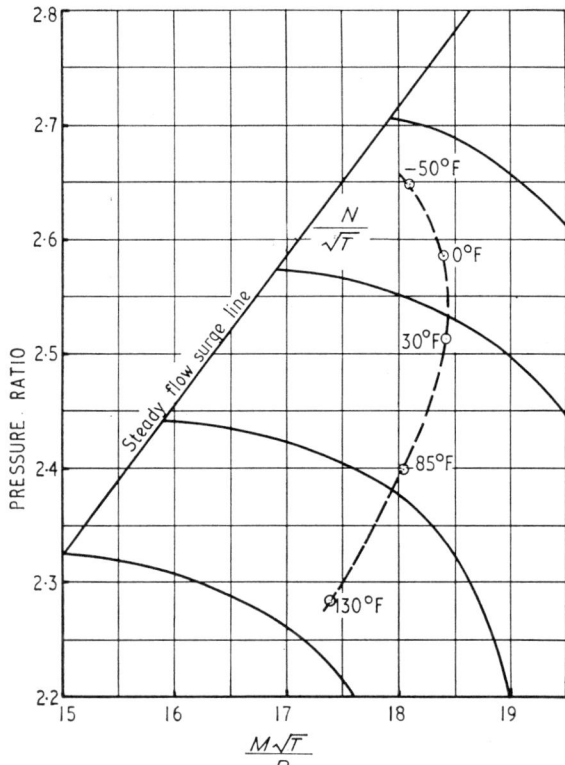

Fig. 3.4. Constant power at fixed match

surge, due to the reduction in mass flow from the lower density of the air entering the engine, accompanied by a rise in pressure ratio from the higher turbine-inlet temperature. This characteristic is often used as a practical check on the margin of surge-free operation by gradually shutting off the water supply to the charge-air cooler and noting the rise in charge-air temperature before surge conditions are reached.

In most cases, the charge-air coolant supply is from a raw-water source which does not vary much with ambient conditions. In the cases where a variation in coolant temperature is to be expected, it is important to match the turbo-charger for the maximum temperature to avoid the surge risk.

This problem is particularly difficult where the raw-water supply is limited or non-existent and radiator cooling is employed, since the engine must then be rated and the turbo-charger matched to suit the coolant temperature at the highest ambient temperature to be encountered.

AIR-TO-AIR COOLING

It has already been shown that charge-air cooling has a large effect on the permissible rating of a turbo-charged engine. With water-cooling of the charge air, the amount of cooling is usually determined by the size of charge cooler which can be conveniently mounted on the engine. The proportions of engine and cooler are such that an effectiveness of about 0·75–0·8 is usually achieved, effectiveness being defined as:

$$\frac{\text{charge-air temperature into cooler} - \text{charge-air temperature out of cooler}}{\text{charge-air temperature into cooler} - \text{coolant temperature into cooler}}$$

To achieve an effectiveness appreciably higher than 0·8, a very large increase in cooler size would be necessary, on a law of diminishing returns. There have been some cases where water-cooled charge-air coolers have been mounted off the engine, but this has not been a common practice.

In recent years there has been great interest shown in the use of air cooling of the charge air, and this is often attractive in locations of high ambient temperature where a cooling-water supply is not available. In such cases, radiator cooling of the engine oil and jacket water is arranged, and then a charge-air cooler is placed in front of the oil and water radiator to cover the same area. The cooling-air temperature rise across the air-to-air cooler is small and has little effect on the proportions of the oil and water radiators. A very large charge-air cooler is thus accommodated with only the small disadvantage of a few extra inches of radiator depth, together with a rather greater disadvantage of air ducting between the engine and the cooler, which must be large in diameter to minimize flow losses.

The very large area available for the charge-air cooler increases the effectiveness to well above the usual 0·8 level of an engine-mounted unit, and an effectiveness of 0·93–0·95 can often be achieved. If such a high effectiveness was achieved at normal temperatures, the engine could be increased in rating by some 6–8 per cent and then derated for increased ambient temperature, as described under 'Turbo-charged engines with charge-air cooling'. However, it is not common practice for manufacturers to allow even theoretical increases above their normal rating, so that the manufacturer generally maintains his standard rating with an air-to-air cooler up to an ambient temperature of around 110°F, and then derates from that temperature. Fig. 3.5 illustrates the dependence of trapped air/fuel ratio and exhaust temperature on intercooler effectiveness over a range of ambient air temperatures. In this figure, the effect of charge-air coolant temperature and that of air-intake temperature have been combined together into a single variable of ambient air temperature; as in Fig. 3.2, the turbo-charger is matched for the site conditions. The broken lines represent a hypothetical permissible increase in rating above the standard rating for high values of intercooler effectiveness. The numerical values of the exponents of equation (3.12) are:

constant trapped air/fuel ratio $b+c = 1·1$

(cf. $0·2+0·8 = 1·0$ in Figs 3.2a and 3.2c),

constant turbine-inlet temperature $b+c = 4·2$

(cf. $1·1+3·3 = 4·4$ in Figs 3.2a and 3.2c).

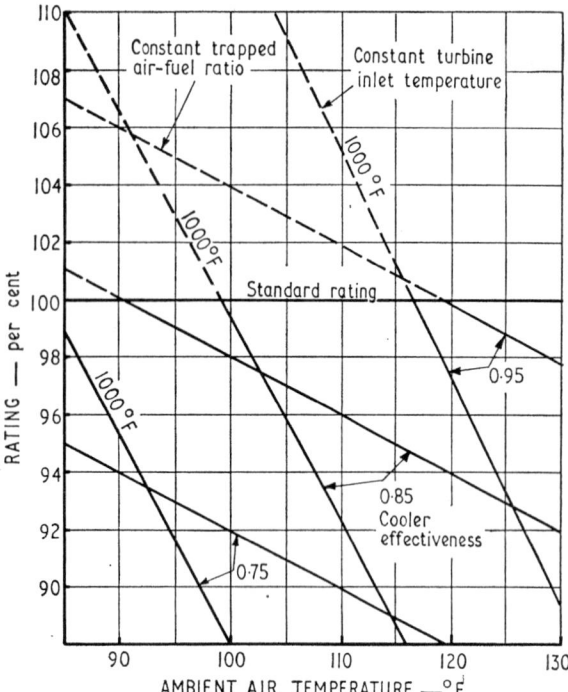

Fig. 3.5. Charge-air cooler effectiveness

INSTALLATION DETAILS

Although it may be stating the obvious, careful assessment of the site ambient conditions for a particular application is of little value if lack of thought in the installation details causes the engine to suffer temperature and pressure conditions far removed from those used in the assessment.

The practice of drawing air from the engine room to feed the engine aspiration system, and hence to use the engine as a ventilation system, has fortunately fallen into deserved disrepute. Most modern installations duct the engine air from outside the engine room, thus avoiding the danger of a false high air-intake temperature. Even with ducted air systems, mistakes have been made in the positioning of the air intakes; cases are known where the intakes have been shrouded by obstructions such as waste heat boilers and tanks, so that air which has been heated by silencers and exhaust pipes has been drawn down into the air intakes. Care must be taken to allow an unobstructed supply of air to reach the air intakes without being artificially heated on the way.

It has also been known for waste steam to be blown off near the air intakes, and in low ambient temperatures for the air filters to be almost completely blocked by a film of ice. Air filters need to be adequately sized, and provision must be made for easy cleaning to prevent an excessive pressure drop across the filter.

CONCLUSIONS

Some of the derating methods published in national standards and others used by individual manufacturers have been described. In the author's opinion, there is no suitable standard for derating modern high-output turbocharged engines with charge-air cooling, and a method is described based on constant trapped air/fuel ratio which could be applied to all such engines up to a point depending on the particular engine design. This point, and the degree of derating beyond it, would be declared by the manufacturer for his engine design.

ACKNOWLEDGEMENT

The author wishes to express his thanks to the Board of Mirrlees-Blackstone Ltd for permission to publish the information contained in this paper.

Paper 4

CRITICAL FACTORS IN THE APPLICATION OF DIESEL ENGINES TO RAIL TRACTION

W. Petrook* W. A. Stewart†

The authors consider that the operation of diesel engines for rail traction is arduous because of the 'on–off' operation of the engine duty cycle in service, and hence thermal cycling is more severe than that in other applications. The joint paper discusses the limitations of the cooling systems, the turbo-blower and its lubrication, and the arrangements made to ensure that fuel input is related to turbo-blower speed during acceleration of the engine. The silencing of diesel engines is of additional importance from the exhaust noise aspect and from the mechanical noise of the engine itself. The rating of diesel engines is discussed briefly, as are overhaul life, ease of maintenance, and reliability.

INTRODUCTION

THE OPERATION of diesel engines for rail traction is very similar to that for any other application of engines of similar powers, but in the opinion of the authors it may be more arduous. This is brought about by the on–off operation of the engine duty cycle, and hence thermal cycling is more severe as extremes of power are of longer duration. Thus, normal critical factors may be made more critical for traction applications, especially in this high-speed high-output era.

One of the first critical factors involved in diesel locomotive design is how to determine which type of diesel engine is the most suitable for the application in mind. The development of diesel engines has now reached the stage where it is possible to build a BB locomotive, i.e. two motored axles on each of two bogies, up to 3000 hp, and a CC locomotive, i.e. three motored axles on each of two bogies, up to 6000 hp without exceeding a static load of 18 ton/axle. The latter case would probably involve the use of two high-speed engines.

In the early days of dieselization, the aim of the locomotive designers was to produce general-purpose 'mixed traffic' locomotives, which could be used to haul heavy freight trains or high-speed passenger trains, as required. As the industry progressed and patterns of new traffic emerged, it became evident that there is now a divergence between the hauling requirements for freight trains and those for high-speed inter-city trains, although both duties require the maximum installation of power. The freight application requires a greater weight of locomotive for greater tractive effort, but at lower speeds. The fast, inter-city application requires the high power for the higher speeds, but lighter trains to reduce track damage at the higher speeds. From existing data and our future designs, it would appear that the total diesel engine weight should be no more than about 35 000 lb to enable a six-axle locomotive to be built under about 110 ton.

Any increase in this engine weight must distort the design parameters of the remainder of the locomotive equipment. However, regardless of application, the authors consider that the most critical factor is the ability of the engine to withstand the effect of rapid power cycling on highly rated turbo-charged and charge-cooled engines.

The criteria for any locomotive design involve the following points: highest possible power, highest possible reliability, low initial capital cost, low maintenance and overhaul cost, and suitable rail loading for the particular locomotive application.

If we take, for example, the criterion of packing as much power as is possible into each design of diesel locomotive, and if we need the lowest axle to rail loadings for a high-speed application, then we must have the lightest possible diesel engine with high rotational speed and high specific power. These factors immediately affect some of the other criteria mentioned. For example, a diesel engine of high rotational speed and high specific power might adversely

---
The MS. of this paper was received at the Institution on 3rd March 1970 and accepted for publication on 25th March 1970. 42
* *Power Equipment Engineer, British Railways, Trent House, Railway Technical Centre, London Road, Derby.*
† *Consultant, Kennedy and Donkin, Brewood, Winter Hill, Cookham Dean, Berks.*

affect the reliability, maintenance, and overhaul costs, and possibly initial costs, compared with an engine of low specific power and lower rotational speed, which has in most cases proved itself to be longer lived and more reliable. Thus, any design must necessarily be a compromise between the five factors listed above.

Another point of major importance is that, within the British Railways (B.R.) loading gauge, the cooling equipment is a limiting factor against increasing power still further. Today, the most modern diesel engines are turbocharged and charge-cooled, and it may be of interest to know that the equipment necessary for the cooling of the air charge is as big as that for the engine cooling system itself.

The factors with which we intend to deal in this paper are cooling systems, turbo-chargers and problems associated with the relation of fuel input to air quantity, exhaust silencing and engine noise, rating of diesel engines for traction, overhaul life of traction diesel engines, ease of maintenance of diesel engines, and reliability.

COOLING SYSTEMS

As mentioned earlier, one of the critical factors in packing as much power as possible into a locomotive, within the B.R. loading gauge, is the cooling system itself. In fact, within this loading gauge the maximum power which can be installed in a locomotive is about 6000 hp, and this is limited by the size of the cooling system. In addition, the size of the cooling system installed in a locomotive is governed by some of the margins which are necessary to compensate for possible high ambient temperatures and for design purposes of the radiator elements, to allow for fouling and possible contamination by dirt. On B.R. it is always assumed for design purposes that the maximum ambient temperature is 85°F—although it is appreciated that this situation arises all too seldom in this country—and we also allow about 10 per cent fouling margin. Despite these precautions, there has been at least one example on B.R. where a diesel engine overheated when the locomotive was travelling in one direction but was fully satisfactory when the locomotive was travelling in the opposite direction. Of course, this became more evident during the rare periods of high ambient temperature, and on many occasions the diesel engine was shut down automatically by the high water-temperature safety device. This deficiency of operation was overcome by immediate attention to the pipework of the engine cooling system, to reduce the system resistance and so increase the volume flow from the water pump. This reduced the maximum coolant temperature when travelling in both directions, but of course did not affect the difference; it merely indicated that some aerodynamic investigations would be necessary for future locomotives, which will probably become essential as higher and higher speeds are anticipated.

Another major factor which governs the size of the cooling system is the operating temperature of the engine itself. Many cooling systems are pressurized, and advantage is taken of this pressurization to ensure that the coolant circulating pump is always filled, or that waterside cavitation erosion on the cylinder liner is suppressed. However, very few cooling systems take advantage of pressurization to increase the operating cooling water temperature by increasing the operating temperature of the engine itself. Most engine manufacturers are reluctant to increase the operating temperatures of their engines, the limiting criterion probably being the piston top gas-ring temperature.

With existing designs of piston, the increase of operating temperature would mean either a derating of the engine itself, to a level where the top gas ring is not increased, or a reduction in engine overhaul life. More attention could be paid to the oil cooling of pistons, to reduce the operating temperature of the ring pack, by careful design of the internal contours of the piston crown and the way in which the oil is allowed to remove the heat.

This, then, is one of the critical factors in diesel engine design applied to rail traction, as an increase in the coolant operating temperature from, say, 190 to 230°F would reduce the volume, and hence weight, of the radiator system by something in excess of 25 per cent, which for a modern locomotive design could be as much as 1000 lb.

The oil companies, too, have some participation here with improved oils, but again we have the compromise between higher available powers with improved oils or a more compact cooling system. It is obvious that from the commercial standpoint, and bearing in mind that traction is but one outlet for diesel engines, engine manufacturers would rather take advantage of improved oils and materials to increase power with atmospheric or near atmospheric cooling systems than sell engines at a lower power, merely to satisfy a rail traction requirement for a light and compact cooling system.

To give some idea of the sources of heat rejection from a diesel engine the following typical heat balance is shown on Fig. 4.1.

From this heat balance the values are as follows:

	Per cent
Heat to useful work	38
Heat to exhaust	37
Heat to engine water jacket	14
Heat to charge-air cooling	7·5
Heat to lubricating oil	3·5
	100·0

For a typical 3000-hp turbo-charged and charge-cooled engine, the actual quantities of heat to be dissipated are quite large, and the following table gives some indication of this:

Heat from engine water jacket
$2 \cdot 9 \times 10^6$ Btu/h = 849·7 kW

Heat from lubricating oil
$0 \cdot 7 \times 10^6$ Btu/h = 205·1 kW

Heat from charge air
$1 \cdot 5 \times 10^6$ Btu/h = 439·5 kW

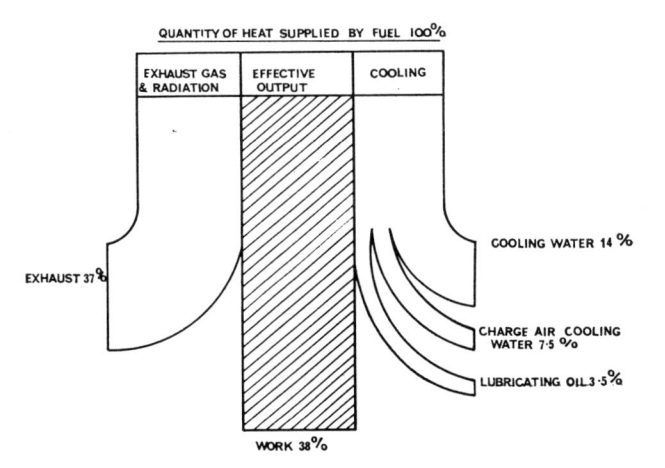

Fig. 4.1. Heat balance of a typical intercooled engine

The modern tendency is to divide the cooling system into two separate circuits, and the operating temperatures involved lend themselves to allowing the charge-air and lubricating-oil systems to be cooled in one circuit, and the main water jacket in the other. In this manner, too, the heat dissipations from each circuit are similar to each other. Fig. 4.2 shows such a divided cooling system similar to that installed in the latest B.R. diesel locomotive.

The authors are agreed that on B.R. the area which gives the greatest trouble with regard to reliability of diesel locomotives is the cooling system. This trouble is usually associated not with highly sophisticated technology or equipment but with the 'simple' problem of leaks. Air-to-air cooling systems could offer some relief to this leakage problem, but with space limitations in modern traction equipment this poses its own problems. It is the lack of any

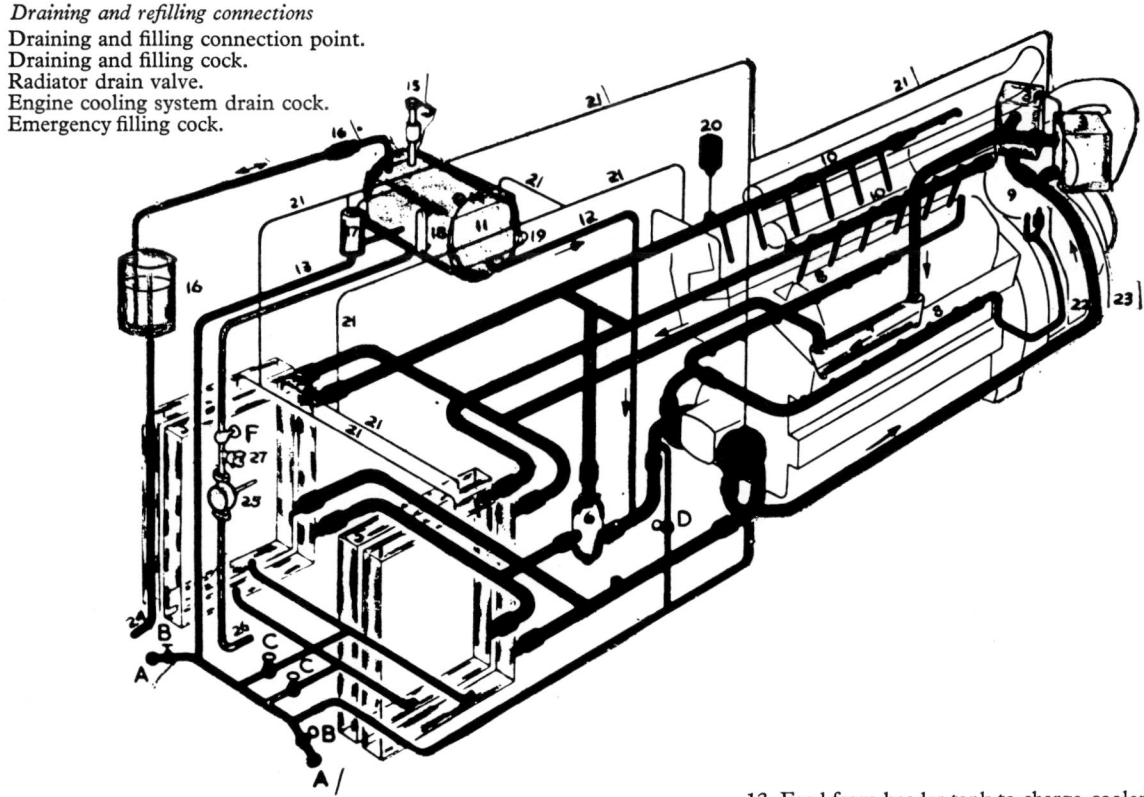

Draining and refilling connections
A Draining and filling connection point.
B Draining and filling cock.
C Radiator drain valve.
D Engine cooling system drain cock.
F Emergency filling cock.

Cooling-water system
1. Radiator for charge-coolers and lubricating-oil cooler–water system.
2. Water pump for charge-coolers and lubricating-oil cooler–water system.
3. Charge-cooler.
4. Lubricating-oil cooler.
5. Radiator for engine cooling-water system.
6. Thermostatic valve.
7. Water pump for engine cooling-water system.
8. Water inlet manifold to crankcase and pipe turbo-blower.
9. Turbo-blower (each side).
10. Water outlet manifold from cylinder heads and turbo-blower.
11. Header tank.
12. Feed from header tank to engine cooling system.
13. Feed from header tank to charge-cooler and lubricating-oil cooler system.
14. Anti-vacuum valve.
15. Water-treatment filler.
16. Pressurizing valve or recuperator.
17. Air separator.
18. Contents gauge.
19. Low-water level switch.
20. High-water temperature switch.
21. Vent pipes.
22. Turbo-blower drain (each side).
23. Condensate drain from charge-cooler.
24. Overflow pipe.
25. Emergency filling hand-pump.
26. Connection to locomotive boiler water tank.
27. Priming cup and cock.

Fig. 4.2. Typical divided cooling system

such cooling system which make the gas turbine concepts so attractive for traction, in spite of their high specific fuel consumption, especially at part load, and their high initial cost.

There is a system which may offer advantages, and that is the 'pressurized' 'sealed-for-life' system which is now common in road transport, incorporating an overflow tank and the ability to draw back the 'spilled' coolant as the temperature falls. This has the advantage of keeping air inhalation to a minimum, so reducing electro-chemical corrosion or erosion with or without vibration excitation. The authors have sketched in on Fig. 4.2 the position of such an overflow tank, should it be used, where the area of exposed air interface can be kept to a minimum.

TURBO-CHARGERS

It has been stated earlier that rail traction engines spend much of their time at idling conditions, and are required to accelerate quickly to full output after idling. Two critical factors apply here. The first is the condition where, at idling, the air manifold pressure is sub-atmospheric and the exhaust will be above atmospheric. This condition could cause exhaust gases to pulse in the wrong direction within the turbo-charger, causing shaft seals to 'carbon up', and the bearings to become contaminated. A method which is adopted on B.R. is to feed these seals with an external air supply to keep them clear under all operating conditions. However, it is surprising how much air is required for this purpose, and it has to be supplied from the locomotive compressed-air system.

The second factor is associated with the fact that although an engine governor can call for full fuel from the injection pumps almost instantly, turbo-chargers take a finite time to reach their maximum speed. This results in overfuelling and black smoke emission from the exhaust during this transition. To overcome this, a 'boost bias control' must be introduced to sense the pressure in the air manifold and limit the fuel accordingly. Another advantage of this control is that when multiple turbo-chargers are employed, it guards against piston seizure due to overfuelling should one turbo-charger fail.

The lubrication of turbo-chargers, too, is a critical factor of diesel engine operation. At present there are several schools of thought with regard to the types of bearing which ought to be used, and their method of lubrication. The question relates to ball and roller bearings versus plain bearings, and whether the turbo-charger should have its own sumps of oil or whether the oil should be fed from the main engine system. The authors themselves lean towards the use of plain bearings, but all they ask is that the overhaul life of the turbo-chargers should be such that it corresponds to the overhaul life of the engine itself. This criterion can be achieved only if the oil is maintained at the correct cleanliness. If the turbo-chargers are of the type having separate oil sumps, then it is imperative that the oil is changed at specified periods. On B.R. this is every 600–700 hours. If, of course, the oil is fed from the main engine oil supply, then it is imperative that it be suitably filtered. There is a case for and against each of these systems, but both appear to work equally well, or equally badly. When operating in a 'hostile' environment of dirt, coal dust, and even brake-block dust, it is possible for turbo-charger impellers to become contaminated with oily dirt, such that their efficiency is impaired with a resultant decrease in air flow, and possible overheating of the exhaust. The method adopted to overcome this shortcoming is to fit special nozzles to each turbo-charger and to spray a mixture of water and paraffin into the eye of the impeller at intervals of between 600 and 800 hours, whilst the engine is running. The effect of this washing is to reduce both smoke and exhaust temperature, in some cases almost in seconds. Of course, the condition of the impeller is also affected by the air-filtration system of the locomotive itself, and how well and how often the filtration system is being maintained.

EXHAUST SILENCING AND ENGINE NOISE

These have become critical factors in railway operation because of the noise nuisance in built-up areas. Both silence and cleanliness of the exhaust require some expenditure of power and fuel. It is well known that turbo-chargers have the facility to absorb some of the sound energy of exhaust pulses as well as the heat energy. In this they act as silencers, but the chargers may make some turbine noise themselves and this is likely to increase with specific power output. Experiments carried out on B.R. have indicated that with medium-speed engines up to about 800 rev/min the turbo-charger alone makes an effective silencer, and in some cases no other exhaust silencer is used. However, with higher speed engines it has been found that only the low-frequency pulses are absorbed by the turbo-charger, and a silencer is still required for the higher frequencies. Hence, unfortunately, exhaust silencers cannot be abandoned, with their attendant costs and back pressures, and have to be allotted space where space is at a premium, viz. near the top of the locomotive above the engine. This space could be better occupied by the radiators and cooling fans, which have been discussed earlier.

The presence of a diesel engine in a locomotive really presents two noise problems. The first is the effect of the mechanical noise of the engine itself, and in many cases extra sound-proof bulkheads have had to be introduced into already full locomotives to protect the driver. However, complete sound proofing is not practicable in a locomotive body, with engines in their present state of reliability. The second problem is related to the noise heard by the public, in stations, and during normal journeys, where the noise is less noticeable. Much experimental work has been carried out on engine noise emanating from the rate of pressure rise in the cylinders and also from the structure and components of the engine itself. Such work is covered in the Proceedings of an Instn Mech. Engrs Symposium on Noise and Noise

Fig. 4.3. Measurement of exhaust noise

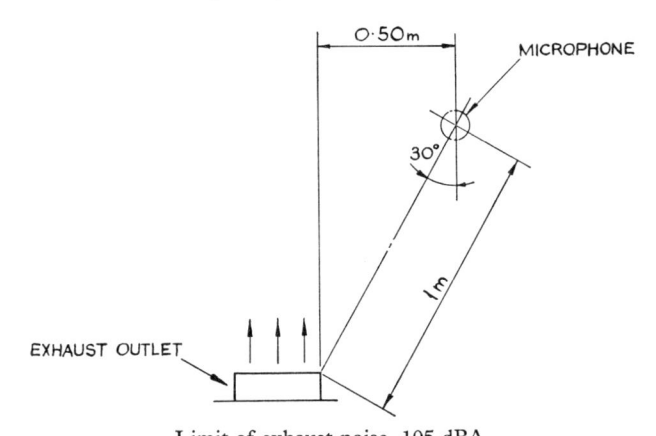

Limit of exhaust noise, 105 dBA.

Suppression, which took place in London in 1958, especially in the paper by Austen and Priede (**1**)*. In addition, some further work has been carried out by Priede (**2**), and Priede, Austen and Grover (**3**).

The authors look forward to some constructive advice being given to educate engine manufacturers so that we, on B.R., may benefit from future designs where reduction of noise has been tackled at the design stage.

In an attempt to assess exhaust noise, there is a standard test using a microphone at a set position which is adopted by most European railways, with values of maximum acceptable noise. This is shown on Fig. 4.3, where the acceptable maximum overall noise level is given as 105 dBA. However, having chosen the diesel engine for

* *References are given in Appendix 4.1.*

one of the criteria mentioned earlier, and with the general situation with space limitation, the problems of exhaust silencing and engine silencing are difficult ones to solve. For the solving of these problems, we lean heavily on the exhaust silencer manufacturers; but in some cases their reaction is to say that little can be done within the parameters which we have allowed them. Here then, too, is a critical factor which has to become a critical compromise.

RATING OF DIESEL ENGINES

The rating by manufacturers of diesel engines for traction purposes is usually higher than the same engine rating for industrial purposes. This is owing to the fact that manufacturers still consider traction as an 'intermittent' operation, related to the time when most traction diesel engines were unsupercharged and built with uncooled pistons, and hence operated at very much lower thermal loading. There is room for some discussion here, as in the opinion of the authors the intermittent operation of traction in this country is more than offset by the thermal cycling of today's turbo-charged engine of high specific power, occasioned by the on–off operation in practice. The normal operation of diesel locomotives on the fast passenger services comprises about 45 per cent of the time on full power and 45 per cent of the time with the engine idling, the remaining 10 per cent being used on all other conditions of power output, acceleration, deceleration, etc. For typical cross-country freight services, the distribution of power is approximately 25 per cent on full power and 65 per cent with the engine idling. A diagram showing this on–off operation of a diesel engine is shown on Fig. 4.4, which is an actual recording taken on a journey from

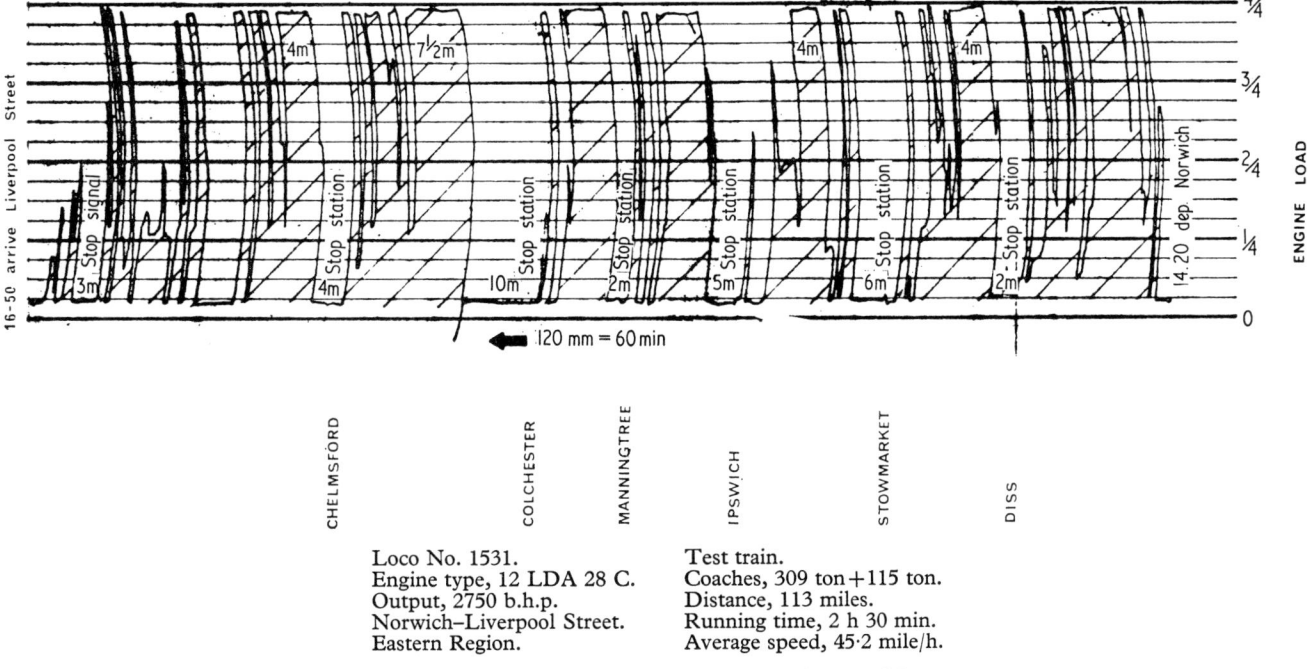

Loco No. 1531.
Engine type, 12 LDA 28 C.
Output, 2750 b.h.p.
Norwich–Liverpool Street.
Eastern Region.

Test train.
Coaches, 309 ton + 115 ton.
Distance, 113 miles.
Running time, 2 h 30 min.
Average speed, 45·2 mile/h.

Fig. 4.4. Typical load cycle, Norwich to Liverpool Street

Norwich to Liverpool Street. The rates of change of power are fast and occur at approximately once every 5 min, giving about six complete cycles per hour, except on one or two specialized places, such as long continuous banks, where it is possible that the locomotive would require full power continuously for up to, perhaps, half an hour. It is the sudden reduction of power rather than the sudden increase of power which is the main problem in thermal cycling, and it is this sudden cooling which subjects the engine to the highest thermal load. If these observations are compared with road transport, where, except for motorway driving, rapid changes of power occur at about 30 cycles/hour, it may be understood why engine manufacturers are tempted to give engines a maximum transport rating higher than that which can be continuously sustained. The authors, then, make a plea that the traction rating of a diesel engine shall be that power which can be continuously subjected to this on–off condition without damage to the engine.

The problem arises when a matter of economics is introduced. Engine manufacturers are reluctant to reduce the power of their engines on any pretext, as the cost per horsepower would rise accordingly. Conversely, B.R. are looking for engines with the highest power that they can obtain with the lowest specific weight, and it is probable, therefore, that the rating of a diesel engine should be related to the operational characteristics or special features of the service for which this engine will be used.

One can possibly foresee in the future the situation with diesel engines of greater power than has been used hitherto, when the maximum horsepower will be used only for a short period in any one journey, but nevertheless has to be available for this short period. Under these circumstances there may be a case for introducing a 'sprint' rating for diesel engines for rail traction in a similar manner to that which was used during the war for military engines in combat aircraft, tanks, and ships. In this manner, the overhaul life of the engine would have to be related to the length of time for which each particular condition of running was used.

OVERHAUL LIFE OF DIESEL ENGINES

The overhaul life of a diesel engine can be measured either by the miles covered by the vehicle in which it is installed, or by the hours run by the engine itself—and it is difficult to choose the right criterion. For example, the same type of locomotive on B.R. today is used either in high-speed passenger service or low-speed freight service. The authors feel that the time between overhauls is dependent, partly, on the total work done, which could be measured in Board of Trade units or kilowatt hours. It is also suggested by the authors that, in order to cover the variation in duty, the overhaul period of an engine could be a function of the total fuel consumed, which, albeit roughly, is related to either ton miles or passenger miles. The determination of the right length of time between overhauls is a problem for the users and not the manufacturers to resolve.

It has been found that top overhauls require to be carried out at about half overhaul life. We are looking forward to a time when all engines will run for, say, 6000 hours before any attention is given, e.g. cylinder heads off for attention to valves; a further 6000 hours before similar attention plus attention to piston rings, etc., is scheduled; and finally, at 24 000 hours a full overhaul. This system is working today for at least some of our medium-speed engines.

EASE OF MAINTENANCE

The reliability of diesel engines on B.R. is measured by the customer by the number of times a delay or a failure occurs on the line, and it is difficult to evaluate whether the cause of such failures is in the design, manufacture, or maintenance. Therefore it is essential from the inception that a diesel engine, together with its installation, must be designed in a manner that will permit easy and simple maintenance. Day-to-day maintenance should be kept to a minimum, as ideally all that is required is to ensure that the fuel, coolant, and lubricating-oil systems are full. In practice, periodic maintenance has to be carried out on a locomotive in a most confined space and in a very hot atmosphere. If the maintenance schedules are difficult to carry out, then in many cases they are not done satisfactorily. For example, with some of the larger diesel engines with a turbo-charger in each corner, it is necessary to fill two sumps with oil on each turbo-charger. Eight sumps to fill or oil to change with drain plugs and fillers in inaccessible places giving rise to eight possibilities of oil spillage. The authors appreciate that locomotive design also plays its part in ease of maintenance, but attention to maintenance detail in the design stage is absolutely essential.

RELIABILITY

It is pertinent to note that defects of diesel engine components themselves amount to only a small part of total locomotive defects which cause breakdowns in service, amounting to only about 8·5 per cent. However, when all the associated engine systems are taken into account this proportion rises to about 35 per cent.

The remainder of defects are due to other components which make up the complicated assembly which is a modern locomotive. However, the authors find it surprising that many 'tried and tested' diesel engine components fail in railway service as soon as they are used under conditions which are only slightly different from those at which the components were 'tried and tested'. These new conditions may not necessarily be more arduous (in theory), but too many failures still occur. Piston rings wear at twice the expected rate, cylinder heads crack, turbo-blower impellers fracture, pistons seize, cylinder blocks, and crankcases fail, etc., etc. These are some of the areas of unreliability which are found during overhaul which may not, in themselves, cause any apparent breakdown in the train service. The authors ask whether these failures are due to changes of manufacturing techniques, insufficient inspection, or just a case of engine manufacturers resting on their laurels.

CONCLUSIONS

The most critical factors in the application of diesel engines to rail traction appear to be in their installation rather than within the engines themselves, the largest installation problem being associated with the cooling system.

Cooling systems

An unpressurized cooling system is the factor which limits the power which can be installed in a diesel locomotive within B.R. loading gauge. This could be overcome by engine manufacturers allowing their engines to run at higher temperatures by the use of pressurizing the cooling systems. The cooling system gives the greatest trouble with regard to reliability of diesel locomotives, and the trouble is usually associated not with highly sophisticated equipment but the simple problem of leaks.

Turbo-chargers

A turbo-charger on a diesel engine presents problems as well as advantages. Water-washing has been shown to increase the overhaul life of a turbo-charger and helps to maintain its performance for longer periods. It is essential, with highly turbo-charged engines, to fit a 'boost bias control' to relate fuel quantity to air quantity when the turbo-charger accelerates. The problem of the type of bearing to be used and its lubrication appears to be unsolved at this moment.

Silencing

There are two problems related to the silencing of locomotives. One is associated with mechanical noise and the other with the exhaust noise. Both sources of noise present problems; the mechanical noise can be alleviated at the design stage, whereas the exhaust noise must be tackled after the engine is manufactured and installed.

Rating of diesel engines

The cyclic operation of modern rail traction engines suggests that a new look should be taken at methods of engine rating for traction purposes, when compared with industrial ratings, both of which must be related to long-life reliability. The problems here may be associated with the economics of manufacture and sales philosophy.

Reliability

The remaining sub-headings of the paper can be summed up under this one heading of reliability. Every effort must be made to improve this factor, and under no circumstances must any design change or installation change be made to a diesel engine which will sacrifice reliability. The most critical factor in the application of diesel engines to rail traction is their ability to withstand the effect of rapid power cycling, especially with highly rated turbo-charged and charge-cooled engines under the hostile traction environment.

ACKNOWLEDGEMENTS

The authors would like to thank Mr T. C. B. Miller, Chief Mechanical and Electrical Engineer, British Railways Board, for his permission to publish this paper, their colleagues on British Railways, and the engine manufacturers who have contributed information to enable the paper to be written.

APPENDIX 4.1

REFERENCES

(1) AUSTEN, A. E. W. and PRIEDE, T. 'Origins of diesel engine noise', *Symp. Engine Noise and Noise Suppression*, 1958, 19 (Instn Mech. Engrs, London).

(2) PRIEDE, T. 'Relation between form of cylinder-pressure diagrams and noise in diesel engines', *Proc. Auto. Div. Instn mech. Engrs* 1960–61, 63.

(3) PRIEDE, T., AUSTEN, A. E. W. and GROVER, E. C. 'Effect of engine structure on noise of diesel engines', *Proc. Auto. Div. Instn mech. Engrs* 1964–65 **179** (Pt 2A), 113.

Paper 5

APPLICATION OF DIESEL ENGINES IN THE ROYAL NAVY

W. H. Sampson*

This paper deals with the application and selection of diesel engines for marine propulsion and auxiliary purposes with particular reference to the special requirements for naval applications. Details are given of the various means that have been developed for dealing with airborne and structure-borne noise, for reducing the effects of external shock, and for checking the quality of lubricating oils during service. Automatic control and protection equipments, including fluidic, are described and the paper finally deals with the means adopted to improve reliability and to reduce maintenance.

INTRODUCTION

THE ADMIRALTY STANDARD RANGE of engines was introduced in 1950. It consisted of five different engine designs covering a range of powers from 3 to 2000 b.h.p. In selecting engines, the following general requirements were considered:

    Reliability
    High power/weight ratio
    Small space, including height
    Ease of operation
    Ease of manufacture
    Ease of maintenance
    Resistance to shock

The largest engine, the ASR1, was designed by the Ministry of Defence (Navy) and was built by H.M. Dockyard, Chatham, and by five commercial firms. The remaining engines were of commercial design, modified as necessary to meet naval requirements.

The Standard Range of engines has been in use for about 20 years and nearly 4000 engines have been completed. When the choice of engine is restricted in this way, and if it is rigidly applied for too long, there is a danger of stagnation in design. Changes have therefore been made as the necessity has arisen. The present list contains six power ranges with a maximum brake horsepower of 8000, and of the original engines selected in 1950 only the ASR1 is included, the others all having been replaced either by improved models of the original engine or by new engines of improved performance. The development of the ASR1 has been such that its brake mean effective pressure in service has been increased from 132 lbf/in^2 to nearly 200 lbf/in^2 and running at the Admiralty Engineering Laboratory (A.E.L.) to 280 lbf/in^2. The advantages of having a restricted range of engines are:

(1) One can be selective.
(2) Logistic support in the form of spare gear and technical documentation is reduced to a minimum.
(3) A policy of refit by replacement of the smaller engines can be introduced with a minimum of expenditure on spare engines.
(4) A modification procedure can be introduced giving the user complete control over the introduction of modifications which may affect interchangeability of complete engines or spare gear.
(5) The requirements for the training of operators and maintainers are reduced to a minimum.
(6) Preparation, issue, and modification of Instruction Books, Spare Gear Identification Lists, etc., are kept to a minimum.
(7) Line overhaul of engines, which reduces overall costs to a minimum, is practical.

TYPE TESTING OF ENGINES

After selection an engine undergoes a type test procedure either at the manufacturer's works or, more usually, at the Admiralty Engineering Laboratory. The test is in two parts. During Part I the engine is run at four or five constant speed loops up to the limit of a number of parameters including fuel consumption, maximum cylinder

---

The MS. of this paper was received at the Institution on 5th February 1970 and accepted for publication on 12th March 1970.
Ship Dept, Ministry of Defence (Navy), Foxhill, Bath, Somerset.

pressure, exhaust temperature and shade, turbo supercharger speed, vibration, noise level, etc., after which the Admiralty Test Rating (ATR) is determined. At the test rating the engine is run for 72 h at 95 per cent rating, 12 h at 100 per cent, ending with 2 h at 110 per cent. This short endurance run is carried out to confirm the results obtained from the loops and to check their repeatability over a reasonable time. It also confirms that the ATR is reasonable and that lubricating oil consumption is acceptable.

Part II is carried out after satisfactory completion of Part I. This consists of an endurance run on a test cycle and because wear rate of components is measured, the engine is stripped and gauged before the test. The test cycle varies with the proposed service application but is normally not less than 1000 h.

Special tests such as cold starting and shock testing, if within the capabilities of existing facilities at A.E.L., are also carried out.

On completion the engine is stripped for examination and gauging.

STRUCTURE-BORNE NOISE AND SHOCK

The introduction of sensitive listening devices, acoustic mines, and homing torpedoes has necessitated a great deal of research into the reduction of noise from machinery in warships. Another problem that arose during the 1939–45 war, and which must be considered, is underwater shock. An explosion in the sea at a distance from a ship may not rupture the hull, but the shock wave can distort the hull sufficiently to cause damage to the machinery and equipment inside.

The vibration levels measured on a diesel engine are probably among the highest that are observed on warship machinery, and all engines must be isolated from the ship's structure by vibration mountings. Unfortunately, vibration mountings on their own are effective only against minor shock forces and will fail or even be amplified under the action of a severe shock, unless additional shock protection is provided. The Royal Navy therefore uses a composite system of shock and vibration mountings beneath diesel engines.

The mountings are designed to function as vibration

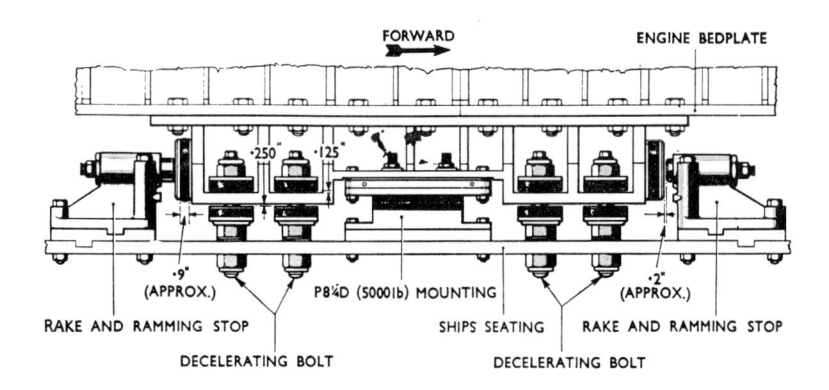

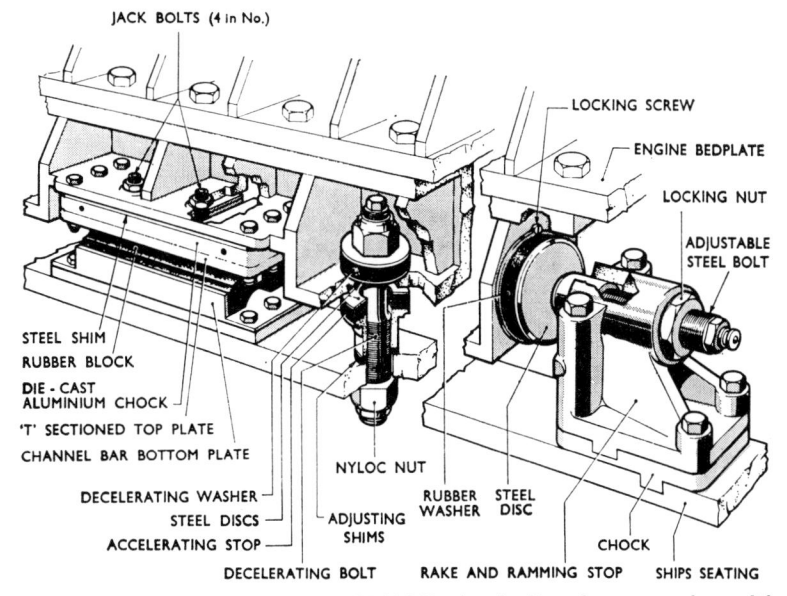

Fig. 5.1. Typical arrangement of 5000 lb shock vibration mounting with accelerating and decelerating washers

mountings over specified frequency ranges and to provide a measure of shock protection for the mounted engine. Until very recently they were invariably of rubber and allowed a considerable relative movement between engine and the ship's seating under shock conditions. Additional shock mountings must be installed to limit this movement and also to limit movement under engine starting and stopping, ship rolling and pitching, rake and ramming. Because the additional shock mountings are too hard to isolate the ship's structure from the vibrations of an engine, they must not be in contact under normal running conditions. A typical arrangement using 5000-lb mountings fitted under an ASR1 16-cylinder engine (38 000 lb) is shown in Fig. 5.1. Fig. 5.2 shows the arrangement of a 600-lb mounting used under smaller engines, and which has the shock mountings built in. Other standard mountings, varying in capacity down to 100 lb, are available.

To obtain the optimum effect from a mounting system it is essential that the mountings should be loaded to their designed load within a tolerance of +5 to −15 per cent, and if this is to be obtained in a long diesel generating set, an accurate assessment of the weight and position of the centre of gravity is required.

A recent development which, at the time of writing, was undergoing evaluation beneath a 12A0 Ruston engine is the Constant Position Mounting System (CPMS) developed by the Yarrow Admiralty Research Department (Y-ARD). The CPMS is a mounting system which has very sensitive position control of the mounted item at low frequencies and exceptionally high damping at the natural frequency. It is capable of limiting vertical relative movement to ±0·050 in under a 10° roll or 10° pitch condition, which makes the system very attractive for propulsion engines. The mount itself (see Fig. 5.3) is an air mount,

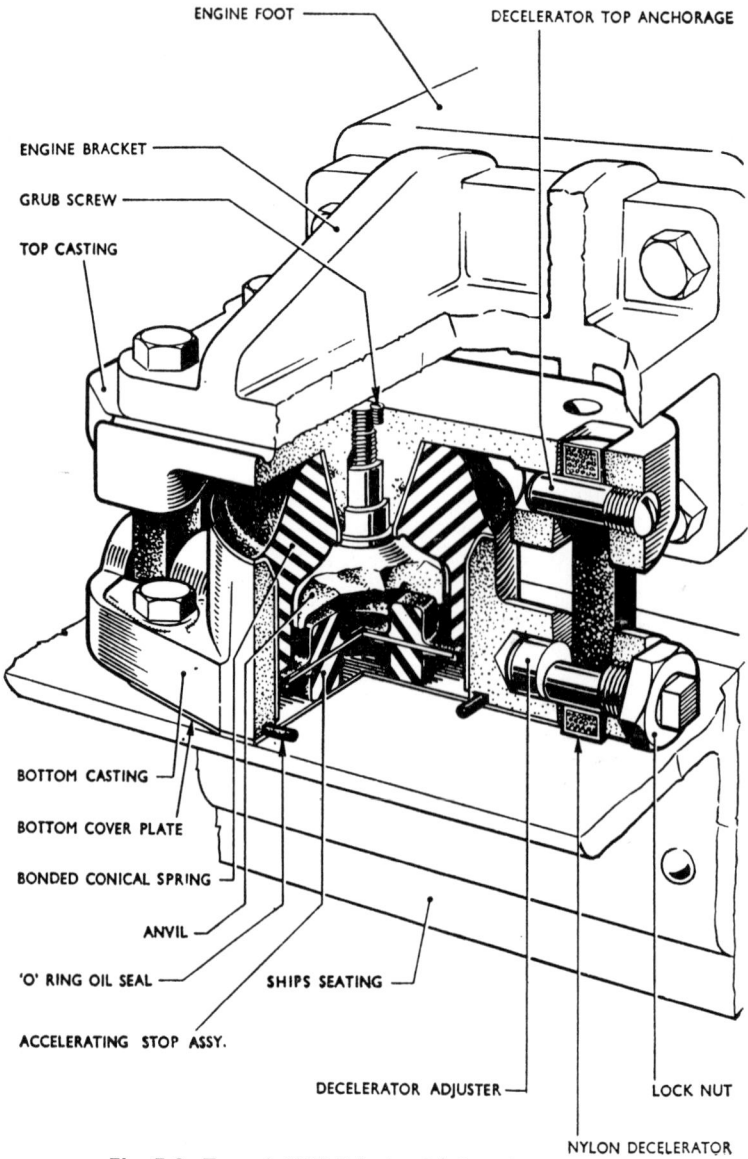

Fig. 5.2. Type S (600-lb) shock/vibration mounting

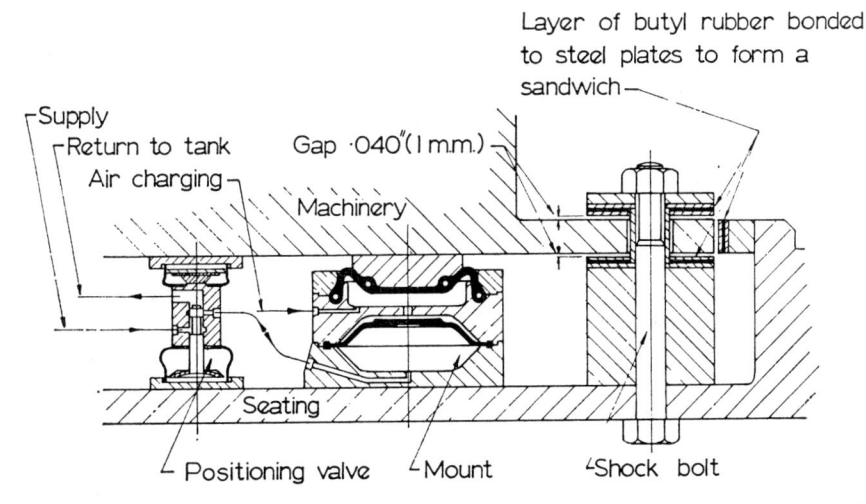

Fig. 5.3. Diagrammatic section through CPMS positioning valve, mount, and shock stop

the air being sealed by a fabric-backed rubber diaphragm. Position control valves sense the position of the mounted item relative to its seating, varying the pressure in the mount such that the force applied by the mount to the mounted item automatically counteracts any tendency to depart from the design relative position. Both hydraulic and pneumatic control systems have been developed. Because first initial cost and maintenance of the CPMS are higher than those of the conventional rubber mountings, it is very unlikely that this CPMS will be used under generating sets.

The arrangements shown in Figs 5.1 and 5.2 provide very little shock protection to the engine. Transmission of shock inwards to an engine may be reduced by fitting a mounting between the engine and the ship's structure that is capable of storing the energy generated by the movement of the structure and releasing it at a relatively slow rate, absorbing as much as possible in the process as internal hysteresis or in the deformation of metal parts.

A great deal of industrial research has been carried out into protection against shock, particularly by the packaging industry where flexible materials are used. Very little of this work is suitable for naval application because the accelerations resulting from the external shock waves are usually far too high ($60g$ minimum) to be wholly absorbed by a flexible material. The only alternative is to build the shock resistance into the diesel engine. The larger diesel engines are usually installed on or near the bottom of a ship where the shock intensity is greatest, and invariably they must have fabricated or cast steel structures. The lighter engines are usually fitted much higher up with a fair amount of ship's structure between them and the outer bottom, and as they are therefore less likely to be subject to high shock loadings, cast aluminium and possibly nodular cast iron structures could be acceptable. Governors, pumps, superchargers, etc., mounted off the main engine structure must be supported by brackets made of a ductile material.

A flexibly mounted piece of equipment cannot rely on the rigidity of the ship's seating to provide the longitudinal rigidity necessary to maintain alignment between important parts of the equipment, and a rigid underbase is therefore essential. The design of this beam must take into account the variations in shock loading throughout the length of the equipment. The provision of the underbase adds to the weight of the equipment and, unless it is very stiff, a flexible coupling must be fitted between the engine and its generator in large generating sets, thus adding further to the length and weight. A logical step forward with large generating sets is to flange-mount the generator to the engine, thus giving a short, very stiff unit of minimum weight. This has been recently employed for three standard generating sets of 500, 700, and 1000 kW capacity (Fig. 5.4).

When an item is installed on flexible mountings, any pipe, shaft, or cable connection to it must be both flexible and a noise insulator. This is not a problem with self-contained equipments, but can be troublesome with a flexibly mounted engine and rigidly mounted gearbox; e.g. with the mounting system shown in Fig. 5.1 the engine can move about 0·5 in under full shock with a 50 per cent possibility that it will become a permanent misalignment. Also, under starting conditions or with the vessel rolling, the amplitude of movement of the top of the engine is about ±2 in. Finally, because rubber slowly deteriorates in use, engines are usually set up above the correct alignment position and the rubber mountings are not changed until the engine has dropped to about 0·125 in below. This means that any connection between the engine and the gearbox must be in the form of a cardan shaft flexibly connected at each end. If the cardan shaft is a flexibly mounted hydraulic coupling, the coupling acts as a dead weight on the end of the crankcase, and problems arise

Fig. 5.4. Standard 1000 kW generating set

with excessive crankweb deflections which cannot be corrected by the normal method of adjusting the height of the engine relative to that of the gearbox. For this reason, static crankweb deflections up to 0·0035 in have had to be accepted on a 9·75 in bore × 10·5 in stroke engine.

The alignment of a flexibly mounted engine can only be checked with the engine in the running condition. This can be a problem if a heavy coupling is fitted between the engine and gearbox. The only satisfactory way is to align the base engine on solid chocks and measure the height between engine and seating at a number of positions. Setting the engine in the running condition to these heights will ensure correct alignment.

AIRBORNE NOISE

Much greater emphasis has recently been given to reducing airborne noise from diesel engines. In a steam driven warship it is logical to use steam turbine driven generators for base load generating duties, using a small number of diesel driven generators for emergency and salvage duties. With the advent of the gas turbine driven ship, as the gas turbine in the generator role is not economical in fuel (particularly at part loads), diesel generators have been selected for base load duties.

The noise level in main machinery spaces adjacent to diesel main engines or high-performance diesel generators invariably exceeds NR85, the generally accepted aural damage limit for continuous exposure (1)*. Octave levels measured on board one of H.M. ships in the compartment containing a Paxman Ventura 16YJCAZ resiliently mounted diesel generator at full power are shown in Fig. 5.5.

Since these high noise levels originate from vibration of the engine surfaces, it is not surprising that this poses particular problems in naval applications, where both airborne noise and seating vibrations are transmitted through the hull into the sea.

Various means have been investigated to reduce the intensity of noise reaching a ship's structure. At first, attempts were made to deal with the various noise

* References are given in Appendix 5.1.

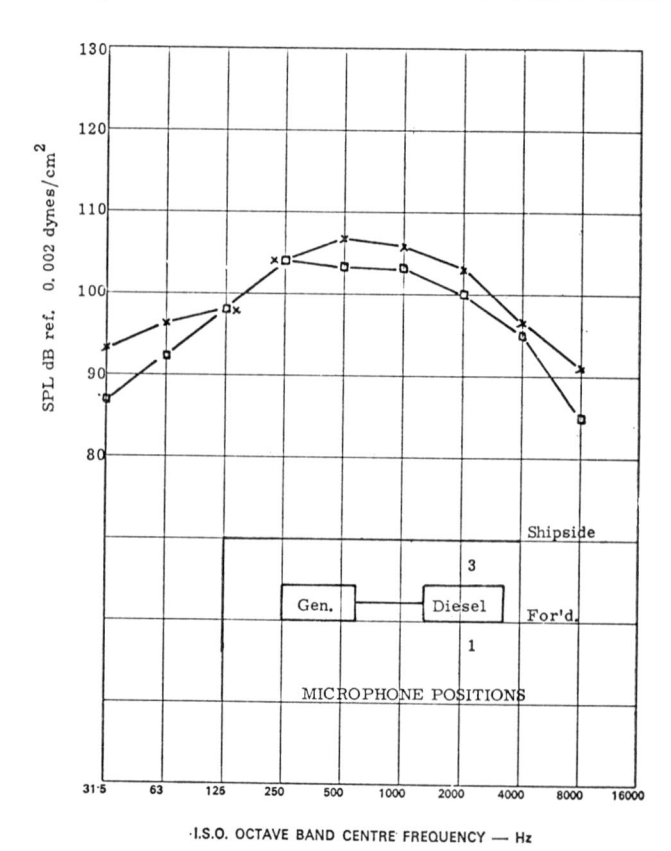

Fig. 5.5. Typical airborne noise levels adjacent to a 1250 kVA diesel generator at full power

emissions at the source. This has been investigated (2) but requires an engine to be designed and manufactured as a 'special', and with a range of engines would require an enormous amount of development. The alternative, which has proved to be easier from a development point of view, is to contain the noise by completely enclosing the engine in such a manner that the sounds radiating from it are attenuated. Two types of acoustic envelope have been developed:

(1) An acoustic enclosure, which is isolated from the engine and is wide enough to permit routine maintenance being carried out inside it.

(2) Acoustic cladding, which is in close contact with the engine.

The vibration attenuation advantages of a well-designed compound mounting system over a conventional single-stage system are well known, the main disadvantage being the weight of the intermediate subframe required for effective isolation. However, when combined with an acoustic enclosure, the intermediate subframe conveniently forms the base of the enclosure, which is sufficiently isolated from the engine by the primary mounts. Fig. 5.6 shows the outline of a 1·5 MW diesel generator module under development by the Navy. This 'modular' approach has the advantage that the set and all auxiliaries are integrated into a 'package unit' with clearly defined boundaries, inspection and maintenance spaces, replacement routes, and built-in top-overhaul facilities.

Acoustic cladding has been developed for Deltic (Fig. 5.7), Foden, and Ventura engines. A number of these installations are being evaluated at present. Acoustically, it is very successful in that reductions of up to 30 dB have been measured, it does not take up much space, and is fairly light, e.g. 2000 lb for the Deltic. From an engineering point of view, however, the first sets of cladding were made in too many pieces and insufficient thought was given to the design of those portions which had to be portable for routine servicing and maintenance. A modified design is under development.

At present it is considered that the acoustic cladding method will be retained for the smaller engines and the

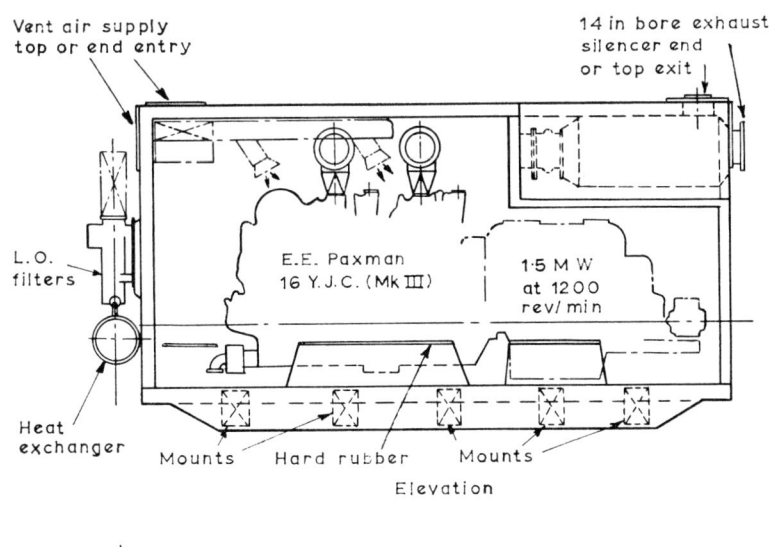

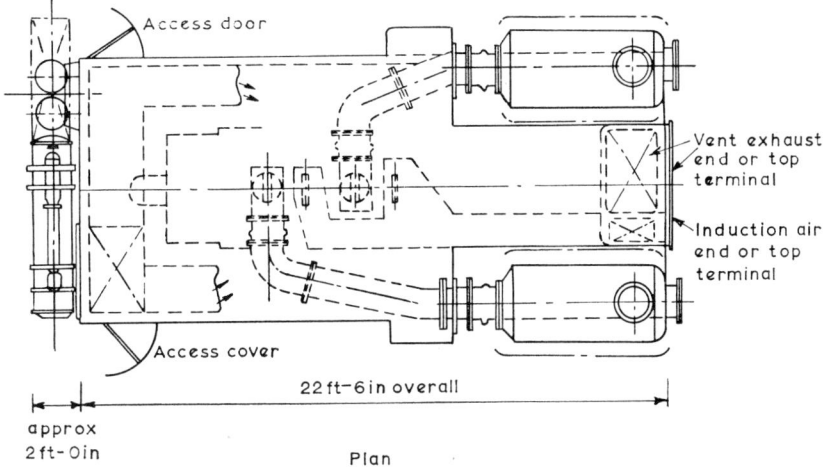

Fig. 5.6. Outline of a 1·5 MW diesel generator module

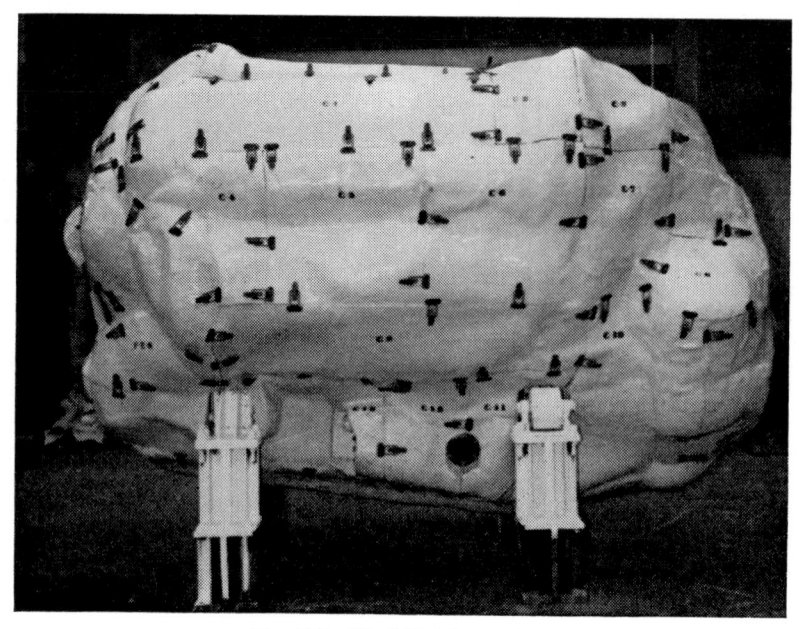

Fig. 5.7. Clad Deltic engine

Deltic, which are refit by replacement engines, and the acoustic enclosure will be used where space permits for the larger engines (Ventura, etc.) which require top and possibly major overhauls *in situ*.

LUBRICATING OIL

The Royal Navy uses oils produced to comply with its own specifications for all diesel engines. The oil in general use is OMD-112, but an improved oil OMD-113, was introduced several years ago to cover highly rated engines. The specification for OMD-113 has recently been revised with the object of improving its availability and it is intended that by the end of 1970 OMD-113 will be used for all engines.

Both OMD-112 and OMD-113 are SAE 30 grade oils and the difference is in the quantity and types of additives needed to comply with the specification performance requirements. The higher quality oil (OMD-113), while not being necessary for some engines, nevertheless gives longer periods between oil changes and the logistic advantages of using only one oil in the Navy are obvious.

Under very cold conditions, for boat and other small engines likely to be exposed to the weather, the Army SAE 10 engine oil OMD-40 is used. Frequency of oil changes would be increased if necessary under these conditions.

It is the Navy's policy not to specify periods for oil change. A test kit has been developed by the Admiralty Oil Laboratory (A.O.L.) which enables operators to determine, by three simple tests, the condition of the lubricating oil in any engine. The test kit contains apparatus for estimating the following properties of the oil:

(1) Development of acidity.
(2) The quantity of free carbon present.
(3) Viscosity changes (fuel dilution, oxidation, presence of carbon).

For some years the limit for item (2) has been set at 1·3 per cent, but recently it has been increased to 3 per cent. Fuel dilution is limited to a 5 per cent maximum.

Additionally, oil can be tested for water contamination using the Speedy Moisture Test Kit which was originally developed for determining the moisture content of foundry core sands.

OIL-CUSHIONED PISTON

The oil-cushioned piston was developed by the Admiralty Research Laboratory (A.R.L.) during a general investigation into the noise produced by the various components in a diesel engine. Fig. 5.8 shows the difference between the standard piston and its oil-cushioned counterpart for the Admiralty Standard Range 1 engine (ASR1). It will be seen that the standard piston has a conventional ring complement of three gas sealing rings and two slotted oil control rings. The oil-cushioned version retains the three gas sealing rings and the upper slotted oil control ring, but has an extra taper-faced down-passing ring between the slotted scraper and the gudgeon pin bore. At the lower end of the skirt, the normal slotted scraper ring is replaced by an up-passing taper-faced ring.

The action of the piston is such that, provided sufficient oil is deposited on the cylinder liner wall after a short period of running, the piston skirt is surrounded by a full oil belt, and this oil acts as a damping medium to stop the piston moving from side to side and to keep it parallel with the axis of the bore. The time taken to fill the annulus and the usual hollow gudgeon pin is from 5 to 15 min, depending on size of engine and speed, but experience has shown that once the annulus is filled, the oil will stay

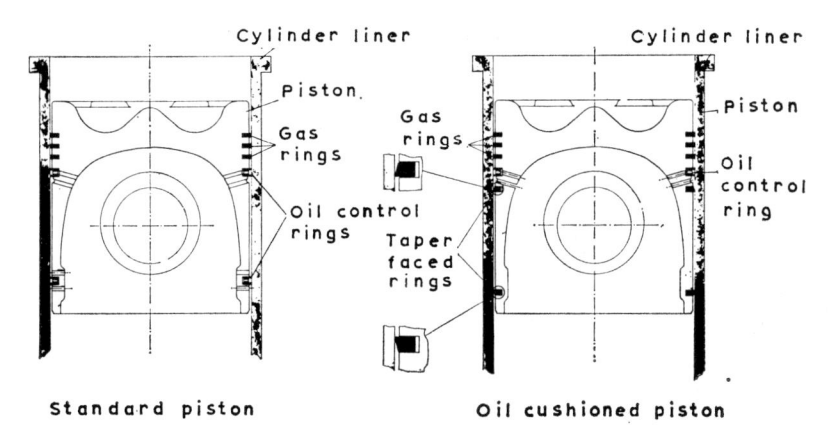

Fig. 5.8. Comparison of standard and oil-cushioned pistons

around the piston for quite long periods after shutdown. It has been found that the oil throw from the big-end bearing is usually not quite sufficient for this purpose. In the ASR1 engine, a $\frac{3}{32}$-in diameter hole drilled in the top of the connecting rod to communicate with the pressure-fed small-end bearing was found to be completely effective.

The oil-cushioned piston was originally developed to counter the effect of waterside attack, but in addition to improving piston lateral control and reducing cylinder liner vibration it also has many other beneficial effects in an engine, namely:

(1) Waterside cavitation of the cylinder liners can be avoided, accompanied by minimal wear on pistons, rings and ring grooves, and liner bores.
(2) Transverse vibratory stresses in the connecting rod, which arise because of piston impacts, are significantly reduced and should lead to a reduction of fretting corrosion sometimes found on the backs of big-end bearing shells.
(3) Gas blow-by is reduced; reductions of 20:1 in volume on a given engine have been measured.
(4) Lubricating oil life is improved.
(5) Oil control is not a problem and may be varied as with standard pistons. Oil consumption of under 1 per cent of the fuel consumption can be obtained without risk of piston seizure.

Many detailed tests and experiments have been carried out, including comparisons between oil-cushioned and standard pistons in the same engine. Fig. 5.9 shows liner vibration records obtained in such a comparison. Each is a double trace record; the upper trace shows crank angle, whilst the lower one shows liner vibration in terms of velocity, using variable reluctance probes inserted through the water jacket.

In the case of the standard piston, the highest peaks of vibration occur just after t.d.c. on the working stroke and during the upward exhaust stroke. With the cushioned piston, the high peaks have disappeared, but there remains a high-frequency residue around the t.d.c. working stroke position. This residue has been found to be water jacket vibration due to combustion, and not that of the liner itself. The probes measure relative movement between the jacket and liner; a small high-frequency vibration is seen on both records at the position of valve closure impact (especially that of the inlet valve on the compression stroke). These impact forces are transmitted via the engine frame.

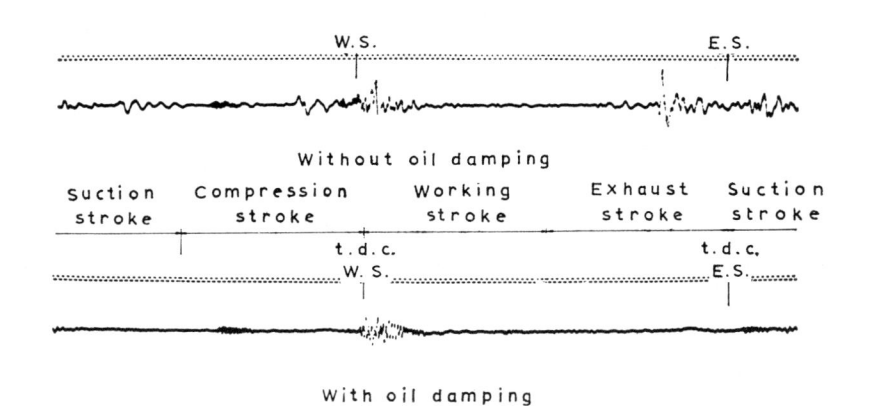

Fig. 5.9. Cylinder liner vibration (ASR1 engine—920 rev/min, 40 b.m.e.p.)

Several thousand hours' development running have been built up in A.E.L., and more by engines in ships and locomotives.

CRANKSHAFT DEVELOPMENT

Problems associated with heavy wear on the ASR1-16 VMS crankshafts have focused attention on the possibility of obtaining surface hardened shafts. Three alternative processes have been investigated, viz.:

(1) Chromium plating.
(2) Induction hardening.
(3) Nitriding.

Using the two rigs described by Barton (**3**), a programme of testing chromium plated and induction hardened fatigue specimens has been in progress at A.E.L.

Four bending test specimens with plating thicknesses of (a) 0·005 in, (b) 0·010 in, (c) 0·020 in, and (d) 0·025 in were first tested and it was found that the fatigue strength tended to fall off with increasing thicknesses of chromium.

It has been estimated that the fatigue stresses at the point of cracking for each of these specimens were (a) $\pm 21 \cdot 4$ ton/in^2, (b) $\pm 18 \cdot 1$ ton/in^2, (c) ± 11 ton/in^2, and (d) $\pm 9 \cdot 7$ ton/in^2. Too much should not be read into the results of such a limited number of tests, especially when subsequent examination revealed that specimens (c) and (d) had numerous weld repair spots where excessive hydrogen evolution during plating had affected the deposition of chromium.

Later bending fatigue tests confirmed that 0·005 in chromium plating of crankpins has no detectable effect on the fatigue strength of the crankshaft, provided the plating is kept clear of the maximum stress points in the crankpin fillets. Torsional fatigue testing commenced in November 1969 on 0·005 in thick chromium plated specimens, and early indications are that this thickness of plating has no adverse effect.

Facilities have recently become available for induction hardening and nitriding of crankshafts of the ASR1 size. The material used for the existing shafts is suitable for induction hardening, and a number of specimens cut from damaged shafts have been treated and will be tested when the chromium plating tests are completed. A nitrided shaft in En 29 has also been ordered.

Subject to satisfactory test results, existing shafts will be removed at a major overhaul and reclaimed either by induction hardening or plating with 0·005 in chromium. New shafts will be nitrided.

ENGINE COOLING SYSTEMS

In all liquid-cooled engines naval practice is to use distilled water containing 20 per cent inhibited ethylene glycol to B.S.S. 3150/59 in arctic, temperate, and tropical waters.

Extensive investigations have been carried out at A.E.L. to find an alternative inhibitor that could be used in distilled water in temperate and tropical waters—and which would be compatible with the inhibited ethylene glycol without requiring extensive flushing when a ship is required at short notice to operate in arctic waters. No alternative has yet been found with the satisfactory protective qualities and latitude in mixing that have been obtained with glycol. Soluble oil inhibitors appear to have satisfactory protective qualities with the wide variety of metals used in the cooling systems of the various types of engines, but their concentration must be controlled to close limits if heat transfer is not to be affected, and this involves methods of testing which the author considers to be unsuitable for shipboard use. Also, service tests on 16 VMS engines have shown a tendency for soluble oil inhibitors to separate out after as little as 50 hours' running. This is being investigated.

ENGINE PROTECTION EQUIPMENT

Engine protective equipment or Automatic Watchkeepers are fitted to all diesel generating sets so that under normal peace-time conditions, human surveillance is limited to periodic visiting. Two types have been developed—(a) electrical, (b) fluidic (**4**)—both of which monitor lubricating oil pressure and temperature, primary coolant temperature (and pressure if required), and generator air temperature.

Should the temperature rise or the pressure fall to an undesirable value, a warning light and audible alarm are actuated. If the condition causing the warning worsens, shutdown and circuit breaker trips are operated. The fluidic equipment also incorporates a remote start sequence.

Both equipments are applicable to propulsion engines without auto shutdown facility. Shutdown must be left to the discretion of the watchkeeper.

MAINTENANCE

Maintenance work on naval diesel engines is carried out at specified intervals depending on the type of engine; i.e. a boat engine of 60–100 b.h.p. will have a top overhaul at 1000 h and a major at 2000 h. An engine of 1500 b.h.p. would have a top overhaul at 5000 h and a major at 10 000 h. Injector, supercharger changes, and other usual maintenance routines are also scheduled. All engines above 350 b.h.p. are, or will be, fitted with running-hour meters to provide accurate records. Development is in hand on a partial load meter which will show the number of hours run in the 0–25, 25–50, 50–75, and 75–100 per cent load ranges. The use of this meter may make it possible to link maintenance periods to the type of service.

Smaller engines are 'refit by replacement' for major overhaul or breakdown, but top overhauls are carried out on board. This policy is being extended to larger engines, e.g. the 16-cylinder Ventura generator engines in the Type 42 frigates. Engines that are too large for refit by replacement must be overhauled *in situ*, but in order to reduce downtime to a minimum, a policy of component refit by replacement (R by R) is being adopted. Reconditioning of R by R engines and components is carried out at nominated dockyards. This will involve initial additional cost in providing spare engines and components but must show a long-term gain in the reduction of maintenance time on board and enhanced reliability of items reconditioned at specialist overhaul centres.

A review of maintenance schedules has highlighted the inclusion of too many items which may be classed as 'Let's have a look and see' inspections. Some of these involve frequent removal of inspection doors which eventually result in complaints of oil leaks, with the result that unnecessary work has made necessary work. Differential pressure indicators are used to indicate when air and oil filter elements need to be changed.

Reliability is not the sole criterion for choosing an engine. Accessibility is equally important when considering maintenance of an engine on board. Too many good engines are cluttered with auxiliary equipment that must be removed by the maintainer before he can carry out his planned maintenance. The man-hours required to do a particular piece of maintenance must be the minimum.

In the design stage of an engine a detailed 'maintenance evaluation' should be carried out to determine the correct procedure for maintenance and the design of any special tools and equipment necessary to enable the maintainer to do his work most effectively.

SERVICE PROBLEMS

ASR1 engines

In 1947 design work was put in hand on a 9·75 in bore × 10·5 in stroke engine designated the ASR1 (Admiralty Standard Range 1). It was a range of engines with 6, 8, 12, and 16 cylinders, turbo supercharged for surface applications and mechanically supercharged for submarines. The first engines entered service in 1954 in frigates, submarines, and a number of other vessels for propulsion and generator duties. About 250 engines have been manufactured.

A number of defects or problems have occurred with these engines.

Liners

Liners are of steel, chromium plated in the bore. Bore wear was not a problem until 1968 when reports were received of serious wear in less than 4000 h running in liners in 6, 8, and 16-cylinder generator and propulsion engines of three frigates. The wear was in the form of a well-defined band at the top of the stroke which extended through the chromium into the steel backing (Fig. 5.10). In addition, a large number of badly corroded Y alloy piston crowns were found in liners with the heaviest wear. Worn liners and corroded pistons were random in any engine, and fewer than half the engines in any one ship were affected. A thick layer of powder discovered on one piston crown was found on analysis to be mainly aluminium oxide with traces of copper, sodium, and sulphur compounds. The two main exhausts on each ship are shared by four propulsion and two generator engines with an isolating valve fitted in each individual engine exhaust, and under cruising conditions only half of the main and generator engines are used. It became evident that the cause of the trouble was exhaust gases from running engines leaking back through the isolating valve, down the exhaust of a stationary engine, then into the engine induction manifolds, allowing sulphur compounds in the gases to attack the copper fins of the air after-coolers and causing severe corrosion. When that engine was run the copper corrosion products plus salt from the atmosphere or incipient leaks in some of the after-cooler elements entered the cylinders and accelerated the corrosion of the piston crowns. The resulting aluminium oxides were deposited on the top compression rings and were hard enough to lap away the chromium plating.

Crankshafts

The ASR1, 16-cylinder, mechanically supercharged engine (16 VMS) has by far the most arduous operating conditions of any diesel engine in naval service. Its net b.m.e.p. is 134 lb/in^2, and with the power required to drive the supercharger the gross b.m.e.p. is 156 lb/in^2. Its speed range is 750–920 rev/min. Under certain operational conditions it must be shut down immediately from full power and, because neither the turning gear nor the lubricating oil priming pump may be used, heat soaking can be a problem. Cases of boiling in the cylinder jackets have been reported and temperatures of 150°C or more have been measured on the outside of the fork rod large end blocks. Severe corrosion of the copper–lead large end and main bearing shells has occurred, and a number of large end bearings have failed completely, resulting in heavy wear of crankpins.

In addition to the arduous operating conditions mentioned above, serious contamination of the lubricating oil has occurred for the following reasons:

(1) Fuel dilution—leaks, etc.
(2) Ethylene glycol from cooling water leaks.
(3) Salt. The air supply to the engines is usually saturated with sea water, and as the engines exhaust beneath the water surface it is possible for it to leak back through the exhaust system.
(4) Large end bearing temperatures of 150°C or more can result in a reduction in the film thickness of the lubricating oil and a breakdown in the quality of the oil left in the bearings.

To ensure that the protective tin–lead flash is retained on the copper–lead bearing surfaces, the finish of the crankpins and journals must not be allowed to deteriorate and the efficiency of the full flow lubricating oil filters must not be allowed to fall off.

The existing crankshaft material is hardened and tempered to a V.p.n. of 250–275, and under the operating condition imposed on them the surface finish of the pins deteriorates. We are therefore faced with changing bearing shells every 4000 h with a possible reclamation of surface

Fig. 5.10. Worn ASR1 engine liner

finish of crankpins every 8000–12 000 h. Experimental work has been carried out on polishing and grinding pins *in situ*, but it is considered that the long-term solution will be to use surface hardened crankshafts with a surface finish of 8 μm maximum in these engines, as already described.

Viscous TV dampers

Random damper failures have occurred due to seizure or due to large changes in the viscosity of the silicone fluid. Why these large changes occur is not at present completely known. It appears that the viscosity falls off first and then increases rapidly and that the increase may be due to particles worn from the bush material when the viscosity is low, promoting a catalytic action on the fluid. Any further evidence on this matter would be welcomed.

All ASR1 engines are fitted with viscous dampers and because in some installations a damper failure, if undetected, could possibly result in a crankshaft failure, regular TV checks are necessary. The conventional methods of carrying out a TV check require the services of a specialist and even when only a routine check is involved are time consuming. The design of a simple tester was investigated by A.E.L. and a prototype instrument was produced and successfully tested on an ASR1 engine at the laboratory. As subsequent service testing in a frigate showed that it could be conveniently used by ship's staff, 100 have been produced and distributed throughout the fleet and naval bases. The unit has since been adapted for use on the Ventura engines in service and has been instrumental in detecting a number of potential damper failures.

Routine tests are made every 1000 h and prior to the ship coming in for a refit.

Fuel injection equipment

Probably the most frequent complaint from ships at sea is the failure of some component in the fuel injection equipment of a diesel engine. Some of these failures may be considered to be of a minor nature by the engine maker or supplier of the fuel equipment, but any defect, however minor, that requires an engine to be stopped before it can be rectified must be classed as an engine failure.

It would appear that the outputs of existing types of block fuel injection pumps have been stretched beyond the limits of reliability, with the result that plungers seize, delivery valves break up, flexible pump drives fail, or the cambox lubricating oil becomes so diluted with fuel that cams and camshaft bearings suffer. To change a block type pump, or even to check the quality of the lubricating oil in the cambox, when the pump assembly is buried between the heads of a Vee engine is not an easy operation and is very unpleasant when one is forced to do it before the engine has cooled down. Fuel injection pipes fail too frequently—often, because the pipe is badly formed, behind a nipple.

The author considers that engine builders and fuel injection equipment manufacturers should institute a programme of research aimed at improving the reliability of fuel injection equipment and suggests that as a target this equipment should be sufficiently robust that, except for injector nozzle changes, it should last the period between top overhauls without failure. Fuel injection pipes should last until a major overhaul, when they could be replaced.

CONCLUSION

Work continues on improving the reliability of engine components and therefore reducing the maintenance work on diesel engines. Development in hand at present includes:

(1) Electrostatic precipitators for crankcase vents (**5**).
(2) Centrifugal superchargers for 16 VMS engines.
(3) Electronic governing.
(4) Rigs for fatigue testing connecting rods.
(5) Wireless telemetry of engine piston and connecting rod parameters (**6**).

ACKNOWLEDGEMENTS

Although Ministry of Defence permission has been granted to present this paper, the opinions expressed are those of the author and do not necessarily constitute naval policy. The author wishes to acknowledge the help given by Y-ARD and colleagues at A.E.L., A.R.L., and A.O.L. during the preparation of the paper.

APPENDIX 5.1

REFERENCES

(**1**) CONN, R. B. 'Survey of noise in merchant ships', *Shipp. Wld* 1969 **162** (No. 3830, February).
(**2**) BERTODO, R. and WORSFOLD, J. H. 'Medium speed diesel engine noise', *Proc. Instn mech. Engrs* 1968–69 **183** (Pt 1), 129.
(**3**) BARTON, J. W. and COX, H. L. 'Fatigue tests on a diesel engine crankshaft', *Proc. Instn mech. Engrs* 1969–70 **184** (Pt 1), 241.
(**4**) LEATHERS, J. W., SAVAGE, A. C. and PEARSON, I. P. Paper B2, *3rd Cranfield Fluidics Conf.*, Cranfield, 1968 (May).
(**5**) *A.E.L. Tech. Note.*
(**6**) *A.E.L. Tech. Note, No. 1783.*

Paper 6

ALIGNMENT INVESTIGATION FOLLOWING A MEDIUM-SPEED MARINE ENGINE CRANKSHAFT FAILURE

D. Castle*

The multiple fracture, apparently without warning, of the crankshaft of one of a pair of medium-speed propulsion engines in a modern cargo ship in mid-Atlantic led to a programme of tests, involving many alignment measurements, chiefly on a sister ship. The damage to the engine is reviewed and an account is given of the investigation into the cause of failure. The measured values of alignment are recorded, together with changes of alignment due to temperature variations of the machinery, the lubricating oil, and the cooling water, and to cargo loading and distribution. Conclusions are stated and a set of rules, which have been supported by subsequent experience, are postulated.

INTRODUCTION

ON THE M.V. *Manchester Port*, a main engine crankshaft failed as the vessel was proceeding at normal speed from Canada to the United Kingdom.

The Chief Engineer stated that the oil mist detector from 2/5 crankcases operated the alarm signal about two weeks before the shaft failed, but at that time an examination of the crankcase revealed no evidence of overheating.

Date of failure, 30th June 1967.

Although the ship proceeded to port safely on one engine and there was no injury or loss of life, such a report causes concern in both commercial and engineering sectors. The cause of the failure is felt to be particularly relevant to the installation of medium-speed engines in modern ships, and the investigation revealed interesting information.

THE SHIP

The M.V. *Manchester Port* is a 470 ft, 12 000 ton cargo liner that was commissioned in November 1966. She is one of a fleet of ships which, in the main, work regularly between the Great Lakes of North America and Manchester. Each round trip involves about 450 engine running hours and some trips involve heavy going in ice for which the hull is strengthened. The ship was on the return voyage of the fifth trip when the failure occurred.

The location of the engine room and the various cargo holds are illustrated in Fig. 6.1.

THE ENGINES

The propulsion engines are two direct reversing 14 cylinder Pielstick type PC2, medium speed diesel engines, built by the author's company.

These engines are four-stroke, single acting trunk-piston vee engines of 400 mm bore and 460 mm stroke, fabricated in steel with cast iron individual water jackets. The crankshaft is underslung and is carried in thin-wall bearings. The construction of this type of engine is described in detail in a paper by Henshall and Gallois (1)†. The engine is designed as an exhaust turbo-charged unit to give a rated brake mean effective pressure of 216 lb/in² (15·2 kg/cm²) at a speed of 520 rev/min rising to 241 lb/in² (17 kg/cm²) at a speed of 400 rev/min. In this case the engines had been supplied at a rating of 5660 b.h.p. at 465 rev/min. The fixed pitch propeller was designed for 11 320 b.h.p., which corresponds to a brake mean effective pressure of 196 lb/in² (13·8 kg/cm²).

In the machinery arrangement (shown later in Fig. 6.7b), each engine drives into the dual input–single output gearbox by means of a rigid crankshaft extension through a hydraulic coupling mounted in the gearbox. The gears have single helical teeth, and a thrust bearing mounted

---

The MS. of this paper was received at the Institution on 3rd February 1970 and accepted for publication on 19th February 1970. 24
* Engineering Manager, Crossley-Premier Engines Limited, Manchester.
† *References are given in Appendix 6.1.*

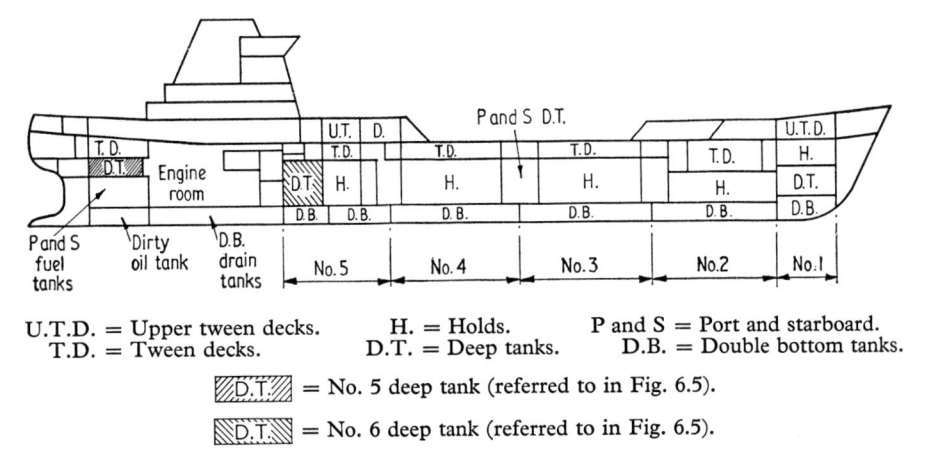

U.T.D. = Upper tween decks. H. = Holds. P and S = Port and starboard.
T.D. = Tween decks. D.T. = Deep tanks. D.B. = Double bottom tanks.

▨D.T.▨ = No. 5 deep tank (referred to in Fig. 6.5).

▧D.T.▧ = No. 6 deep tank (referred to in Fig. 6.5).

Fig. 6.1. General layout of cargo holds on M.V. *Manchester Port* and M.V. *Manchester Progress*

forward of the coupling absorbs thrust from the pinion and locates the engine crankshaft. Cylinders are numbered from the driving (after) end, as are main bearings, crank-webs, and balance weights. There is an additional main bearing at the driving end, beyond the gear train to the camshaft, which is not numbered, and because it can be used to locate the crankshaft, it is referred to as the 'anchor' bearing.

THE FAILURE

Multiple fracture of the crankshaft occurred with impact damage to the crankcase, pistons, and liners local to the fractures. Balance weights were damaged and weight number 7 (from the after web of crank 4/11) was thrown from the shaft and was found lying on the floor plates between the main engines (Fig. 6.2a). Identical balance weights are fitted to all 14 crank-webs.

At crank 4/11 both webs were fractured and the crankpin, with attached big-end bearings and connecting rods, was embedded in the wall of the crankcase (Fig. 6.2b).

Crankpin 5/12 (Fig. 6.2c) was not wholly detached but fractures had penetrated most of the web sections on either side.

At the time of failure both engines were going ahead at full speed, the direction of rotation, looking forward as in Fig. 6.2a, was counter-clockwise. The main bearing shells were examined in detail. Bearings 4 and 5 were so mutilated as to be wholly uninformative. In fact, No. 5 shells had rotated within their housing and No. 4 bottom half had ridden under its top half, the pair then jamming in the bearing cap.

It was evident from the marking of the bearings that the shaft had been in contact with the top halves of bearings 2, 3, and 4 and the outboard 'horns' of bearings 1, 2, and 3, and had been in heavy contact with the top halves of 1 and 8 (Fig. 6.3).

POSSIBLE CAUSE OF FAILURE

The cause of the failure was considered to be one or more of the following:

(1) The balance weight had come adrift.
(2) Lubrication had failed, with consequent bearing seizure.
(3) Overstressing of the crankshaft in fatigue had occurred because of (*a*) unsatisfactory material, or (*b*) excessive torsional vibration stresses, or (*c*) misalignment of the crankshaft in its bearings.

INVESTIGATION

Cause number (1) was a popular theory at the beginning of the investigation. The balance weights are mounted on the shaft webs on a dovetail, and the mating faces are held in contact by a central screw loaded in compression. The head of the screw is fitted with a locking plate and no locking plate has ever been found loose.

Examination of the faces showed that fretting had occurred on webs 1 to 10. Fretting could lead, in turn, to slackness and a detectable wear step was found on weight No. 9, which is one of the webs on the throw 5/12.

However, impact marks on the roof and wall of the crank chamber bay of cylinders 4/11 clearly showed that the shaft had already been displaced when the balance weight No. 7 struck and rubbed the structure. It was clear, therefore, that the jettisoning of the balance weight was a consequence and not a cause of failure.

Item (2) was quickly disposed of. All oilways were checked, even in the damaged pieces of shaft, and found clear. The Chief Engineer of the ship reported that at the time of failure he had carried out an immediate check that the lubricating oil pump was running and was able, subsequently, to trace oil emerging from all points in the crankshaft.

To investigate cause (3*a*), extensive metallurgical tests were performed on pieces of the damaged shaft and the material was found satisfactory in physical properties and in quality. Concerning (3*b*), when a crankshaft breaks, the immediate suspect is torsional vibration. In this case, examination of the fractures showed clear bending fatigue with no sign of torsion failure.

It was evident, therefore, that (3*c*) must be investigated.

Fig. 6.2a. Between main engines looking forward

Fig. 6.2b. View through inboard crankcase door, cylinders 4/11

Fig. 6.2c. View through inboard crankcase door, cylinders 5/12

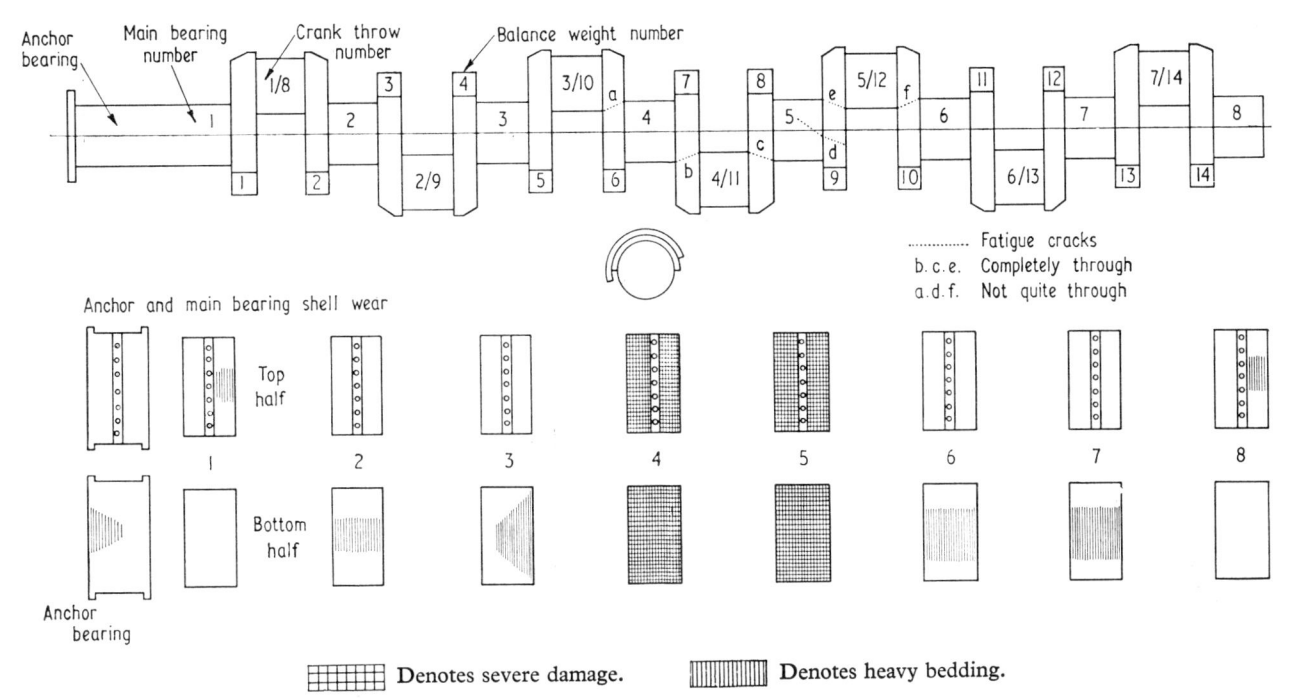

Fig. 6.3. 14-cylinder PC2V crankshaft and main bearing condition

In the work which followed—co-operatively between Lloyd's Register, the shipowners, the shipbuilders, and the engine builder—it was fortunate that there was an identical sister ship, the *Manchester Progress*, which was at the time returning, towards Manchester, from her third voyage. Engine running hours were 1350 compared with 2500 for the engines of the *Manchester Port*.

A programme of alignment checks was planned to be carried out on the sister ship, leaving the *Manchester Port* to have the damaged starboard engine removed and the ship prepared for operating on her remaining engine. The observations on the starboard engine were completed during its dismantling.

TELESCOPIC SIGHTINGS ON THE *MANCHESTER PROGRESS*

Telescopic sightings were made along each side of both engines using five target positions equally spaced along their length. Each line of graduated targets was mounted on the machined levelling pads provided on the upper side of the engine mounting flange.

The machining of these pads is referred to the same datum as the main bearing bores when the engine is manufactured. In addition, two target points were established on each outboard side of the gearcase to allow relative movement between the engines and gearbox to be observed. The telescope was a Taylor-Hobson instrument with optical vernier. This equipment was set up while the ship was in transit from Liverpool to Manchester so that the first sightings were able to be taken with the machinery and its surroundings in the hot condition and with the ship loaded.

The measurements were repeated when the machinery had cooled to ambient temperature, and again at various stages of unloading.

It should be noted that the target positions on the gearbox casing are purely arbitrary and no plotted values of height for positions X and Y relate to any original manufacturing or installation reference.

The effect of cooling from operating temperatures

Fig. 6.4 shows the measuring method diagrammatically. It will be appreciated that there is no fixed datum in a ship and the graphs are plotted relative to a line of sight joining the targets at each end of the engine, i.e. points 1 and 5.

For the port engine, it will be observed that the maximum change of alignment within the engine from hot to cold is 0·0035 in, and for the starboard engine 0·0065 in.

However, since the crankshafts are rigidly coupled to the gearbox pinion shafts, it is important also to consider the change of alignment taking a datum line for the hot condition from point 1 to points X and Y, which is the condition at the time the target points were established on the gearbox. This shows the same magnitudes of movement, but for the port engine point 3 does not now cross the datum line and this point's movement of 0·005 in now represents the maximum change of alignment.

Results led to a series of checks being made on other engines of the same type. Measurements of the effect of the change from cold to running temperatures were made on 16 PC2 engines in ships. Similar checks were made on two further engines on test berths. In every case 'hogging' was found to occur with rising temperatures and all

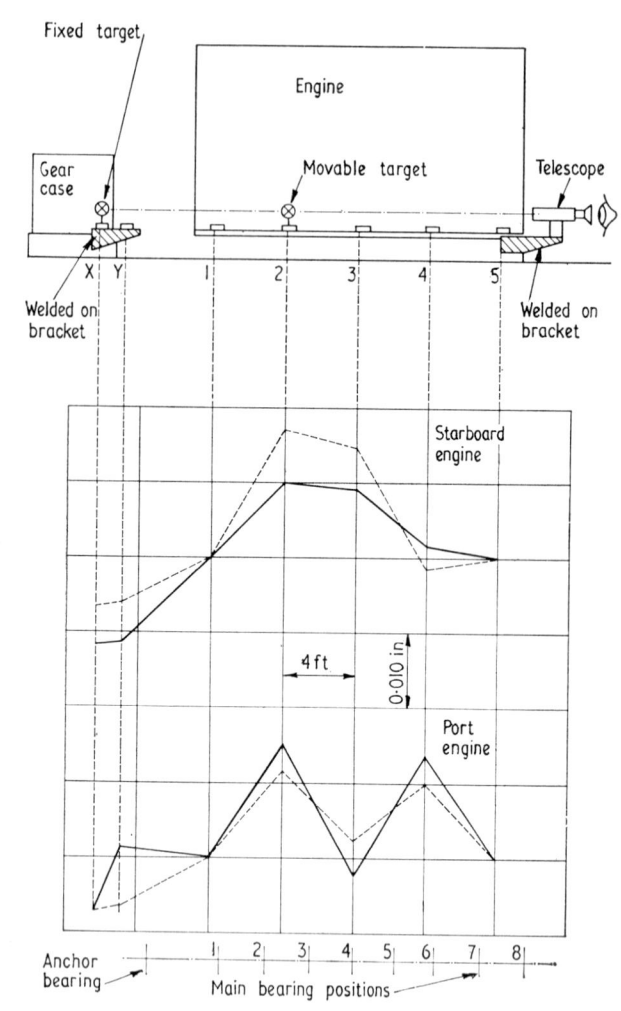

Fig. 6.4. Alignment tests on both engines of the M.V. *Manchester Progress*

Test 1 - - - - Engine hot.   Test 2 ——— Engine cold.
All sightings taken on the outboard side of each engine.

Fig. 6.5. Alignment tests on both engines of the M.V. *Manchester Progress*

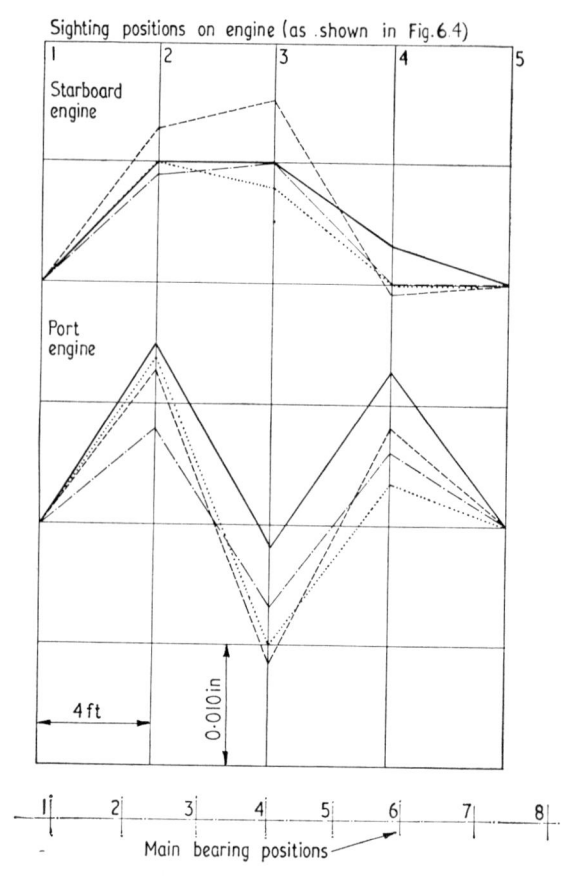

Test 2 ——— Vessel fully loaded (engine cold).
Draught: 18 ft 7 in forward, 19 ft 5 in aft.

Test 3 —·— Discharging cargo. Nos. 5 and 6 deep tanks full.
Draught: 13 ft 8 in forward, 19 ft 4 in aft.

Test 4 - - - - Discharging cargo.
No. 5 deep tank empty; No. 6 deep tank full.
Draught: 13 ft 11 in forward, 16 ft aft.

Test 5 ······ Cargo fully discharged.
No. 5 deep tank empty; No. 6 deep tank empty.
Draught: 12 ft 2 in forward, 14 ft 1 in aft.

Note: No. 5 deep tank is just ahead of engine room and No. 6 deep tank is just astern of engine room (see Fig. 6.1).

All sightings taken on outboard side of each engine.

movements were of the same order as for the engine of the *Manchester Progress*.

Other effects

Fig. 6.5 shows the movement of the engine seatings during cargo discharging. For the port engine the maximum change of alignment is 0·0102 in at point 2 and for the starboard engine, 0·0030 in at point 4.

Finally, from this graph it will be seen that when the ship is fully discharged and, of course, in the cold condition, there is a departure from datum of up to 0·0137 in for the port engine and up to 0·0100 in for the starboard engine.

A notable fact about these plots, however, is that the port seating has a different shape from the starboard. It is not difficult to imagine the effect on main bearing loads of a seating such as the starboard, which rises from the ends to a peak in the centre. It was, in fact, discovered that No. 4 main bearing in the starboard engine was on the verge of total failure. It had been running hot for some time and small areas of copper–lead were missing. Damage to the crankshaft journal was fairly severe but was rectified within the wear limit for a standard size bearing. All other main bearings in this pair of engines were in good condition.

The bore of the bottom half of No. 4 bearing was found to be curved, the ends of the bore being depressed by 0·004–0·005 in. This is a clear indication that this bearing was substantially higher than the remainder.

TELESCOPIC SIGHTINGS ON THE
MANCHESTER PORT

Sightings were made along both sides of the port engine. The results are shown in Fig. 6.6a. The movement from cold to hot is of the same order as for the engines of the *Manchester Progress*, but the cold alignment, although showing a slight hog, is good.

This engine was re-chocked to give an exact cold alignment, and sightings taken after heating up are shown in Fig. 6.6b.

All the main bearings of this engine, although heavily worn, were in good order.

It was obviously desirable to take sightings to confirm that the starboard engine seatings were of a comparable shape to the starboard seatings of the *Manchester Progress*, but the seatings were unmachined and no sightings could be made without the engine.

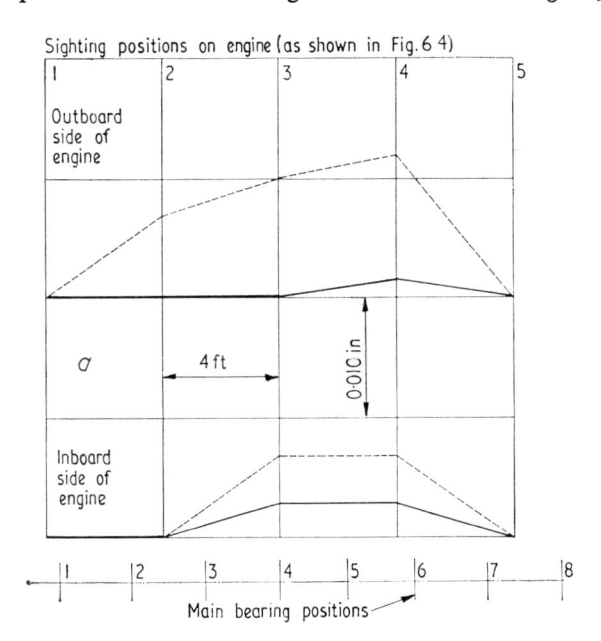

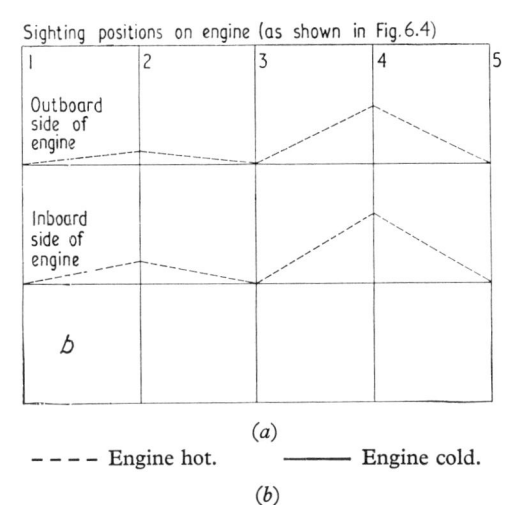

FURTHER INVESTIGATION
Original alignment

When the installation of the machinery was made, a supporting jig was used to centralize the input shaft of the hydraulic coupling, alignment then being made at the flanges forming the coupling between this shaft and the engine crankshaft in each case.

The jig was manufactured by the gear maker at the same time as the first gearbox and was now brought from the shipyard so that the coupling alignment could be checked. Before using it, however, the truth of the jig itself was checked.

It was found to have an error of approximately 0·031 in (0·79 mm) in the horizontal location of its bore, which meant that all four engines had an initial sideways misalignment. This had actually produced lateral crankshaft web deflections of relatively high magnitude in the first place, but within tolerable limits. Owing to lack of experience of this type of engine at the time, the engine builders had not attached significance to the abnormality.

Engine holding-down bolts

The engine had been installed with a combination o clearance and fitted holding-down bolts, the fitted bolts being evenly distributed throughout the length of the engine. The question had to be considered of whether the restraint of fitted bolts was causing an increased tendency to thermal hogging. It has long been the practice for some engine builders to use fitted bolts only towards the driving end and to assume that sliding on the seating would occur.

Measurements were made of the longitudinal movement of one of the engines of the *Manchester Progress* due to temperature change and it was found that the engine frame moved 0·07 in (1·8 mm) towards the gearbox during heating to working temperature. The test was repeated with two fitted bolts each side at the driving end and precisely the same movement was recorded.

A similar test was carried out on another 14-cylinder engine on a fabricated test berth of similar form to the ship's seatings, although of heavier section. The total extension of the engine (length 19 ft or 5795 mm) was 0·115 in (2·9 mm) and the movement of the driving end of the frame towards the dynamometer was 0·075 in (1·9 mm).

These results tended to show that the position of fitted bolts does not influence the movement due to thermal expansion.

Thermal hogging was observed by Bureau Veritas surveyors in large slow-speed engines and reported by Bourceau and Wojcik in 1966 (2). They attributed the effect to the large temperature difference which exists on such engines (56 ft or 17 m long and 26 ft or 8 m high) between the cylinder heads and the engine mounting.

(a)
---- Engine hot.   —— Engine cold.
(b)
---- Sightings taken on hot engine after rechocking, such that the cold alignment was level.

Fig. 6.6. Alignment tests on the port engine of the M.V. *Manchester Port*

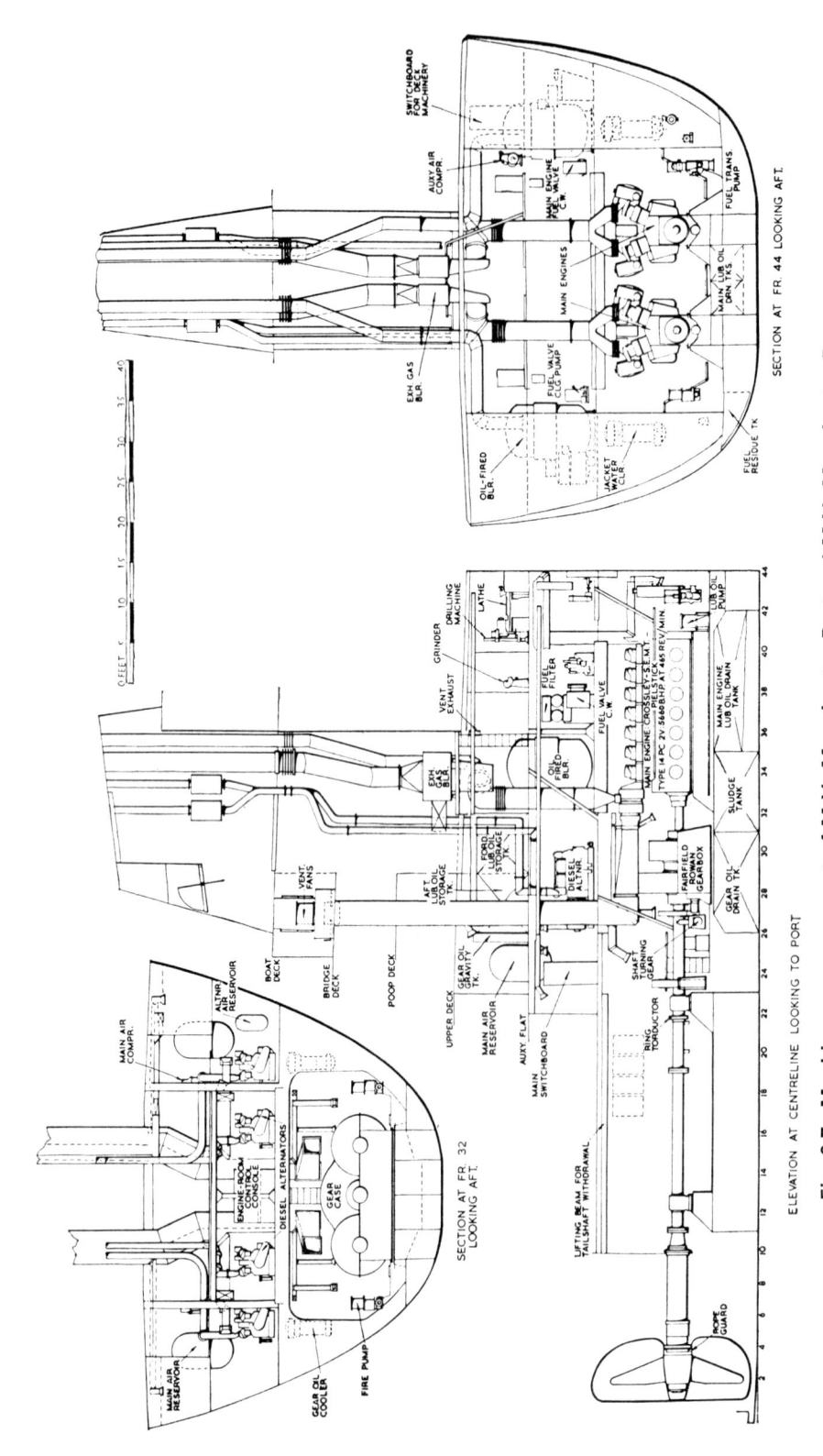

Fig. 6.7. Machinery arrangement of M.V. *Manchester Port* and M.V. *Manchester Progress*

However, measured values of temperature on a PC2 engine show a difference of only 5 degF between cylinder head temperature and the frame temperature at crankshaft level, and the cause of hogging, certainly in this type of installation, would seem more likely to be failure to slide.

Movement of the crankshaft to gear input shaft flanges

During the series of rechocking experiments on the port engine of the *Manchester Port*, which were directed at ensuring the safety of the ship to put to sea again on one engine, measurements were made at the coupling flanges.

There were marked changes in alignment in the various stages of developing heat in the engine and in the double-bottom lubricating oil drain tank. The location of the drain tanks can be seen in Fig. 6.7. The pattern of events is disclosed by the following series of results measured with the coupling bolts removed (Fig. 6.8 and Table 6.1).

The engine is tilted outboard when the drain tank is heated and inexplicably moves when hot water is circulated through it such that lateral alignment is partially restored but vertical alignment is considerably worsened. When

Table 6.1. Changes of alignment at crankshaft coupling flange

X Vertical coupling misalignment.
Y Lateral coupling misalignment.

Dimensions at 1, 2, 3, and 4 are the gaps between the faces of the coupling at the top and every 90° clockwise, looking ahead.

Condition	Measurements		Measurements at positions			
	X	Y	1	2	3	4
Engine cold (19th August 1967)	0	0	1	0	0	0
Engine oil tank at 160°F (21st August 1967)	+12	+25	5	4	0	0
Jacket water at 160°F (22nd August 1967)	+25	+15	5	4	0	0
Engine run not coupled (22nd August 1967)	+20	*	*	*	*	*

* Readings could not be made because engine had moved about 0·020 in axially towards the gearcase, closing the coupling gap.

the engine is run light, vertical expansion begins to restore the vertical alignment. It was expected that when power was transmitted, further vertical expansion of the engine and expansion of the gearcase would be almost matched, but in practice it was found that the engine side of the coupling lifted in relation to the gear side almost the whole of the 0·020 in vertical misalignment. This must be regarded as purely fortuitous.

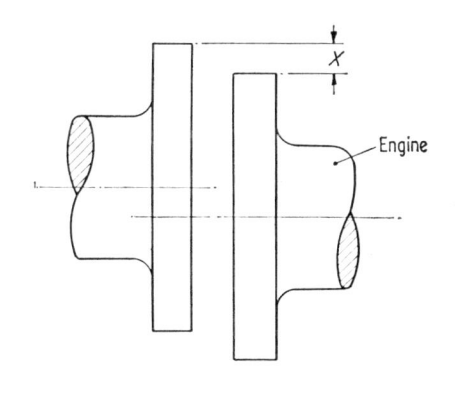

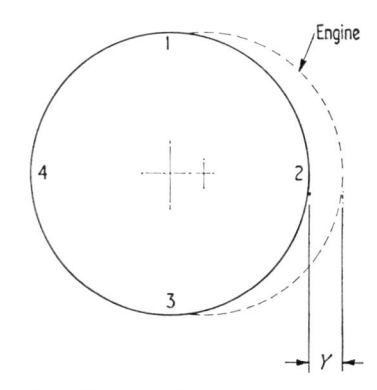

Further checks on the relative movements of the empty starboard seating and the gearcase in athwartships and vertical directions were planned and measuring equipment installed by Lloyd's. Readings taken during the next voyage showed relative movement of up to 0·006 in both vertically and laterally. No movement was recorded when the ship was not pitching.

Record of ships since the failure

Rechocking was carried out with fastidious care to ensure the most satisfactory alignment achievable. Crankshaft web deflections were taken at the end of each voyage and, in the case of each of the four engines, only small variations have been recorded. Original values of deflections were recorded from time to time, which was reassuring. The need to monitor the alignment remains, however, and design and building practice incorporating such a need cannot be regarded as satisfactory.

The degree of discontinuity in the seatings of these ships can be seen in Fig. 6.9, and Fig. 6.10 shows the heavier section and better continuity of the seating in a later installation where the engine builder was consulted at the design stage.

DISCUSSION OF RESULTS

The primary cause of failure of the crankshaft was overloading of the main bearings in the centre of the engine,

X positive: Engine below gears.
Y positive: Engine to port.

Fig. 6.8. Measurements at crankshaft coupling flange

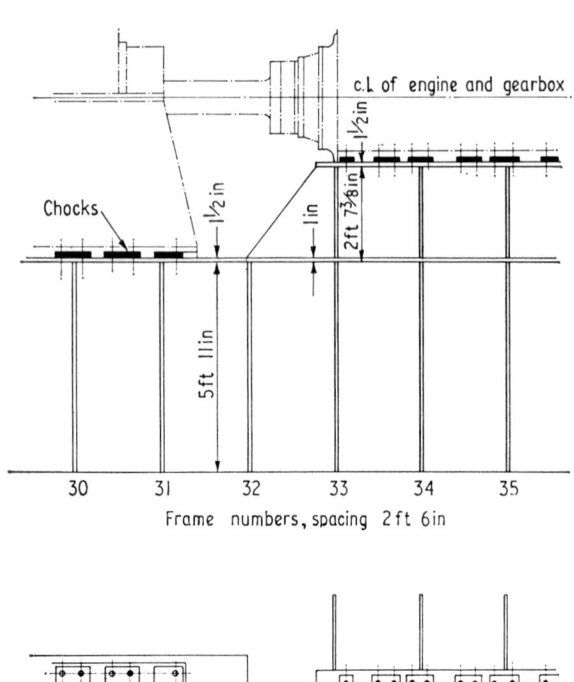

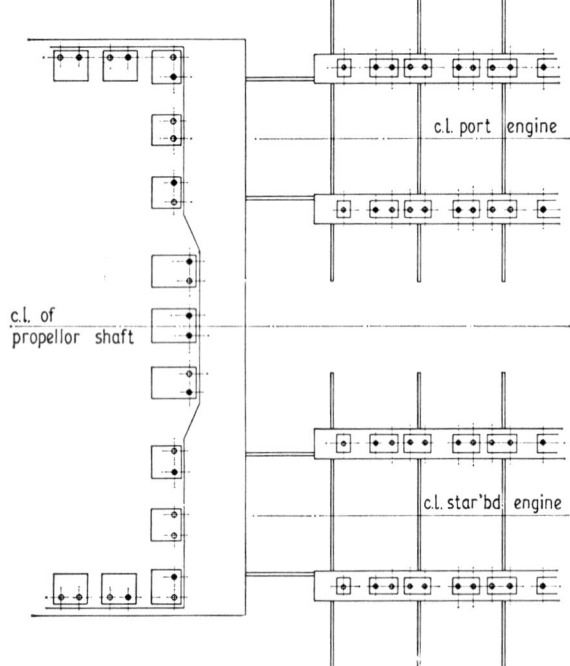

Fig. 6.9. Lightness of sections and seating discontinuity M.V. *Manchester Progress* and M.V. *Manchester Port*

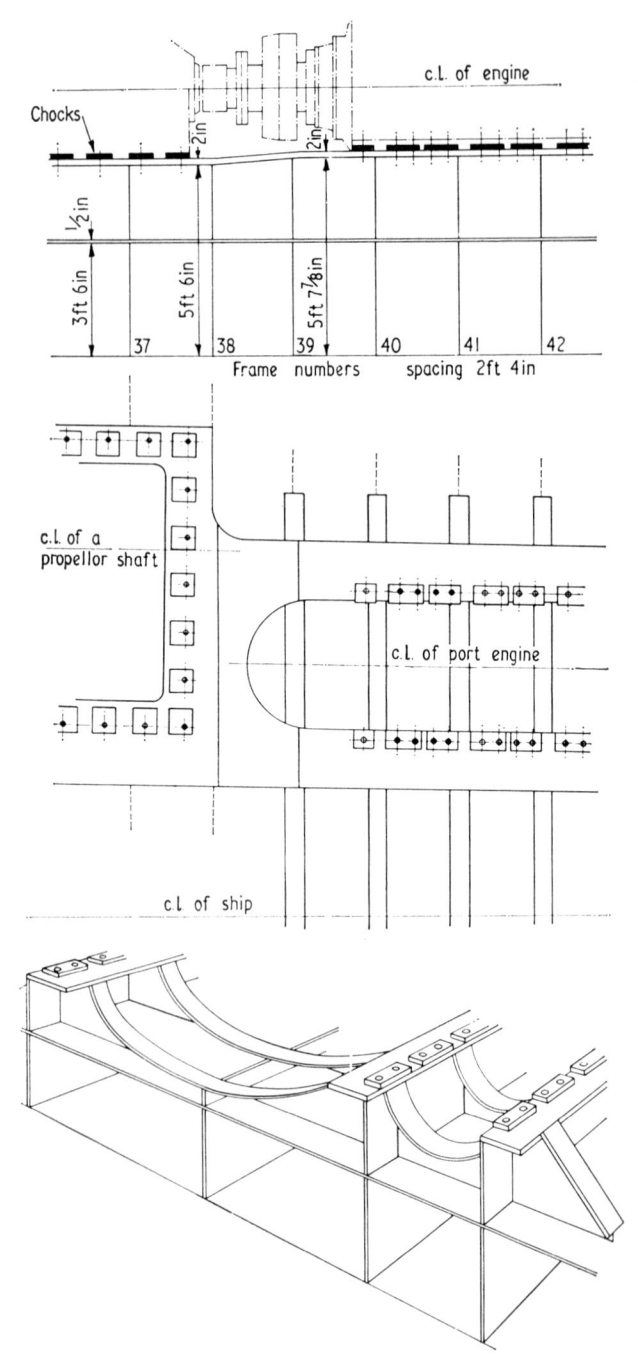

Fig. 6.10. A later seating design incorporating heavier sections and better continuity

which in turn led to bearing lining collapse and hence to a loss of support of the crankshaft. The shaft then failed in bending fatigue.

(1) The excessive main bearing loads came about as a result of deformation of the engine seatings to the hogged shaped assumed when the machinery was hot. The addition of the thermal deflection of the engine to this permanent deformation caused the main bearings in the centre of the engine to be overloaded. The magnitude and profile of the hogging of an engine must depend upon the stiffness of the seatings and their supporting structure. For all PC2 engines examined, the loss of alignment of main bearing housing falls within the main bearing clearance of 0·012–0·014 in (0·30–0·35 mm). The fatigue lines on the fracture surfaces showed no interruption fronts, which implied that crack initiation to failure occurred in one running period.

(2) According to bending fatigue tests performed on the material of the failed shaft, a bending stress of $\pm 18\cdot7$ ton/in^2 (29·5 kg/mm^2) is necessary to initiate fatigue;

stress of ± 18.5 ton/in^2 (29.5 kg/mm)^2 did not cause failure after 50×10^6 cycles.

Although some work was done on the effect of loss of bearing support on a PC2 engine crankshaft, it was not economic to pursue this beyond the removal of a single main bearing. Tests were made in which strain gauges measured journal and crankpin fillet stress in the plane of symmetry of the crankpin when load was applied in this plane to this crankpin both towards and away from the shaft centre.

The measurements were made on No. 4 journal and the adjacent crankpin of a 12-cylinder vee-engine shaft, first with all bearings in position and then with No. 4 bearing removed. It is thus possible to relate radial load on the crankpin to stress in the fillets for any given clearance under the journal. The cyclic variation of the radial component of load in the connecting rod was calculated and peak values were determined. It was found that with no bearing the stress range produced at full engine power was 16.45 ton/in^2 (26 kg/mm^2) compressive to 5.7 ton/in^2 (9.0 kg/mm^2) tensile in the journal fillet. The corresponding stress range in the crankpin fillet was 15.2 ton/in^2 (24.0 kg/mm^2) tensile to 5.4 ton/in^2 (8.6 kg/mm^2) compressive.

This means that, after the failure of No. 4 bearing, No. 5 bearing must fail and develop increased clearance before fatigue cracks would be initiated. It would be possible to extend the work to determine the point at which fatigue stresses would reach failure level, but it is impossible to estimate accurately.

The effect of applying a bending moment at the coupling flange is not measurable beyond the second throw when all bearings are in position, and unless there is a whirling tendency in the long span created by the absence of two central bearings, the misalignment at the coupling seems unlikely to have been a contributory cause of the failure.

(3) The coupling of the engines to the gearbox by rigid crankshaft extensions, through hydraulic couplings which give only torsional flexibility, allowed bending moments to be applied to the crankshafts and to the fluid coupling which varied in magnitude according to (*a*) the state of loading of the ship, (*b*) the relative temperatures of the units of machinery and oil drain tanks, and (*c*) whether the ship was in port or at sea, and the state of the sea. The total discontinuity between engine and gearbox seatings served to concentrate deflections into this region and to allow each seating to move independently under the above influences.

(4) Initial misalignment took place between the engine and gearbox at the installation stage due to the use of a faulty jig. The precision of the alignment was also influenced by the fact that the engine was mounted on chocks which rested on unmachined surfaces.

(5) The permanent set taken by the engine seatings could well result from a relaxation of locked-in stress in the structure during the early operation of the ship. The stability of this structure during the subsequent operation of both ships points to such an effect as the only feasible cause of the early deformation. There can be no doubt that considerable stress is accumulated in a welded ship during construction and that most of it will be relieved fairly quickly in service.

(6) By chocking the engines so that the main bearing bores assume a 'sagging' curve when cold, and this curve is readily deduced from crankshaft web deflections, it is possible entirely to compensate for the hogging effect of thermal expansion in any given installation. With sufficient experience from installations it is possible to cater for the varying degrees of seating restraint with a standard curve to which the main bearing housings are bored by the engine builder. This practice was subsequently adopted by the author's company.

RULES OF PROCEDURE

The investigation discloses a situation in which the uncoordinated efforts of the engine builder and the shipbuilder produced a machinery installation which suffered from a number of features that were individually acceptable but unsatisfactory in combination. There appears to be no code of practice in existence and design can proceed on an *ad hoc* basis.

As a result of the investigation a set of rules was drawn up for observation within the author's company which may appear obvious to the experienced but which serve to avoid the inherent pitfalls. These rules are as follows:

When medium-speed engines are to be applied to ship propulsion:

(1) It is important to ensure that the alignment of the main bearings is such when the engine is cold that thermal expansion will produce a near-straight line at operating temperature.

(2) The seating supporting the engine(s) and gearbox should be designed as a single and continuous structure with no sudden changes in section at any point. This applies equally to members below the double-bottom tank top as to those above. It is then important to ensure that propeller thrust is distributed to underside longitudinal girders other than through the engine bearers. A drawing of the proposed engine(s) and gearbox seating showing double bottom tanks and structure should be supplied by the shipyard to the engine builder for comment before the design is finalized.

(3) Although it appears that no sliding of an engine on its seating normally occurs during thermal expansion, it is desirable to provide for this possibility by locating any fitted holding-down bolts towards the driving end of an engine. To prevent any possibility of swinging of the engine resulting from shock due to collision or running aground, fixed chocks should be arranged at each side of the free end of the engine.

(4) Precision of alignment during installation will not be achieved in a reasonable time unless the top of the engine seating has a machined face. Such machining also makes

it a simple matter to assess the linearity of the faces before installation and, if necessary, subsequently.

(5) An engine seating must be sufficiently stiff to accept the 'hogging' of the engine due to thermal expansion, as well as out-of-balance forces and couples and the reaction torque, and it must hold its geometry in the face of any relaxation of locked-in stress in the hull and engine room structure. A specimen seating arrangement drawing should be provided by the engine builder to the shipyard, giving typical basic dimensions.

(6) A double-bottom lubricating oil drain tank, if located other than symmetrically about the longitudinal centre-line of an engine, will exert a tilting force when hot. Similar effects about a transverse axis will occur if the tank lies under the engine but is shorter than the engine itself, or protrudes beyond the end of the seating. The tank should therefore be located so as to lie symmetrically beneath the engine and within its length.

(7) A shaft coupling should be provided as close to the engine as possible which has capacity for angular and parallel misalignment. Provision for linear expansion must, of course, be made, taking account of the possible movement of entablatures as well as of shafting.

(8) Consultation on alignment and methods of alignment being used must take place between shipbuilder and engine builder during the period of this operation.

APPENDIX 6.1

REFERENCES

(1) HENSHALL, S. H. and GALLOIS, J. 'Service performance of S.E.M.T.–Pielstick engines', *Trans. Inst. Mar. Engrs* 1964 **76** (No. 12, December).
(2) BOURCEAU, G. and WOJCIK, Z. 'Some considerations of marine engine crankshafts', Association Technique et Aeronautique Meeting, Paris, 1966.

Paper 7

THE TURBO-CHARGED DIESEL AS A ROAD TRANSPORT POWER UNIT

E. Holmér*   B. Häggh†

The lowest possible cost of transport determines the choice of vehicle engine. Installation space, weight, and fuel consumption favour turbo-charged diesels which also have good smoke and noise characteristics. Much development work has been done at Volvo to increase reliability as regards the turbo-charger unit, cylinder head gaskets, valve wear, and bore scuffing. In this connection a decreased compression ratio is an important factor. However, cold starting and white smoke under part load must be solved in a way that does not decrease reliability.

INTRODUCTION

THE RATED OUTPUT of a combustion engine is related to the weight of air per unit of time available for combustion in the cylinders. By compressing the induction air in a turbo-compressor the density of the air is increased and a greater weight of induction air is available while retaining the same cylinder displacement. The source of power for the compressor consists of a turbine unit in the exhaust system which utilizes a part of the exhaust energy that would otherwise be lost. This results in a lower fuel consumption. The higher output available from the same engine size reduces the amount of mechanical loss, and this also helps to improve the fuel consumption.

Thus, an increased output by means of turbo-charging is theoretically very simple and the development costs for this increase are relatively low. However, the sacrifices necessary to ensure acceptable reliability are apparently very great.

In this article we describe the turbo-charged diesel engine and the measures that were necessary to achieve the high standard of reliability that is offered by Volvo engines. The possibilities of further development are also discussed.

THE CHARACTERISTICS OF THE TURBO-CHARGED DIESEL ENGINE

A truck owner's primary interest is to obtain the largest possible pay-off for the lowest possible capital investment and at the lowest possible running cost. Table 7.1 shows the calculated relative pay-off potential for various sizes of vehicle.

The calculations take account of annual milage, typical load schedule, average speed, and fuel and service costs. The output which gives the best pay-off potential for each size of vehicle—i.e. optimum output—has been calculated for operation on normal Swedish roads. Here, the increase in average speed resulting from a higher output has been compared with the increase in fuel consumption, purchasing price, and engine weight, which reduces the payload. The calculations assume the use of turbo-charged diesel engines.

With a 50 ton gross vehicle weight the pay-off potential is 62 per cent higher than a 38 ton vehicle, provided optimum brake outputs are used. A comparison between optimum and lower output shows that the pay-off potential is not significantly influenced. For example, the pay-off potential is only 11 per cent higher on the 50 ton vehicle if the power is raised from 250 to 430 ps, i.e. an increase in output of 70 per cent. On the other hand, if the gross weight is raised from 38 to 50 ton using the same 250 ps engine, the pay-off potential increases by 48 per cent.

As can be seen, it is primarily the size of the vehicle (the payload) which influences the pay-off potential, while the output—up to the optimum—has only a slight, but favourable, influence on this factor. Against the background of the economic utilization of road networks, their capacity, and safety, West Germany will bring a law into force on 1st January 1972 which demands that a truck engine must have a minimum output of 8 ps/ton

---

The MS. of this paper was received at the Institution on 2nd February 1970 and accepted for publication on 26th February 1970. 32
* *Engine Laboratory Manager, AB Volvo, Truck Division, Gothenburg, Sweden.*
† *Chief Design Engineer (Truck Engines), AB Volvo, Truck Division, Gothenburg, Sweden.*

Table 7.1. Relative pay-off potential for various sizes of vehicle and output with turbo-charged engines. Optional, legislative, and required output

Gross vehicle weight (G), ton	38	44	50
(A) Relative maximum pay-off potential, G/G_{38}	1·00	1·37	1·62
Optimum brake output (N_{opt}), ps	320	390	430
Corresponding brake output per vehicle weight, ps/ton	8·4	8·8	8·6
Legislative brake output N_L at 8 ps/ton vehicle weight, ps	304	356	400
The influence of brake horsepower on relative pay-off potential ($N_{opt}/N = 250$), ps	1·02	1·07	1·11
Relative pay-off potential with brake output 250 ps, G/G_{38}	1·00	1·31	1·48
(B) Required brake output at 70 km/h with gradient $s = 0$ per cent, ps	155	172	188
Corresponding fraction of legislative brake output	0·51	0·48	0·47
Required brake output at 50 km/h with $s = 3$ per cent, ps	325	380	420

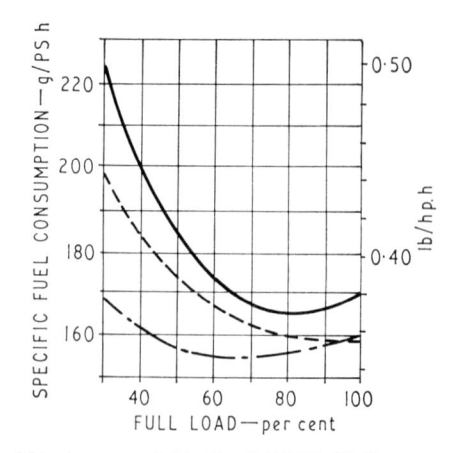

——— NA; b.m.e.p. 8·6 kgf/cm² (122 lbf/in²).
- - - - TC; b.m.e.p. 11·1 kgf/cm² (158 lbf/in²).
·—·— TC+after cooler; b.m.e.p. 18·2 kgf/cm² (260 lbf/in²).

Fig. 7.1. Fuel consumption as function of load for different types of diesel engines at 90 per cent of rated speed (2000 rev/min)

vehicle total weight. Other European countries will probably adopt the German norms or even a higher figure (e.g. Switzerland or Austria, 10 ps/ton). In this case, a turbo-charged engine is an interesting alternative because of the lower cost per horsepower and also because of the lower fuel consumption, especially with part loads (Fig. 7.1), which is an important factor with higher outputs relative to vehicle weight since a large part of transport work will be carried out with only partly loaded engines.

Since the size of the payload is decisive for a high pay-off potential, a low engine weight per horsepower is imperative. The weight of the naturally aspirated Volvo engine D 100B is 4·4 kg/b.h.p. (9·8 lb/b.h.p.) compared with the turbo-charged version TD 100A which is 3·6 kg/b.h.p. (7·9 lb/b.h.p.). Experimental engines with a lower compression ratio have a weight of 2·8 kg/b.h.p. (6·2 lb/b.h.p.), including after-cooling 2·5 kg/b.h.p. (5·5 lb/b.h.p.). The naturally aspirated engine is comparatively heavy. It is basically the same engine as the turbo-charged version which is designed to work with 140 atm cylinder peak pressure. Modern vee-type engines designed for natural aspiration have weights of 2·9–3·5 kg/b.h.p. (6·3–7·6 lb/b.h.p.).

The robust construction, the low maximum engine speed, and the six-cylinder in-line design of Volvo turbo-charged engines favourably influence noise level, and this may prove to be a vital factor since a legislative proposal concerning maximum noise has been announced. The noise level in the cab of the TD 100A is 80 dB, and with visco-fan at 80 km/h, 74 dB.

The characteristically harsh noise of the diesel engine, which is caused by the instantaneous combustion of a large part of the fuel injected during the ignition delay, is considerably reduced with a higher brake mean effective pressure, since the higher pressure and temperature brought about by turbo-charging reduces the ignition delay. A higher brake mean effective pressure gives a smaller engine for a given output, and this in turn means that there is a smaller area of noise-transmitting surfaces which simplifies the insulation of noise in the design of the vehicle. A considerable dampening of induction and exhaust noise is obtained in turbo-charging and the turbo unit itself does not result in any noise problems.

A further vital aspect in the choice of an engine intended as a road transport power unit is the limitation of exhaust smoke to 45 Hartridge units for trucks and 35 for buses, which has been stipulated by the authorities in Sweden. For a naturally aspirated engine the exhaust smoke is the output limiting factor. This is not the case with a turbo-charged engine, except at the lowest utilized speed, which depends upon the deviation between the amount of air supplied by the turbo-charger and the breathing requirements of the engine. This means that within the main work range a turbo-charged engine provides a considerable surplus of air (Fig. 7.2). In addition, the degree of charge increases at higher altitudes, thus compensating a great deal for the reduction in density of the air, and in general, there is no need to limit output because of exhaust smoke. When load is quickly applied, however, the amount of fuel injected must be reduced, to avoid smoke puffs, until the turbo has picked up sufficient speed. This need is met by an acceleration control unit which senses charging pressure and which simultaneously limits full load performance in the low speed range (below 50–60 per cent of maximum engine speed). When moving away from rest under difficult conditions this pick-up time and the relatively low output of a turbo-charged engine at low engine speeds is a disadvantage. The best method of solving this problem is by use of a torque converter. This unit, in combination with the effect caused by engine pick-up time, gives smooth starts which reduce shock load on the transmission and make moving away from rest much easier.

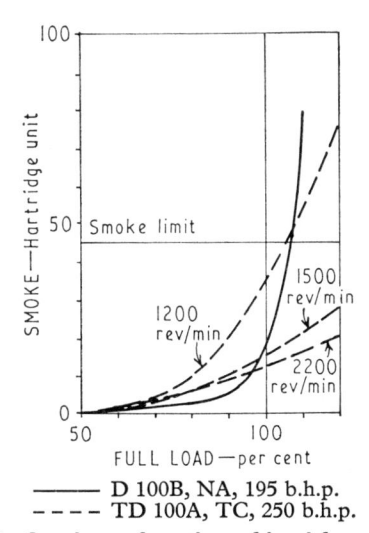

Fig. 7.2. Smoke as function of load for naturally aspirated and turbo-charged engine

—— D 100B, NA, 195 b.h.p.
- - - - TD 100A, TC, 250 b.h.p.

The turbo-charged engine has thus every chance of providing a truck owner with the best economic returns together with a very low noise level and small amounts of exhaust smoke. One absolute necessity, however, is that such an engine must be reliable.

RELIABILITY

We will now describe the growth of development concerning the reliability of Volvo's TD 96 engine which has now been redesigned and designated TD 100 (1)*. This is a six-cylinder in-line direct-injection turbo-charged diesel engine with a bore of 120·7 mm (4·75 in), stroke of 140 mm (5·5 in), total swept volume of 9·6 litre (568 in^3), and dry weight of 890 kg (1960 lb), rated 250 b.h.p. at 2200 rev/min.

The turbo-charger

In order to gain experience concerning the reliability of the turbo-charger, it was replaced free of charge after 50 000 km for detailed study. No serious turbo-charger damage resulting from its design was found, but certain secondary faults resulted from worn injection pump couplings of Bakelite cross type, which could not withstand the forces caused by the high injection rate. The delay in injection caused by this wear gave a higher exhaust temperature, resulting in too high a turbine speed.

The first turbo-chargers were not designed to withstand the high exhaust back-pressure caused by the exhaust brake which had, therefore, to be mounted in front of the turbine. The exhaust brake required a throttle valve for each manifold port which resulted in too small a volume of gas being locked between the exhaust brake and the exhaust valves. This resulted in heavy pressure shocks which gave a good exhaust brake effect, but which simultaneously gave a considerable exhaust valve bounce, with consequent fractures in the valve head or stems. The problem was partly solved by using better valve material,

References are given in Appendix 7.1.

but a fully reliable solution was not obtained until the exhaust brake had been mounted behind the turbine so that the pressure absorbing volume became sufficient to reduce uncontrolled valve movement to an acceptable level.

The disadvantage of placing the exhaust brake unit behind the turbine was that the sealing rings on the turbine shaft wore, thus allowing oil to enter the exhaust system where it was burnt up, causing smoke emission and carbon deposits. A great deal of work has taken place to solve this problem. Various types of material for the piston sealing rings and turbine shaft have been used. The best solution was reached with an improved material which gave a tighter fit of the sealing rings in the bearing housing. After a very short running time the rings centre themselves in the ring grooves through the movement of the rotor caused by axial play, and their position is not altered by subsequent pressure impulses in the exhaust system. In this manner metal-to-metal contact is prevented, while at the same time the clearance is sufficiently small to ensure the correct degree of gas and oil sealing.

After a few years Volvo introduced service inspections for the turbo-charger every 100 000 km (63 000 miles), and this service inspection interval has now been increased to 160 000 km (100 000 miles). Volvo recommends the exchange of the turbo unit for a Volvo reconditioned unit, but some customers do not follow these recommendations since, under the conditions their trucks work, they have found that the recommended milage can be at least doubled before the turbo-charger needs replacing. Providing that blow-by is not sucked in with the air, and that paper air filters are used, no reduction in turbo-compressor output is caused by dirt. Before the above-described piston ring assembly was introduced there were cases where the turbo-charger decreased in speed because of carbon deposits which built up between the turbine wheel and the housing. These deposits had resulted from oil leakage and necessitated more frequent cleaning.

Valve wear

One of the biggest problems in the application of turbo-charging was wear on the inlet valves. The primary reason for this wear was high peak pressure in the cylinders. Attempts to solve this problem by using various new materials did not give the required result. A reduction in wear was achieved by a modification of valve seat angle from 45° to 30°. A more flexible valve, reinforced valve mechanism, and a modified cam curve gave the final solution.

The exhaust valve seats have never given any wear problems. In this case there is good compatibility between wear and the build-up of deposits (seat angle 45°).

Cylinder head gaskets

Another problem which immediately made itself felt in turbo-charging was blown cylinder head gaskets. The higher maximum pressure resulted in movement in the cylinder head which fatigued the copper–asbestos gaskets then in use. A solid steel gasket, with Viton rings around water and oil holes, was therefore introduced.

However, market research showed clearly that blown gaskets were more frequent during the winter months. They were often located in areas where intense cold prevailed and where the amount of heat dissipated to the coolant was not sufficient to hold a normal operating temperature, especially with the engine idling or under light load. At the same time it was noted that under constant high loads (marine applications) there was no occurrence of blown gaskets. In laboratory tests it soon became apparent that the quickest testing method was to fluctuate loading from maximum speed and output to idling at the same time as coolant temperature was varied from 90 to 30°C. At this stage of development it was concluded that thermal deformation was the main culprit, and this was verified by an appraisal of stress measurements.

In order to reduce the effect of thermal deformation the cylinder head bolt load was doubled. A defined part of the bolt load was transferred to the cylinder block face so that the variations in surface pressure on the sealing surface resulting from gas impulses and heat fluctuations were reduced.

Piston and ring scuffing

In the early stages, scuffing was not a serious problem on our turbo-charged engines. At one stage, however, ring scuffing was experienced and resulted from excessively fine liner surfaces. In connection with the introduction of our TD 100A power unit, scuffing appeared but was eliminated by modifications to the piston and liner and by the removal of the lower oil ring (4-ring piston).

The type of oil recommended by Volvo (DS/Series 3) is necessary when running the engine at full load, and has totally eliminated the problem of sticking rings. Oil spray cooling of the piston is not utilized for the present output rates. The temperature in the upper piston ring groove does not, as shown by various methods of measurement, exceed the critical temperature for carbon deposits of 220°C.

DECREASED COMPRESSION RATIO

A contributing factor to the good reliability is that the peak cylinder pressure has been kept relatively low by low compression ratio. It has been reduced from 17/1 to 15/1.

In order to improve cold starting, inlet valve closing was altered from 44° to 23° after b.d.c. This meant that the same degree of compression was obtained when turning the engine over with the starter motor. Satisfactory starting was achieved down to the same temperatures as with the higher compression ratio ($-18°C$), but a little longer time is needed before the engine runs smoothly and without white smoke because of the low compression pressure at higher speed (2).

Various attempts to improve cold starting have been made. Glow plugs give good ignition but after a short time the engine misfires and emits white smoke because the glow plugs cool down. Flame heaters somewhat reduce white smoke after starting, but the improvement seems to be insufficient for engines with a low compression ratio.

The best results have been achieved by induction air preheating (electrical or by heat exchanger and burner) in the induction manifold close to the cylinders, combined with a loading of the engine immediately after starting by raising the back-pressure in the exhaust system. However, this pressure must not exceed the opening pressure of the exhaust valve (Fig. 7.3). With this design immediate starts are made possible to $-18°C$ with $\epsilon = 12.5/1$ without misfiring, and the engine quickly (within 10 min) reaches full working temperature (Fig. 7.4). White smoke disappears after 1 min when the exhaust system has been heated so that the exhaust gases are not cooled below dewpoint. This seems to illustrate that the white smoke consists of water.

With the standard battery and starter motor as fitted to our trucks, starting is possible down to $-25°C$. However, below $-18°C$, one or more cylinders misfire for 3 or 4 min. White smoke emitted in this case also includes unburnt fuel, giving an irritating exhaust.

Theoretical calculations (3) and practical tests show that when the peak pressure is kept constant and is limited

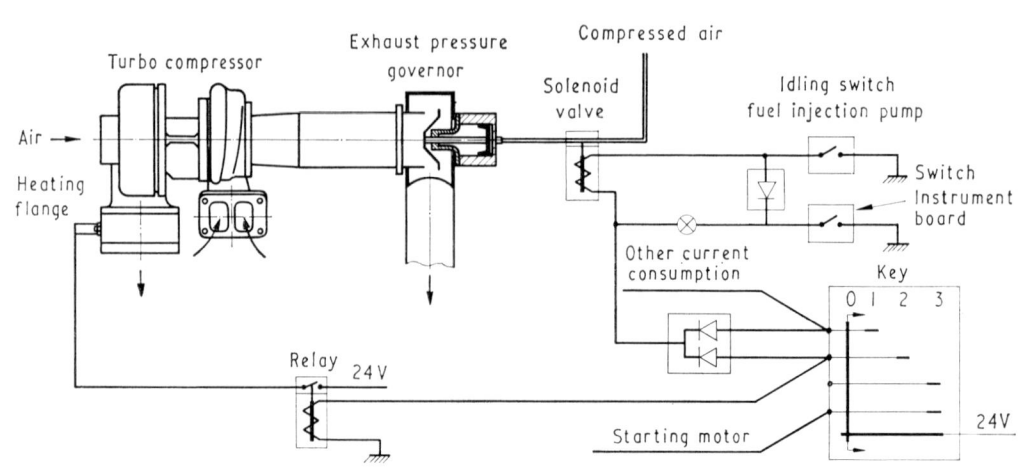

Fig. 7.3

Engine exp. TD 100A, $\epsilon = 12\cdot 5/1$.

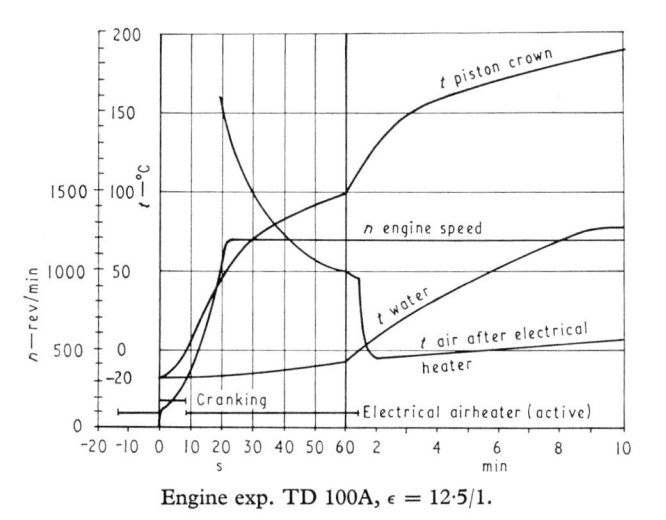

Fig. 7.4. Start at −18°C with electrical air heating and exhaust back-pressure

because of, for example, mechanical stress, then the total degree of efficiency is, within a wide range, independent of the compression ratio. On the other hand, the combustion efficiency, which is the possibility of emitting smokeless exhaust gases with a low amount of excess air, is highly dependent upon the compression ratio. A low compression ratio results in a greater ratio between the maximum pressure in the cylinder during combustion and the compression pressure. The larger pressure ratio means that a greater amount of fuel can be combusted at t.d.c. when the larger part of air in the cylinder is collected in the piston bowl and the air turbulence has reached its maximum. Combustion is thus concluded earlier during the expansion stroke and a relatively low amount of air rotation in the cylinders is required for an acceptable quality of exhaust gases, with a low amount of excess air.

MATCHING TURBO-CHARGER TO ENGINE

The turbo-charger is powered by energy derived from the exhaust gases. This energy consists of pulse energy, produced when the exhaust valves open, and the pumping work carried out by the piston during the exhaust stroke. It is the latter which, in the matching of a turbo-charger to an engine, is controlled by the choice of a nozzle ring or flow area of turbine housing; in other words, the pumping work of the piston is influenced by throttling the turbine. This throttling greatly increases with engine speed, which means that an increase in engine speed gives a corresponding increase in charging effect.

The pulse energy depends on the energy contents of the gas in the cylinder when the exhaust valve opens, and is thus a function of the energy supplied by the fuel minus the indicated work energy and cooling losses, and is not a function of engine speed.

In order to build an engine with a wide speed range and a good torque rise it is vital that the turbo-charger be driven primarily by pulse energy. To achieve this, it is important to have the high utilization of air that a lower compression ratio offers. The loss due to throttling past the exhaust valves should be kept low by providing the quickest opening of the valves and the resulting pulse should be maintained by a narrow throated exhaust manifold and well-sealed ducting in the turbine housing.

It is essential to try to obtain the lowest possible cooling losses. On the basis of Eichelberg's equation for heat dissipation in an engine

$$\alpha = 2\cdot 1 \cdot \sqrt[3]{C_m} \cdot \sqrt{(PT)}$$

one is able to see what action is necessary. A low value of C_m, which is a denomination of air turbulence in a cylinder, favours direct-injected diesel engines with a low air rotation. Low pressure P can be reached with a low compression ratio. A reduction of temperature level T is achieved with an after-cooler.

As shown by Fig. 7.5 an engine with low compression ratio (12·5/1) and after-cooler has a considerable degree of charging even at low engine speed.

CONSTANT HORSEPOWER ENGINE

An engine providing constant output throughout a wide engine speed range is the ideal source of power for a road vehicle since, in this way, a degree of tractive effort is assured which facilitates a favourable driving condition by reducing the number of gear changes necessary and the number of gearbox ratios. For highly turbo-charged engines, which always have a certain output delay in connection with rapid increases in load because of the turbo unit's pick-up time, this reduction in gear changes is a favourable factor which eliminates the effects of this delay. With constant output one can gear an engine to give lower engine speed at the maximum permissible vehicle speed, which results in lower fuel consumption and lower noise level.

As early as 1954 Volvo carried out tests with constant output engines utilizing a differential coupled supercharger powered from the output shaft of the engine, thus giving a degree of charge related to the degree of load and engine speed (**4**).

The test illustrated in Fig. 7.6 shows that with an after-cooler and a low compression ratio it is possible to achieve a constant horsepower engine. Various lines of limitation which are of considerable importance for the reliability of the engine are shown. Provided that an exhaust gas temperature of 700°C is the limiting factor, then a constant output down to 75 per cent of maximum engine speed represents an increased utilization of the available engine resources.

FUTURE ASPECTS

The continuous endeavour to provide increased economical gains will bring about larger vehicles. Vehicles of 100 tons gross vehicle weight are under discussion in Sweden and practical tests have been carried out. With the output

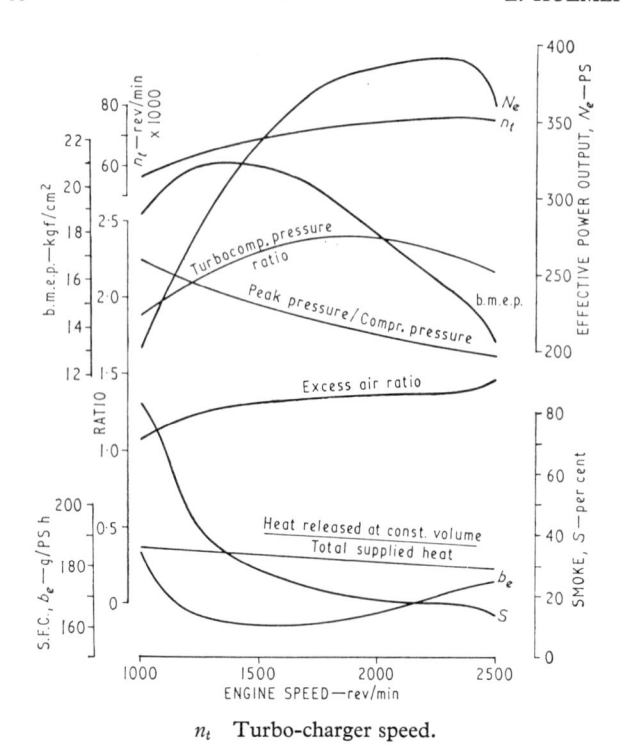

Fig. 7.5. Operating data for experimental engine TD 96 with $\epsilon = 12.5/1$ and after-cooler

n_t Turbo-charger speed.

requirement of 8 ps/ton, 800 ps would be necessary. It is possible to have a division of power source utilizing two diesel engines or one diesel engine mated with a gas turbine which would only be used under certain conditions.

On the basis of Fig. 7.6 we would like to outline a possible design for an engine giving 800 ps. The diagram is based on tests with a direct-injection diesel engine with a compression ratio of 12·5/1 and after-cooling to 60°C of air temperature in the air manifold, which would appear to be possible with an air-to-air after-cooler. In order to make the diagram more generally applicable the Y-axis is designated b.m.e.p. and the X-axis shows the mean piston speed. Output curves of interest have been numbered and the required 800 ps output is represented by number 3 on the graph.

As can be seen, a 700°C exhaust temperature gives curve number 1. With 20 per cent torque rise, peak pressure in the cylinder can be kept below 110 atm—a value which is acceptable today. By increasing the permissible exhaust temperature to 800°C it is possible to increase the output available, as shown by curve 2. A requirement here is, however, better materials for exhaust valves, exhaust manifold, and the exhaust turbine. Forced cooling of exhaust valves and/or exhaust valve seats is also feasible. The peak pressure in the cylinders of 130–140 atm, depending upon the torque graph, can be controlled with aluminium pistons without the risk of crack formation in the gudgeon pin holes, but improved cylinder head

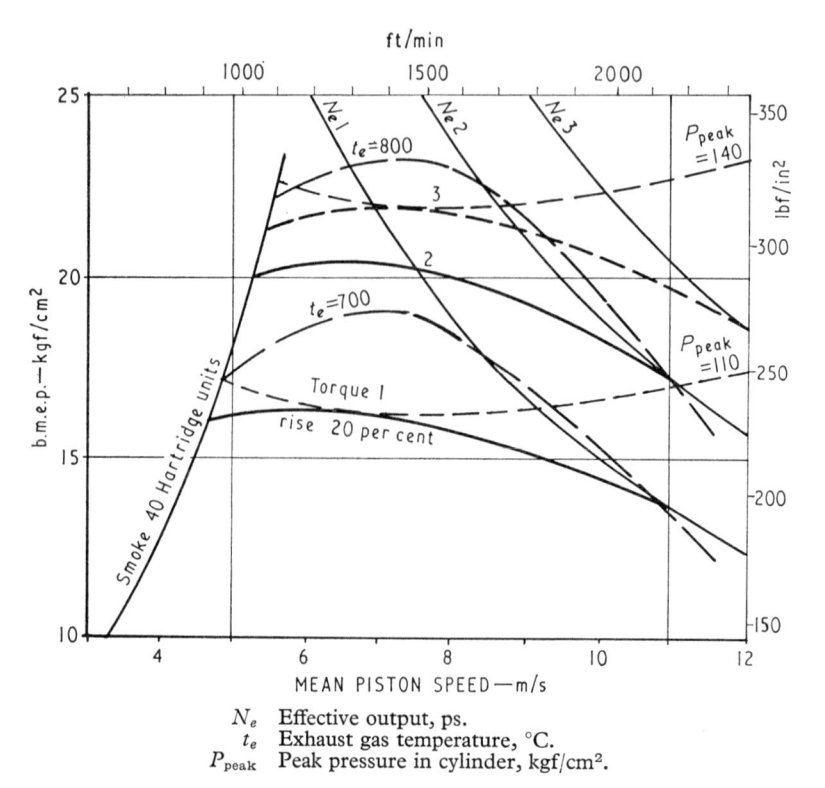

N_e Effective output, ps.
t_e Exhaust gas temperature, °C.
P_{peak} Peak pressure in cylinder, kgf/cm².

Fig. 7.6. Output limiting factors for turbo-charged diesel with after-cooling and compression ratio of 12·5/1

gaskets would be necessary. However, it seems to be possible to solve these problems with conventional and easily serviced designs. Cooling of the piston around the ring zone would be necessary. One reason for the fall in the limitation graph for exhaust temperature at high engine speeds is that the compressor efficiency declines since it has too narrow a working range. If a compressor giving a greater volume of air is chosen, then the compressor's surging limitations are exceeded at low engine speeds and thus limit the working range of the unit, which is something to be avoided in engines for road vehicles. With a compressor giving a wider working range the output shown by curve 3 should be reached if the mean piston speed is increased to 12 m/s. This piston speed should be possible with acceptable oil consumption and blow-by. With earlier injection at higher engine speeds an acceptable fuel consumption is achieved without the maximum pressure being exceeded, there being room within design margins for this.

The compressor pressure ratio in the proposed engine would be about 3/1. A higher pressure ratio would give a higher maximum cylinder pressure and a narrower working range for the compressor. A series coupling of turbo-chargers with inter- and after-cooling would widen the working range but is complicated, expensive, and requires more space. The problems experienced with 110 atm of peak pressure make it inadvisable to use peak pressure of more than 130–140 atm. A lower compression ratio would allow the degree of turbo-charging to be increased. If cold starting and combustion without white smoke and smell under low loading can be solved, then this alternative would seem promising. Variable compression ratio is a solution which may come into use if sufficient reliability can be achieved.

A turbo-charger with after-cooling and a fixed low engine compression ratio combined with starting and low load aids are therefore the most probable alternatives for commercial diesel engines in road transport service. With a 10 per cent safety margin for high peak pressures at low ambient temperatures and high exhaust temperatures in high ambient temperatures and at high altitude, then brake mean effective pressures of 17 kgf/cm² (242 lbf/in²) can be realistic. Engine size could then be calculated for 800 ps.

Various types of engine design are acceptable. When turbo-charging it is vital that the pulse energy of the exhaust gases is well utilized if the results shown in Fig. 7.6 are to be achieved. This is most easily done with a 6- or 12-cylinder unit. The Brown-Boveri developed Comprex method gives a freer choice. Some alternatives are shown in Table 7.2.

Each option has its advantages and disadvantages but all would appear achievable. Decisive for the choice of engine would be the possibility of fitting the engine and its auxiliaries in the vehicle, the weight of the engine plus transmission, the conditions of noise emission, and future requirements concerning the toxic content of exhaust gases.

These engines give 560 ps with values acceptable today corresponding to curve 1 in Fig. 7.6 (b.m.e.p. = 13 kgf/cm², C_m = 11 m/s). Output corresponding to curve 2 with 10 per cent safety margin, possible in the near future, will be 670 ps (b.m.e.p. = 15·5 kgf/cm², C_m = 11 m/s).

SUMMARY

The turbo-charged diesel engine is an economical source of power for road transport. The relatively low price per horsepower and the low fuel consumption, especially under part load, means that the requirements of higher output per ton vehicle weight will not represent a loss for a truck owner.

A higher mean effective pressure would necessitate a more robust design, but would facilitate the provision of required output with a lower displacement and at a lower engine speed. This, together with the turbo-charger's noise-absorbing quality as regards induction and exhaust noise, would give a favourable noise level.

The turbo-charged diesel engine has, at all times, except under full load combined with lowest utilized engine speed, a generous excess of air for combustion. Requirements concerning maximum permissible exhaust emission can therefore be met with a built-in margin for the variations in air conditions and altitude.

Lower compression ratios, which could be achieved with a method developed for cold starting, and after-cooling of air behind the turbo-compressor make it possible to raise the mean effective pressure without exceeding the limits which are permitted today for the mechanical and thermal loads. Through continued development work these limits can be raised.

A turbo-charger with good efficiency within a wider working range at a pressure ratio of 3/1 and with good reliability up to temperatures of 800°C is a vital link in the continued development towards higher mean effective pressures and mean piston speeds.

On the basis of results already achieved, the feasible future development, brake mean effective pressure of

Table 7.2. Optional turbo-charged and after-cooled diesel engines for 800 ps with b.m.e.p. = 17 kgf/cm² (242 lbf/in²) and mean average piston speed of 12 m/s (2360 ft/min)

Optional	Number of cylinders	Bore, mm	(in)	Stroke, mm	(in)	Swept volume, litres	(in³)	Engine speed, rev/min
1	6	158	(6·22)	158	(6·22)	18·6	(1130)	2280
2	12	112	(4·41)	112	(4·41)	13·2	(805)	3220
3	12	112	(4·41)	134	(5·28)	15·8	(965)	2690

17 kgf/cm² (242 lbf/in²) and a mean piston speed of 12 m/s (2360 ft/min) should be achieved. The proposed 800 ps engine incorporating the above data is no larger than will enable it to be installed in the space available in present-day trucks. The problem of installation of these high-output engines will be to find room for a radiator with sufficient capacity.

The 60 000 turbo-charged diesel engines produced by Volvo during the last 16 years have given experience of how to make this type of engine reliable. The continuously increasing share of turbo-charged engines of different sizes has shown that customers freely choose this type of diesel engine more and more.

APPENDIX 7.1

REFERENCES

(1) HÄGGH, B. and HOLMÉR, E. 'Experience and development of turbo-charged diesel engines in Scandinavia', S.A.E. Paper No. 690746, 1969 (October).
(2) BIDDULPH, T. W. and LYN, W-T. 'Unaided starting of diesel engines', *Proc. Instn mech. Engrs* 1966-67 **181** (Pt 2A), 17.
(2) LYN, W-T. 'Calculation of the effect of rate of heat release on the shape of cylinder-pressure diagram and cycle efficiency', *Proc. Auto. Div. Instn mech. Engrs* 1960-61, 34.
(4) LARBORN, Å. and STÅLBLAD, J. 'Some experiments with high-pressure supercharging in high-speed diesel engines with special torque requirements', CIMAC Colloquium, Wiesbaden, 1959.

Paper 8

THE DIESEL ENGINE AS A SOURCE OF COMMERCIAL VEHICLE NOISE

P. E. Waters*   N. Lalor†   T. Priede‡

The drive-past noise of diesel-engined commercial vehicles is controlled by engine speed rather than road speed. The principal noise source related to the power unit is the noise radiated by the engine structure, which has a level of 85–91 dBA during the ISO test. The combined intensity of all the other such sources is, at the most, equal to the noise radiated by the structure. It has been found that the parameters controlling engine noise are the bore, speed, and combustion systems. This has led to an empirical formula which predicts engine noise to within ±2 dBA.

INTRODUCTION

THE DIESEL ENGINE is now almost exclusively used to power the commercial vehicle because of its economic advantages of low fuel consumption, long life, and reliability. In achieving this position it has produced a severe noise problem.

The automotive diesel engine was introduced in the U.K. in the early 1930s, and at that stage in its development it had a slow speed and did not have a particularly good specific weight or specific bulk volume.

A Royal Commission on Transport was set up in 1928 to investigate, among other things, the regulation of vehicle maximum weights, dimensions, and mechanical condition. As a result of the work of this Commission, Motor Vehicle (Construction and Use) Regulations were introduced in 1931 under the 1930 Road Traffic Act. These regulations imposed a gross weight limit of 22 tons, a maximum width of 7 ft 6 in, a maximum length of 30 ft, and a maximum speed of 20 mile/h for vehicles over 2½ tons (later raised to 3 tons) unladen weight. The Salter Committee Report of 1932 recommended a taxation system based on unladen weight which was implemented by the 1933 Road and Rail Traffic Act. These two Acts, in spite of protests from parts of the vehicle-building industry, brought about a revolution in commercial vehicle design, which Cornwell (1)§ suggests put the British commercial vehicle far ahead of its Continental rivals. These weight and dimensional restrictions set the shape of the commercial vehicle with forward control cab and compact engine, which has not significantly changed to the present day.

Initially these regulations and the technology of the 1930s produced engines with a stroke/bore ratio of 1·2/1·4 and maximum speeds of some 1800–2000 rev/min. Recent vehicle weight increases (to 32 tons gross for an articulated vehicle in 1964), a general speed limit of 40 mile/h, and a reasonably high speed (70 mile/h) on motorways have resulted in a demand for more engine power. This has been achieved without significant change in the bulk of the engine by an increase of engine speeds to 2800–3000 rev/min and a reduction of stroke/bore ratio to as low as 0·8. These changes have produced an increase in engine noise of some 10 dBA. Dimensional increases to 36 ft long for a rigid vehicle and 42 ft 7 in for an articulated vehicle with a width of 8 ft 2½ in were granted in 1964, but the opportunity afforded by these was not taken to alter the now traditional design of commercial vehicles or to increase engine shielding.

It seems that the only way to stimulate improved vehicle design is by legislation. To this end the Wilson Committee Final Report of 1963 recommended the introduction of vehicle noise limits. A construction limit of 89 dBA for commercial vehicles has been in force since April 1970. Currently, consideration is being given to the increase of gross weight to between 40 and 44 tons and to the introduction of a minimum power/weight ratio,

---

The MS. of this paper was received at the Institution on 23rd February 1970 and accepted for publication on 6th March 1970. 22
* *Research Fellow, Institute of Sound and Vibration Research, Southampton University.*
† *Research Fellow, Institute of Sound and Vibration Research, Southampton University.*
‡ *Professor of Automobile Engineering, Institute of Sound and Vibration Research, Southampton University.*

§ *References are given in Appendix 8.1.*

probably of 8 b.h.p./ton. The combination of these changes will, no doubt, bring about as valuable an advance in vehicle design as did the changes in 1930 and 1933.

CHARACTERISTICS OF THE NOISE OF DIESEL-ENGINED COMMERCIAL VEHICLES

Fig. 8.1 shows the noise produced by an unladen 8·0-litre diesel-engined truck when driven at various constant speeds and in various gears past a microphone situated 7·5 m from the centre-line of the truck. Lines of constant gear ratio and constant engine speed are shown for the top five of the six gears. The results obtained cover a road speed range of 3–50 mile/h and an engine speed range of 700–2700 rev/min. Also shown is the rolling noise obtained by coasting past the microphone with the engine stopped and the clutch disengaged. Measurements of a similar form for constant speed drive past noise have been reported previously for a 9½ ton, six-cylinder, 6-litre engined truck (2).

It can be seen that, although there is a considerable difference in road speed between the various gears, there is little difference in noise level. If these results are plotted against engine speed, as has been done in Fig. 8.2b, they reduce almost to one line. Fig. 8.2a shows the noise emitted by a 12-litre diesel-engined truck and Fig. 8.2c that emitted by a 6-litre diesel-engined truck.

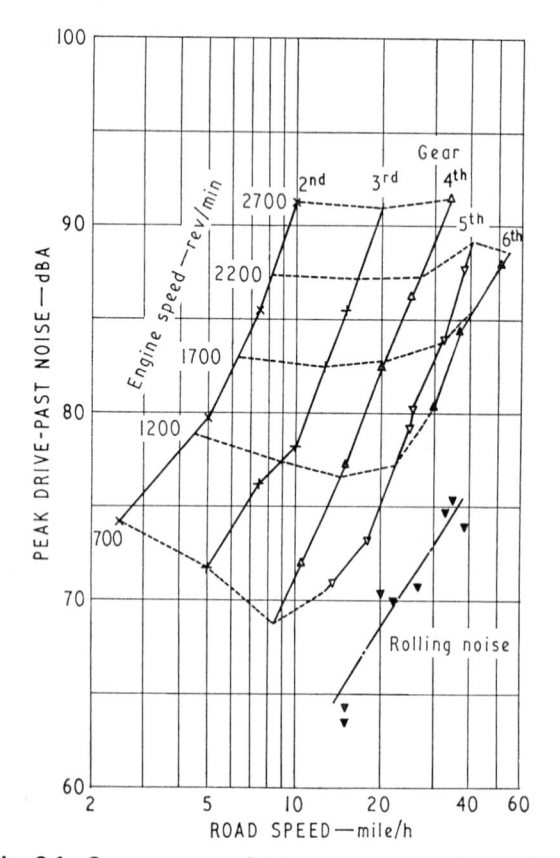

Fig. 8.1. Constant speed drive-past noise of an 8-litre diesel truck

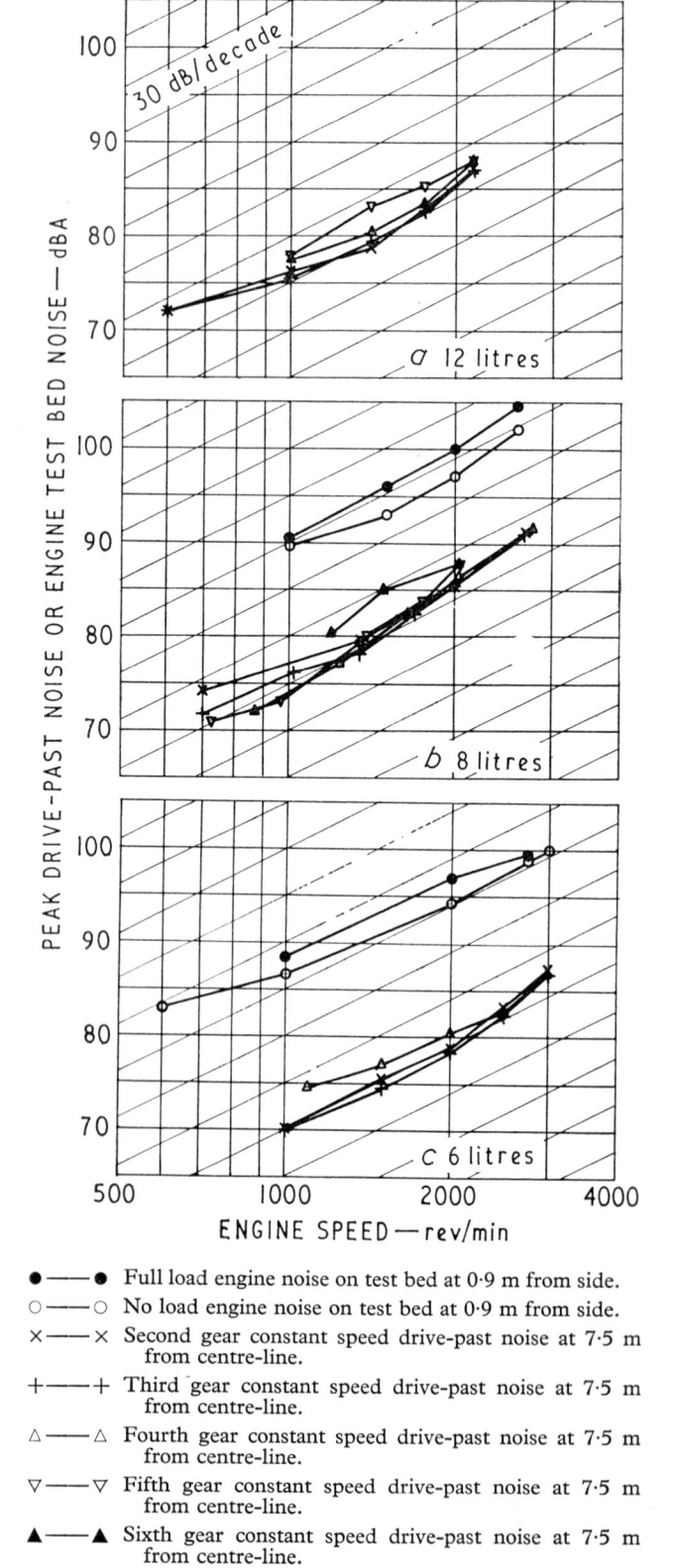

Fig. 8.2. Constant speed drive-past and engine test-bed noise for three typical normally aspirated diesel-engined commercial vehicles

● —— ● Full load engine noise on test bed at 0·9 m from side.
○ —— ○ No load engine noise on test bed at 0·9 m from side.
× —— × Second gear constant speed drive-past noise at 7·5 m from centre-line.
+ —— + Third gear constant speed drive-past noise at 7·5 m from centre-line.
△ —— △ Fourth gear constant speed drive-past noise at 7·5 m from centre-line.
▽ —— ▽ Fifth gear constant speed drive-past noise at 7·5 m from centre-line.
▲ —— ▲ Sixth gear constant speed drive-past noise at 7·5 m from centre-line.

The noise of the engine and gearbox, measured at 0·9 m on a test bed with the intake and exhaust fully silenced and with no cooling fan, is also shown for both full load and no load for the two vehicles in Fig. 8.2*b* and *c*. It can be seen from these two examples that the principal noise source must be the engine. The noise emitted by the truck is predominantly engine-speed controlled, and the rate of change of noise with speed is similar for both the vehicle noise and the engine noise.

Assuming hemispherical radiation, the distance correction between test bed and vehicle noise measurements is

$$20 \log_{10} \left(\frac{7 \cdot 5}{0 \cdot 9 + a} \right)$$

where a, half the width of the engine, varies from about 0·1 m for an in-line engine to 0·4 m for a large vee engine. This gives distance correction values of 17·5–15·2 dB. However, the attenuation of the engine noise apparently varies from some 13 dB in the higher gears and at high engine speeds to some 16 dB in the lower gears at low engine speeds.

This suggests that the engine noise is modified under some conditions by secondary noise sources. These secondary sources can include such things as the cooling fan, exhaust, air intake, and body vibration. However, these are mostly engine related systems and at their loudest are no louder than the engine structural noise.

It has been shown (3) from a large sample of engines that (for engines of similar configuration) where V is the swept volume the test bed noise is given by

$$\text{dBA} \propto 17 \cdot 5 \log_{10} V$$

Fig. 8.3 shows the mean levels of a number of trucks plotted against cylinder capacity for two different engine speeds. The engines of these trucks are all normally aspirated, in-line, six-cylinder and have a stroke/bore ratio in the range 1·05–1·09. It can be seen that they basically fit the above relationship, which confirms that the effect of engine capacity on vehicle noise is the same as it is on engine test-bed noise. Although the 6- and 12-litre trucks have similar noise levels at their maximum speeds, at their minimum speeds the 6-litre truck is some 5 dB quieter than the 12-litre truck.

STANDARD TEST METHOD FOR MAXIMUM VEHICLE NOISE

The examples discussed above indicate that the maximum noise of current generation diesel trucks is produced at maximum engine speed and that, in many examples, the engine is noisier on full load than on no load. There is also reason to expect that some engines will be noisier under transient conditions than under steady-state conditions. Therefore, it can be expected that the maximum noise of the vehicle will occur during maximum acceleration close to the maximum engine speed. Such a vehicle test has been recommended by the International Standards Organization (ISO). See references (4) and (5).

The ISO test involves accelerating the vehicle at maximum rate in a low gear over a distance of 20 m from three-quarters of the engine maximum power speed past a microphone situated 10 m from the start of the acceleration and 7·5 m from the centre-line of the vehicle. The use of a low gear, second or third depending on the vehicle, ensures that maximum engine speed is reached at or shortly after the microphone.

This test procedure is the basis of vehicle construction regulations in several countries, and for this reason most investigations of vehicle noise have been limited to this test. The drive-past test is frequently supplemented by a stationary test at maximum governed engine speed. It should be noted that these tests are only justifiable when the total vehicle noise is engine controlled.

All of the vehicle test data presented in this paper have been measured to the ISO procedure, with the exception of the constant speed tests, which were conducted on a site conforming to the ISO test site recommendation.

THE NOISE OF THE POWER UNIT

It has been shown by the previous examples that the noise of current generation commercial vehicles is controlled by the noise of the power unit. In this context, the power unit is defined (Fig. 8.4) as the engine/gearbox assembly plus its auxiliaries, such as cooling fan, generator, air and hydraulic pumps, inlet and exhaust systems. Of these the principal sources of noise are the air intake and exhaust systems, the cooling fan, and the noise radiated from the surfaces of the engine/gearbox assembly.

It is often difficult to determine the characteristics and effect of each of these sources. The only practical method is to eliminate a particular source and to determine the effect of this change on the overall vehicle noise. The differences between the noise obtained on the ISO drive-past test and on the static test can also give a useful indication of the characteristics of individual sources. For this reason it was decided to measure the static noise at 7·5 m from the vehicle centre-line opposite the engine instead of

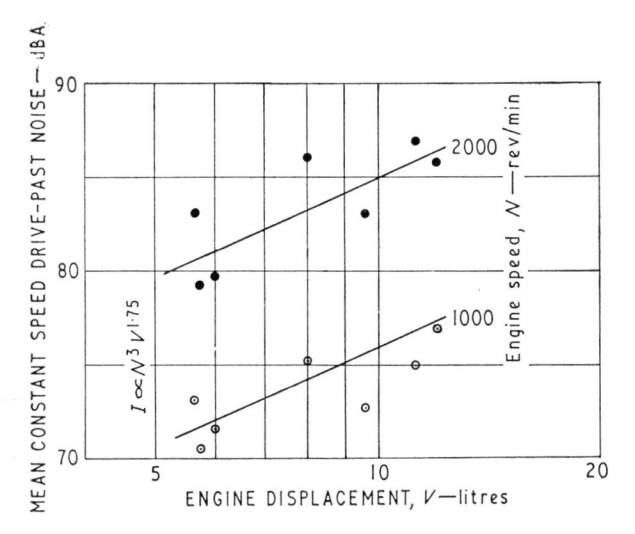

Fig. 8.3. Effect of engine capacity on drive-past noise

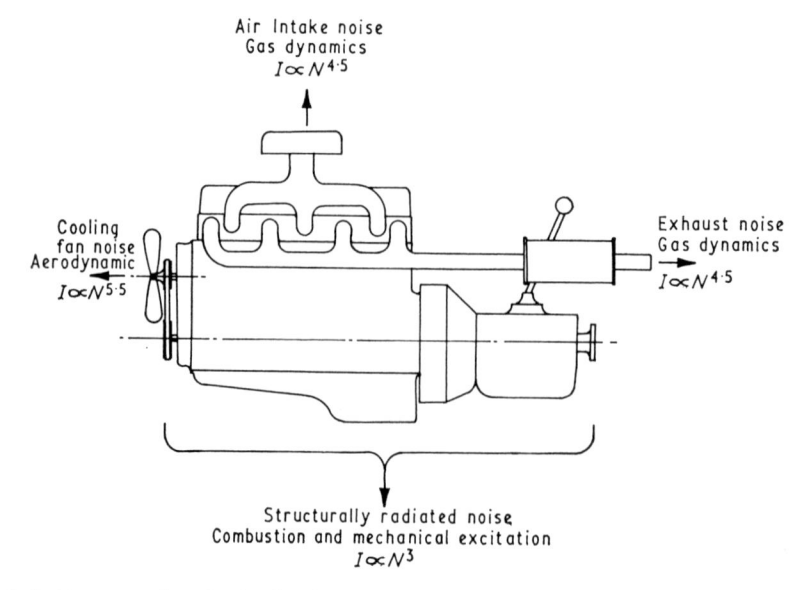

Fig. 8.4. Power unit principal noise sources with normally aspirated diesel engine

at 7·0 m from the side of the vehicle as recommended by the ISO.

Inlet and exhaust noise

Both inlet and exhaust noise of an internal combustion engine are due to gas columns vibrating at high pressure amplitudes which directly communicate with the atmosphere. The common principle of controlling these sources of noise is to use silencers which reduce the induced pressure fluctuations. Without silencing, exhaust and inlet would be the predominant source of noise (**6**). Both inlet and exhaust noise are found to have the same relationship with engine speed (**7**):

$$\text{dBA} \propto 45 \log_{10} N \quad \text{or} \quad I \propto N^{4 \cdot 5}$$

where N is the rotational speed and I is the intensity.

Fig. 8.5 shows the ISO drive-past and static noise spectra for a 7¾-litre V-8 truck. The standard truck produced an ISO level of 90·7 dBA and a static level of 90 dBA. The first modification tried was the addition of an extra exhaust silencer. This made no change to the drive-past noise, but reduced the static noise to 88·5 dBA, which indicated an exhaust noise level of 84·7 dBA in the static test. The addition of an extra air intake silencer still did not change the ISO level, but reduced the static level to 86·5 dBA, indicating an air intake noise level of 84·2 dBA during the static test.

Fig. 8.6 shows the ISO drive-past spectra for the 8-litre in-line six-cylinder engined truck of Figs 8.1 and 8.2*b*. Silencing the air intake in addition to removing the fan reduced the total noise from 89·5 to 88·3 dBA, which gives a level of 83·3 dBA for the air intake.

These examples and measurements on other vehicles fitted with their standard inlet and exhaust silencers show maximum levels of inlet and exhaust noise of 80–86 dBA.

Cooling fan

The axial flow type of cooling fan is usually used with water-cooled engines. The mechanism of noise generation by this type of fan has been investigated (**8**) and the relationship of noise and speed has been shown to be

$$\text{dBA} \propto A \log_{10} N$$

where A is of the order of 55–60 and N is the rotational speed.

Fig. 8.7 shows the ISO drive-past and static noise spectra for an 8½-litre V-8 diesel-engined truck. The second modification tried on this was the removal of the cooling fan. This reduced the total noise from 89 dBA on the drive-past test and 85·5 on the static test with only the gearbox covered to 88·5 dBA on the drive-past test and 83 dBA on the static test. The fan noise can be estimated from these measurements as 79·3 dBA on the ISO test and 81·9 dBA on the static test. The engine speed on the static test was 2850 rev/min. Therefore, assuming a fan noise to speed relationship of dBA $\propto$ 55 $\log_{10} N$, the engine speed corresponding to the fan noise in the ISO drive-past test can be estimated at 2320 rev/min. The ISO test approach speed was 2100 rev/min in third gear.

The third modification to the 7¾-litre V-8 engined truck of Fig. 8.5 was to remove the cooling fan. This gave no change to the ISO drive-past level, but reduced the static noise to 84 dBA. This is equivalent to a fan noise level of 82·9 dBA.

In the case of the 8-litre truck of Fig. 8.6, removal of the fan reduced the total noise from 91·5 to 89·5 dBA. Therefore, the fan noise was 87·2 dBA. Measurement of the noise of these and of several other vehicles with and without fans has enabled the level of fan noise to be deduced as 80–90 dBA during ISO test conditions. Fan noise is usually more troublesome in the static test than in the

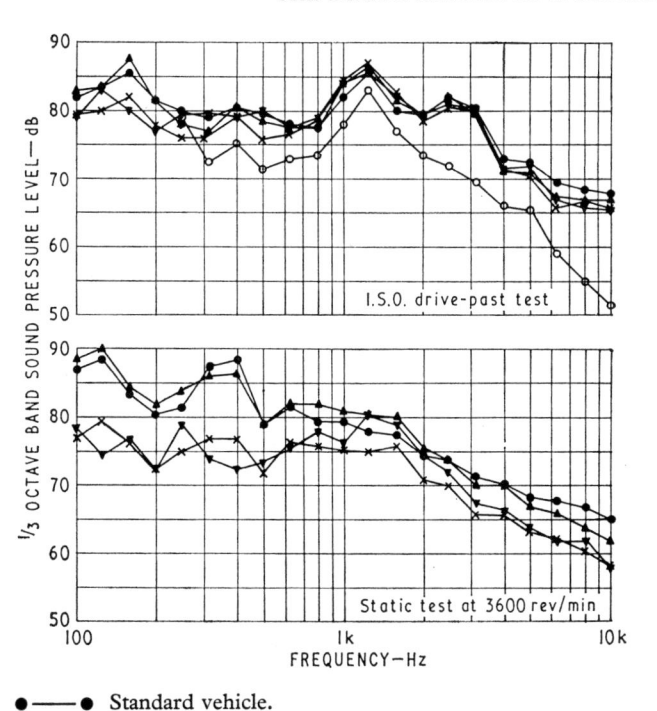

THE DIESEL ENGINE AS A SOURCE OF COMMERCIAL VEHICLE NOISE

- ●——● Standard vehicle.
- ▲——▲ Extra exhaust silencer.
- ▼——▼ Extra exhaust silencer and extra intake silencer.
- ×——× Extra exhaust and intake silencers and fan removed.
- ○——○ Extra exhaust and intake silencers, fan removed, and sides and front of engine covered with lead sheet.

Source	ISO drive-past test, dBA	Static test, dBA
Exhaust noise level . . .	?	84·7
Intake noise level . . .	?	84·2
Fan noise level . . .	?	82·9
Engine noise level . . .	90·7	84·0
Other sources . . .	Negligible	Negligible
Total noise level . . .	90·7	90·0

Fig. 8.5. Instantaneous spectra (at peak dBA) for a 7¾-litre diesel-engined truck showing the effect of exhaust, air intake, fan, and engine structure

drive-past test as the fan noise rises rapidly with engine speed. This leads to the conclusion that fan noise can be controlled easily by the use of a special fan drive to limit fan speed. This has the added advantage of reducing the power absorbed by the fan.

Gearbox noise

The source of excitation for the gearbox is not altogether clear. As the gearbox, in most installations, is closely coupled to the engine, it is likely to be excited by vibrating forces transmitted from the engine or by vibratory forces caused by gear meshing. Levels of gearbox noise under ISO test conditions are of the order of 75–85 dBA in the worst cases.

The 8½-litre engined truck of Fig. 8.7 produced an ISO drive-past test level of 90·2 dBA and a static level of 86 dBA in standard form. The first modification carried

- ●——● Standard vehicle.
- ▲——▲ Fan removed.
- ×——× Fan removed and extra air intake silencer.

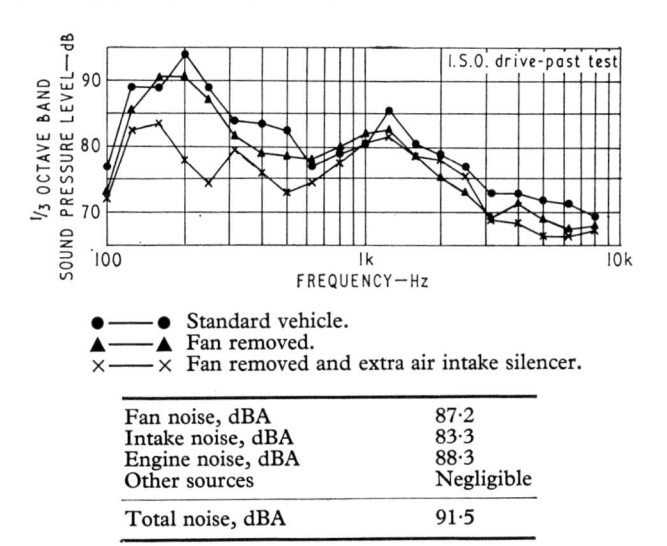

Fan noise, dBA	87·2
Intake noise, dBA	83·3
Engine noise, dBA	88·3
Other sources	Negligible
Total noise, dBA	91·5

Fig. 8.6. Instantaneous spectra (at peak dBA) for an 8-litre diesel-engined truck showing effect of fan and air intake silencer

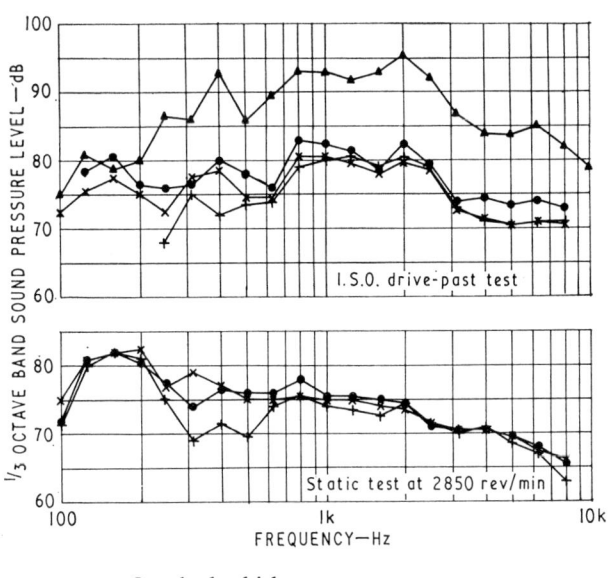

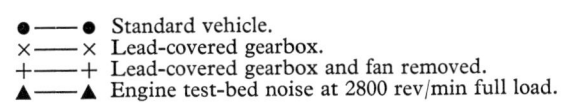

- ●——● Standard vehicle.
- ×——× Lead-covered gearbox.
- +——+ Lead-covered gearbox and fan removed.
- ▲——▲ Engine test-bed noise at 2800 rev/min full load.

Source	ISO drive-past test, dBA	Static test, dBA
Gearbox noise level . . .	84·0	76·3
Fan noise level . . .	79·3	81·9
Engine noise level . . .	88·5	83·0
Other sources . . .	Negligible	Negligible
Total noise level . . .	90·2	86·0

Fig. 8.7. Instantaneous spectra (at peak dBA) for an 8½-litre diesel-engined truck showing the effect of the gearbox and cooling fan

out was to encase the gearbox in lead sheet lined with glass fibre mat. This covering technique has been found to give some 12–15 dB reduction of noise, so that the noise from the item covered is effectively eliminated. This led to reductions of noise to 89 dBA on the ISO test and 85·5 dBA on the static test. From these results the level of noise from the gearbox can be estimated as 84·0 dBA in the ISO test and 76·3 dBA in the static test. This reduction in gearbox noise of some 8 dBA from the drive-past to the static test suggests that gear noise could be a significant source of excitation in the drive-past test, as the change in combustion induced noise for this power unit from no load to full load is only of the order of 3 dBA.

Engine noise

For the $7\frac{3}{4}$-litre V-8 engined truck of Fig. 8.5 the final drive-past level of 90·7 dBA represents a reduction of 15·3 dBA from the full-load test-bed noise at the rated speed of 3300 rev/min, and the final static test level of 84 dBA represents a reduction of 14·1 dBA from the no-load test-bed noise at 3600 rev/min. That the drive-past noise is predominantly engine noise is confirmed by the final test which involved covering the sides of the engine block, crankcase, and sump and the front of the crankshaft with glass fibre lined lead sheet. This reduced the ISO drive-past level to 86·2 dBA after modification to the other sources had had no effect.

The final level of 88·3 dBA for the 8-litre truck (Fig. 8.6) represents a reduction of 16·2 dB from the test-bed noise at rated load and speed. The 3 dBA contribution from the fan and gearbox accounts for the apparent increase of noise at high engine speed in Fig. 8.2b.

In the case of the $8\frac{1}{2}$-litre V-8 engined truck of Fig. 8.7 the final levels of 88·5 dBA for the drive-past test and 83 dBA for the static test represent reductions from the full-load and no-load test-bed noise levels at the rated speed of 2800 rev/min of 15 and 18 dB respectively. Fig. 8.7 also shows the test-bed engine noise spectrum at full load at 2800 rev/min with fully silenced intake and exhaust and with gearbox attached. From 630 Hz upwards there is fairly close agreement between this spectrum and the drive-past spectra allowing for some 15–16 dB attenuation with distance. Below 630 Hz the level of the engine spectrum decreases with decreasing frequency, but the spectrum of the vehicle noise remains relatively flat. This is typical of the results obtained with other trucks and is an indication that engine noise is much less important in the low-frequency region than it is around 1–10 kHz.

The results presented above for a range of trucks from 6 to 12 litres and both in-line 6 and V-8 cylinder engines indicate that engine noise itself on the ISO test is around 88–91 dBA. The engine noise is then a limiting factor to achieving current legislated noise levels of 88–89 dBA.

It is useful to be able to correlate engine test-bed data with the vehicle noise. It has been shown above that in the absence of any effective body shielding, the engine noise measured at 0·9 m on the test bed with fully silenced exhaust and air intake is reduced by some 15–17·5 dBA when measured in the vehicle at 7·5 m. Therefore, we can say that if the engine itself gives 104 dBA on the test bed, then in the absence of any other noise source on the vehicle, the vehicle would just comply with present British legislation. However, it is recommended that to avoid any serious problems in a practical installation, and in meeting Continental legislation for engines under 200 DIN hp, a test-bed level of 101 dBA should be the aim.

FACTORS AFFECTING ENGINE NOISE

Investigations have been carried out to determine the effects of design parameters on the noise emitted by the engine. It will be shown in the following discussion that the two basic parameters controlling engine noise are (a) the cylinder bore, and (b) the rated engine speed. Other parameters such as engine load, number of cylinders, cylinder configuration, details of engine design, combustion, and injection systems as currently employed, although significant, have only a secondary effect on noise with the result that it has been possible to obtain an empirical formula to predict engine noise from just these two basic design parameters.

Effect of basic design parameters

In the recent past, automotive engines were generally designed with a stroke/bore ratio of approximately 1·2, and it has been established from theoretical considerations and from measurements on a number of these engines that for a constant speed there was an increase of noise of the order of 13·3–17·5 dBA per ten-fold increase in cylinder volume (3).

Recently, however, the bore sizes of many automotive engines have been increased without change in the stroke and rated speed. This has enabled a considerable increase in power output to be obtained without an increase of

● ——— ● Stroke/bore ratio 0·8, 106 dBA.
× ——— × Stroke/bore ratio 1·06, 103·5 dBA.

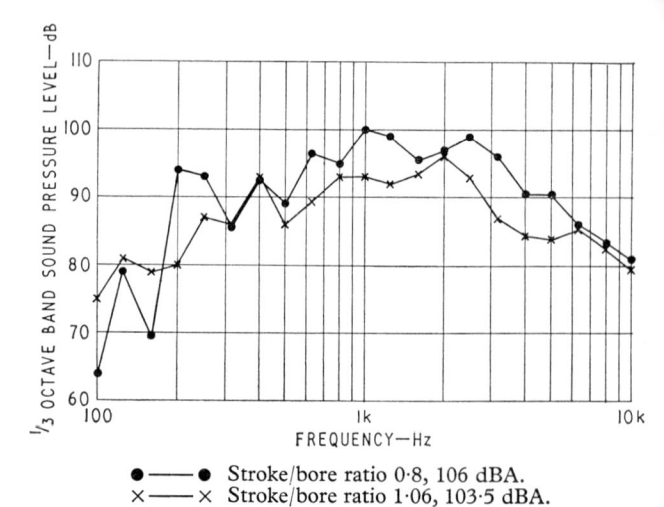

Fig. 8.8. Test-bed noise spectra at 2800 rev/min full load for two 1-litre/cylinder V-8 engines showing effect of stroke/bore ratio

mean piston speed and with an improvement in the specific bulk volume. An example is quoted by Herschmann (**9**) in which the bulk volume of a 90° V-8 engine was reduced by 16 per cent by attention to a number of design factors, including a change in the stroke/bore ratio from 1·06 to 0·94. The power output was maintained by an increase in engine speed with only a 4 per cent increase in mean piston speed.

Fig. 8.8 shows the spectra at 2800 rev/min of two 8-litre, eight-cylinder engines with stroke/bore ratios of 1·06 and 0·8. It can be seen that the short stroke engine is considerably noisier over the part of the frequency range which determines the dBA level. The cylinder pressure diagrams in the two engines were identical.

From these considerations Grover (**10**) suggested that the relation found between the cylinder displacement and noise is no longer adequate as the stroke/bore ratio must also be taken into account. The previous relationship, $I \propto V^n$ where $1·33 \leqslant n \leqslant 1·75$, was found for engines of similar stroke/bore ratio, i.e.

$$S \propto B$$
Now
$$V \propto B^2 S$$

Therefore, combining these three relations gives
$$I \propto B^m$$
where $4·00 \leqslant m \leqslant 5·25$.

The rate of increase of noise with speed is generally determined by the form of the cylinder pressure diagram (**11**). The general slope, in the frequency region containing the maximum engine structure response, of the spectrum of the cylinder pressure considered as a repetitive wave determines the rate of increase of noise with increasing speed if the cylinder pressure diagram remains of constant form and size on a degree basis. With the normal abrupt form of the cylinder pressure diagram of a normally aspirated diesel engine, the spectra have a slope of about 30 dB per ten-fold increase in frequency which gives an increase of noise of 30 dBA per ten-fold increase of engine speed:
$$I \propto N^3$$

Since the load (**11**) has negligible effect, the noise of the engine can be defined by the relation
$$I \propto N^3 B^m$$

It has been found empirically, as will be shown later, that the best value for m is 5·0 for the range of engines for which data are available. Thus,
$$I \propto N^3 B^5$$

Effect of number of cylinders and cylinder configuration

Comparison of a number of in-line engines of the same make has shown that there is no major difference in the overall, A-weighted, noise emitted between a four-cylinder and a six-cylinder engine, all other design parameters being the same. There are, however, differences in the characteristics of the noise, as is shown in Fig. 8.9.

The characteristics of the noise from in-line engines

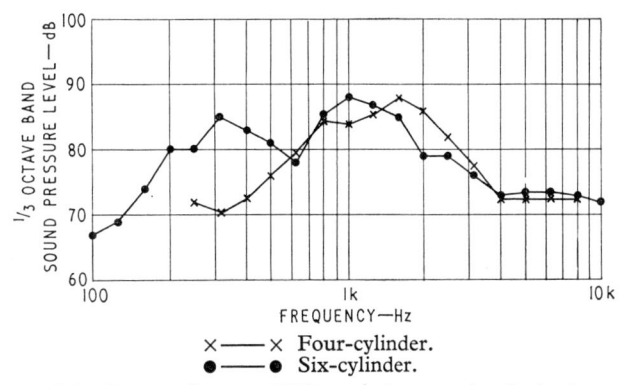

Fig. 8.9. Comparison at 1500 rev/min, part load of a four-cylinder engine and a six-cylinder engine which are otherwise identical and of 1 litre/cylinder capacity

× —— × Four-cylinder.
● —— ● Six-cylinder.

are determined by horizontal bending vibrations of the engine crankcase and cylinder block. The relation for the frequency of the fundamental mode of these vibrations is of the same form as that for a simple free-free beam:

$$f = \frac{c}{l^2}\sqrt{\frac{EI}{\mu}}$$

where l is the length of the cylinder block and μ the mass per unit length of the cylinder block.

In the example shown in Fig. 8.9 the quantity inside the square root is common to both engines, resulting in the frequency of the mode being inversely proportional to the square of the length. The four-cylinder engine has a peak at 800 Hz and the six-cylinder engine has a peak at 613 Hz. These frequencies are approximately in the ratio of 6^2 to 4^2.

In general, I and μ are functions of l and it has been found empirically (Fig. 8.10) that for engines with dry liners and main bearings between each cylinder:

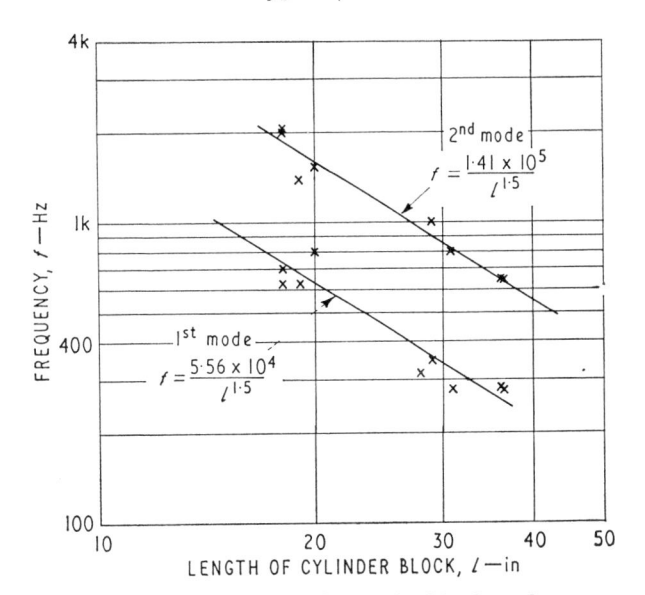

$$f_b = c/l^{1·5}$$

Fig. 8.10. Effect of length of cylinder block on frequency of lateral bending modes of in-line engines

where l is the block length (in), and c is equal to $5 \cdot 56 \times 10^4$ for the fundamental mode and $1 \cdot 41 \times 10^5$ for the second mode.

The noise of vee-form engines has been examined in some detail (12). On these engines the crankcase constitutes only a small part of the total surface and the total engine noise is much greater than would be expected from this surface alone. This is because, in addition to the basic modes of vibration identified in in-line engines, pronounced cylinder bank-to-bank vibrations are exhibited similar to a large tuning fork. On some engines this mode is the source of the predominant noise. The bank-to-bank frequency can also be determined by a relevant dimension of the engine, in this case the distance, h, between the centre of the crankshaft and the top deck of the cylinder block:

$$f_{\text{btb}} = (173/h)^{2 \cdot 5}$$

where h is in inches.

Fig. 8.11 shows a comparison of noise spectra between a V-8 and an in-line 6 diesel engine of the same bore. As can be seen, the overall levels of noise are the same. Therefore, the effect of cylinder arrangement on overall noise is insignificant and the V-8 engine, which contributes to more compact design, allows greater specific bulk volume to be obtained without an increase of noise.

Effect of engine design detail

The more important variations of in-line engine design are as follows:

(a) Engine cylinder design (wet or dry liner);
(b) Crankcase design (underslung or skirted crankcase);
(c) Crankshaft and main bearing design (number of main bearings).

The effect of these features on the noise produced by the engine has been investigated (12) and the basic conclusions drawn are as follows:

(1) Dry liner engines have higher bending stiffnesses and higher natural frequencies than wet liner engines and, therefore, emit lower levels of noise in certain frequency ranges.

(2) The skirted crankcase engine exhibits, in addition to a horizontal bending mode, a crankcase 'flapping' mode similar to that of the prongs of a tuning fork.

(3) Main bearings between each cylinder increase the horizontal bending stiffness of the engine over other bearing arrangements.

Fig. 8.12 shows the spectra at full load, 2800 rev/min, of three 6-litre in-line six-cylinder engines of similar stroke/bore ratio produced by three different manufacturers. There are many detail differences in their designs, including one engine with underslung crankshaft and one engine with a thin sheet steel sump, and there are also differences in the design of the cylinder head and rocker cover. However, their total noise levels vary by only 1 dBA.

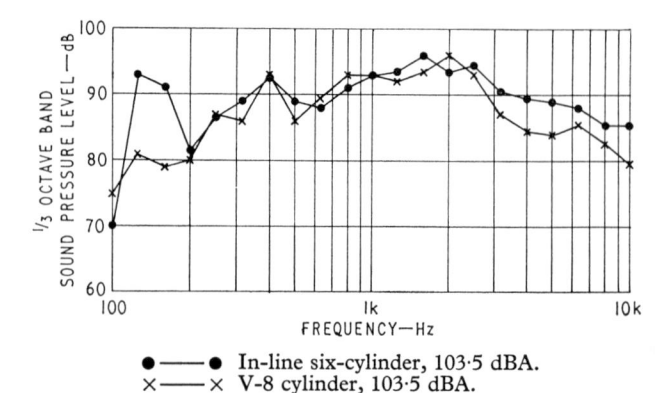

●———● In-line six-cylinder, 103·5 dBA.
×———× V-8 cylinder, 103·5 dBA.

Fig. 8.11. Comparison of in-line 6 and V-8 engines of same bore at 2800 rev/min, full load

Effect of combustion and injection systems

The rate of increase of noise with speed was previously given as 30 dB per ten-fold increase in speed, corresponding to the majority of commercial vehicle engines, which have an abrupt form of combustion. Now that engine speed ranges for a given cylinder displacement are increasing, the fixed timing injection systems of these engines are beginning to be found inadequate to maintain optimum performance and to comply with smoke regulations. Therefore, injection systems with speed and load advance characteristics have been introduced. The injection timing for minimum smoke is generally some 8° to 10° in advance of the timing for optimum performance. In addition, the timing needs to be progressively advanced with increasing speed up to 10–15° to achieve optimum fuel consumption. The result of such timing is that the noise no longer corresponds to the simple relation of dBA $\propto$ 30 log N. On high-speed direct-injection engines an appreciable steepening of the pressure diagram with increasing speed is observed with a corresponding effect on the rate of increase of noise with speed. On several indirect-injection engines of 1·5–3·0 litres, with maximum speeds of 4000–4500 rev/min, it is possible to control the noise at high speeds by gradually retarding the

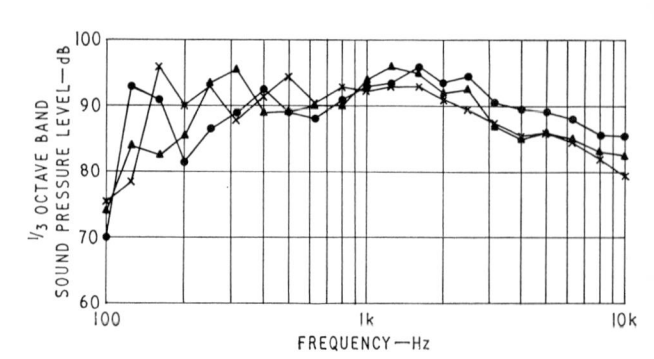

●———● Skirted crankcase, aluminium sump, 103·5 dBA.
▲———▲ Skirted crankcase, sheet steel sump, 103 dBA.
×———× Underslung crankshaft, aluminium sump, 102·5 dBA.

Fig. 8.12. Full load spectra of three in-line six-cylinder engines of the same bore showing effects of design details

injection with speed until combustion occurs so far past t.d.c. that the peak cylinder pressure becomes the compression pressure. In such cases the mean rate of increase of noise with speed over the whole speed range may be as low as 25 dB per ten-fold increase in speed.

The noise at rated engine speed can, in any case, be controlled by some ±3 dB by varying the injection timing without significant loss of performance.

On turbo-charged engines the cylinder pressure diagrams are smoother and on such engines $I \propto N^4$ has been found to give good agreement with observed noise levels.

PREDICTION OF AUTOMOTIVE ENGINE NOISE

It has been shown above that of all the current variables in engine design only the bore, speed and, to some extent, the combustion system have any significant effect on noise. It is therefore worthwhile to consider establishing an empirical formula to predict engine noise.

The following relations can be advanced for the noise at rated speed and load measured on a test bed at 0·9 m for four-stroke cycle diesel engines, and as far as they are affected by combustion relate approximately to optimum performance settings:

$$dBA = D \log_{10} N + 50 \log_{10} B - (3 \cdot 5 D - 73 \cdot 5)$$

where N is in revolutions per minute, B is in inches, and D is the predominant slope of the cylinder pressure spectrum in decibels per decade.

For a normally aspirated engine $D = 30$, and for a turbo-charged engine $D = 40$.

Table 8.1 shows the calculated and measured values for a range of engines.

CONCLUSIONS

The following conclusions can be drawn:

(1) The overall level of noise of the diesel commercial vehicle results from a number of sources related to the power unit. The predominant source is noise radiated from the engine surfaces, and at present this source alone produces a level of around 88–91 dBA under ISO vehicle test conditions. Other major sources of noise—inlet, exhaust, fan, and gearbox—are lower, approximately 82–86 dBA, but their combined effect can raise the maximum vehicle noise by some 3–4 dBA above that produced by the engine alone.

(2) Over the rest of the operating range of the vehicle the power unit is the controlling factor and vehicle noise is determined largely by engine speed. The noise of the vehicle shows little dependence on road speed.

(3) A definite relation has been established between the noise of the engine on the test bed at rated speed and load and the ISO test level obtained when the engine is fitted in a vehicle. For a normal control truck, which is a type of vehicle giving negligible shielding of engine noise, the difference between the test bed and vehicle noise is some 15–17 dB. An apparently lower value, down to 13 dB, is usually observed, but this is due to the effect of the other secondary sources such as the inlet, exhaust, and fan.

(4) No significant effect on the overall noise has been found from various features of engine design such as number and configuration of cylinders, wet or dry liners, number of main bearings, and crankcase design. These features only affect the characteristics of the engine noise and of the resultant vehicle noise. The factors that determine the overall level of noise are the basic design parameters of bore size and engine speed and, to some extent, the combustion and fuel systems. The intensity of noise has been found to be related to these basic parameters by $I \propto N^3 B^5$. This has been found to predict levels of noise of current generation diesel engines to ±2 dB.

ACKNOWLEDGEMENTS

The work presented in this paper is the result of several projects that have been undertaken at the Institute of Sound and Vibration Research over the past five years, and acknowledgement must be given to the following organizations who have sponsored and assisted in this work: Science Research Council, Ministry of Technology, Road Research Laboratory, Ford Motor Company Ltd, Perkins Engines Company, Cummins Engine Company, and British Leyland Motor Corporation. In addition, acknowledgement is due to Mr N. Ross and Mr G. Harland of the Road Research Laboratory, and members of the staff of I.S.V.R., particularly Mr L. Pullen and Mr F. Cummins for their assistance in the experimental work. Finally, thanks are due to Mrs G. Pullen for the preparation of the manuscript.

APPENDIX 8.1

REFERENCES

(1) CORNWELL, E. L. *Commercial Road Vehicles* 1960 (Batsford).
(2) WATERS, P. E. 'Control of road noise by vehicle operation', *Conf. Road and Environmental Planning and the Reduction of Noise* 1969 (British Road Federation).

Table 8.1

Engine type	Induction	Capacity, litres	Overall noise level, dBA Calculated	Overall noise level, dBA Measured
V-8	N.A.	8	107·2	108
,,	,,	8	107·2	109
,,	,,	8	103·3	103·5
,,	T.C.	15	107·3	107
In-line 6	,,	14	103·3	102
,,	,,	9	101	100
,,	N.A.	8	104·4	102
,,	,,	8	104·9	105
,,	,,	6	101	102
,,	,,	6	102·7	102·5
,,	,,	6	101·3	103
In-line 4	,,	4	101·3	103
,,	,,	2·5	104·9	107
,,	,,	2	101·3	101·5

(3) AUSTEN, A. E. W. and PRIEDE, T. 'Noise of automotive diesel engine: its cause and reduction', S.A.E. Paper No. 1000A, 1965.

(4) ISO Recommendation R362:1964. *Methods of measurement of noise emitted by vehicles* 1964.

(5) B.S. 3425:1966. *Method for the measurement of noise emitted by motor vehicles.*

(6) AUSTEN, A. E. W. and PRIEDE, T. 'Origins of diesel engine noise', *Symp. Engine Noise and Noise Suppression* 1958, 19 (Instn Mech. Engrs, London).

(7) PRIEDE, T. 'Origins of automotive engine and vehicle noise', lecture presented to the Engine Manufacturers Association, Chicago, U.S.A. 1968 (28th June).

(8) SHARLAND, I. J. 'Sources of noise in axial flow fans', *J. Sound. Vib.* 1964 **1** (3), 302.

(9) HERSCHMANN, O. 'Daimler-Benz high output engines—A study in compact design', S.A.E. Paper No. 670519.

(10) GROVER, E. C. 'Control of road noise by engine design', *Conf. Road and Environmental Planning and the Reduction of Noise* 1969 (British Road Federation).

(11) PRIEDE, T. 'Relation between form of cylinder-pressure diagrams and noise in diesel engines', *Proc. Auto. Div Instn mech. Engrs* 1960–61, 63.

(12) PRIEDE, T., GROVER, E. C. and LALOR, N. 'Relation between noise and basic structural vibration of diesel engine', S.A.E. Paper No. 690450, 1969.

Paper 9

FUEL-INJECTION SYSTEM REQUIREMENTS FOR DIFFERENT ENGINE APPLICATIONS

P. Howes*

Modern diesel engines are used in a wider variety of applications, installations, and environments than probably any other power unit. It is quite common for one engine type to be used for tractors, automotive, industrial, and marine purposes, with very little change to the basic engine. Considerable changes are necessary in the fuel-injection system, however, to meet the varying torque curve, speed range, and governing requirements. Operation over a wider range of temperatures and viscosities is often needed, and noise and smoke legislation have added further requirements. The paper outlines these problems and the compromises that must sometimes be made. Direct-injection and indirect-injection engine cases have been considered.

The paper concludes with some comments on pre-setting fuel pumps for high-volume production.

INTRODUCTION

THE PAPER HIGHLIGHTS some of the problems facing the supplier of fuel-injection equipment (F.I.E.) when meeting the varying demands of the modern high-speed diesel engine. The engines of today are used in a wider variety of applications, installations, and environments than ever before, with very little change to the engine. However, there need to be major changes to the fuel system, as it largely controls the performance of the engine.

The environmental conditions under which the engine is expected to work also change considerably. Temperature down to −40°C and up to 100°C is a typical range, with fuel viscosities which can vary from 1 cS to 7 cS. Water and other impurities in the fuel are additional hazards. The fuel-injection system must operate under these conditions, must be capable of being tailored to meet the application needs, and to be pre-set in high-volume production.

The most favoured choice in the appropriate engine size range is the Distributor Pump A (D.P.A.), the production numbers of which far exceed those of any other type. This paper, therefore, deals mainly with application experience with the D.P.A.

APPLICATION CLASSIFICATION

Applications can be divided broadly into the following groups:

The MS. of this paper was received at the Institution on 9th February 1970 and accepted for publication on 6th March 1970. 23
* *Chief Engineer (Applications), C.A.V. Ltd, Warple Way, Acton, London, W.3.*

(1) *Automotive*—covering commercial vehicles, taxis, and private cars.

(2) *Tractor*—covering, in the main, agricultural tractors.

(3) *Industrial and marine*—covering combine engines, industrial units, generator applications, and marine engines.

These applications require different torque curves, and an example of the extremes are shown in Fig. 9.1. One engine type is required both as a high-speed vehicle on the one hand, and a low-speed tractor on the other. To achieve these differences, two distinctly different fuel-delivery speed curves are required from the F.I.E.

Whilst this is an extreme case, the same applies to nearly all engine sizes. It is quite common for a 2800-rev/min truck engine to be developed to suit a 2000-rev/min tractor or a 1500-rev/min generator set, the F.I.E. being changed accordingly to give the required torque–speed curve.

In addition to changes to the speed-output characteristic, governor changes are also required. The percentage governor run-out for each classification is broadly as follows:

(1) Vehicle, 10–15 per cent.
(2) Tractors, 6–8 per cent.
(3) Industrial and combines, 4 per cent or less.

There are, however, many other governing conditions to be met, such as a particularly low idle speed or a special speed setting for power take-off, etc.

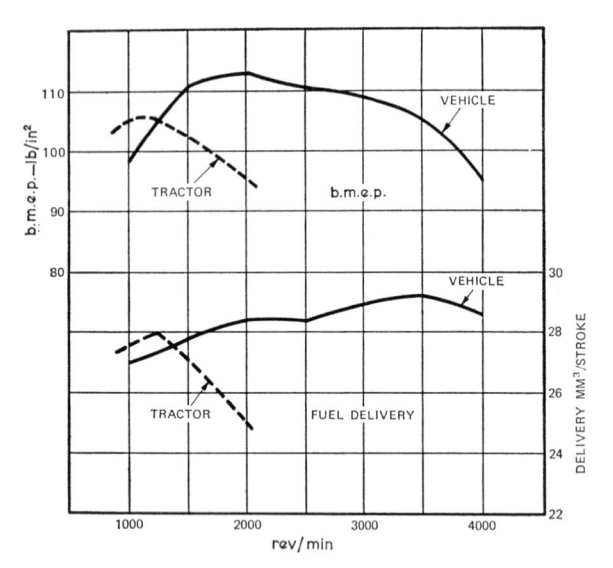

Fig. 9.1. Torque and fuel delivery requirements for different applications of the same engine

OPERATING PARAMETERS

Irrespective of the final application of the engine, it is possible to list the requirements which must be satisfied if the engine is to be successful. A complete check-list has been presented elsewhere (1)* but the more important are listed here.

Performance

The F.I.E. must be developed to enable the engine to give the required torque curve, and maximum power, with the most economical fuel consumption.

Smoke

This is a critical factor, especially now that legal limits are in force.

Noise

Again critical because of legal limits.

Governing and stability

Sometimes with legal limits on governing.

Starting; Durability; Pre-setting and Production

It is not possible to discuss these requirements exhaustively here, but some of the more critical aspects are discussed below.

PERFORMANCE, SMOKE, AND NOISE

The specific output of a diesel engine is very dependent upon the performance of the F.I.E., and considerable tailoring is required to make an engine perform in a satisfactory manner. As power output, smoke, and noise are closely linked it is convenient to consider them

* *References are given in Appendix 9.1.*

together, their relationship differing for direct- and indirect-injection engines.

Indirect-injection engines

Engines with a separate or pre-combustion chamber are generally used where a wide speed range is required. This is partly because pintle-type nozzles can be used—having a variable nozzle orifice area, they are suitable for a wider speed range than the fixed orifice or hole-type nozzle—and partly due to the better air mixing which can be achieved over a wide speed range. On the whole, however, these engines are more difficult to get right than direct-injection (D.I.) engines.

Combustion noise

A common problem with indirect-injection (I.D.I.) engines is, however, part-load combustion noise. Some of this is owing to a reduction in the combustion chamber temperature, causing an increase in the ignition lag. An illustration of this is given in Fig. 9.2, which shows cylinder pressure and nozzle opening at full load, half load, and quarter load. It will be observed that, although the dynamic timing has been held constant, the ignition delay increased with reduction of load.

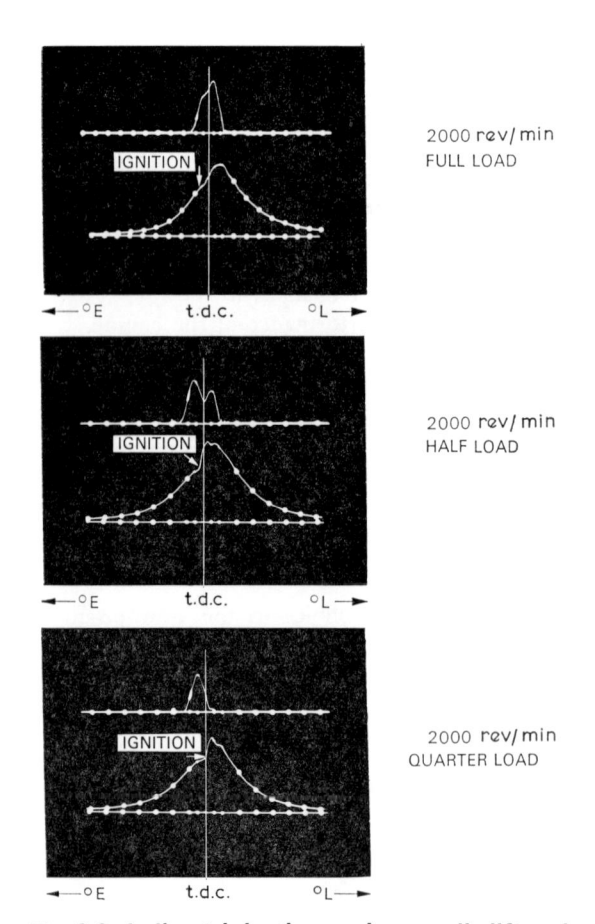

Fig. 9.2. Indirect-injection engine, needle lift and cylinder pressure

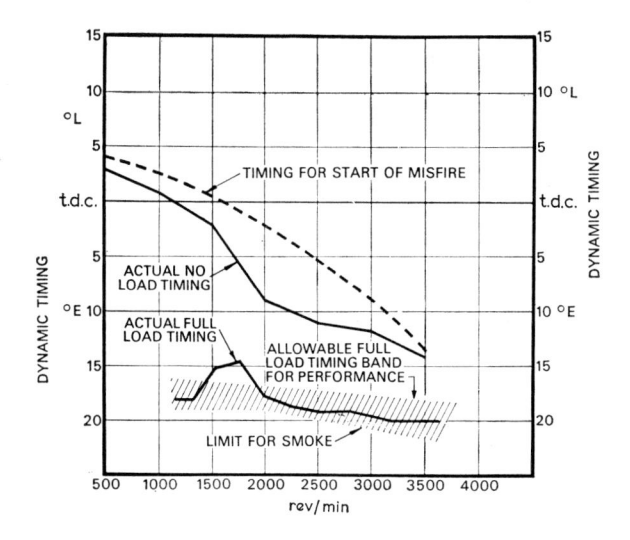

Fig. 9.3. Actual and required timing for indirect-injection engine

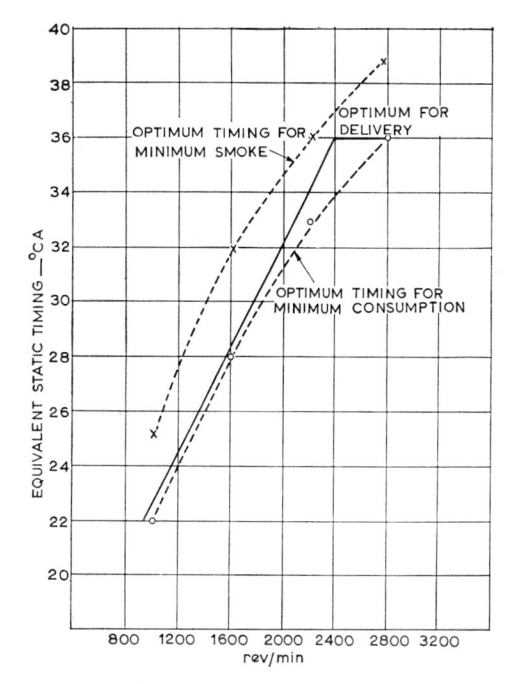

Fig. 9.4. Engine and pump advance compromise, direct-injection engine

To overcome this, means have to be found to reduce the fuel injected during the initial injection period. This reduction can sometimes be obtained by changes to the pump cam profile and nozzle-opening delay period; but much effort is required to find the optimum system. Clearly, some part-load retard can be an advantage, within the limitations determined by misfire. Modification to the engine to increase the compression ratio is also beneficial.

However, the problem, as always, is how to achieve a compromise solution between noise (or cylinder pressure), performance, and smoke. Some of the difficulties have been presented before (2), but by a suitable choice of F.I.E. it is possible to achieve a practical solution. This is illustrated in Fig. 9.3, in which the dynamic timing requirements to avoid misfire, and the tolerance band for performance are shown. The actual dynamic timings for full and no load are shown, and it will be seen that the specification is critical on timing to avoid misfire at idling and maximum speeds.

The direct-injection engine

Until very recently, the D.I. engine had not, as a rule, a noise problem, and development work was therefore simplified. In some instances, however, maximum cylinder pressures have been a limitation, particularly in turbocharged applications, and performance has had to be sacrificed accordingly.

The timing requirements for a D.I. engine are usually a compromise between the requirements for smoke and consumption. Usually, the optimum timing for minimum smoke is in advance of that for minimum consumption by about 4–8°. With a D.P.A. pump having a built-in speed advance, it is usually possible to arrange the timing to be a reasonable compromise.

It so happens that the delivery curve shape is also, to some extent, under the control of the pump advance. This is owing to the phasing of the cam relative to the delivery ports, which is changed when the cam is moved by the advance device. See under 'The production test plan'.

Fig. 9.4 shows a typical example of a compromise solution for an engine running up to 2800 rev/min. It will be seen that the advance movement chosen for delivery curve shape has been pitched to strike a balance between that required for smoke and consumption.

The introduction of noise and smoke legislation has already had an effect on engine performance. Noise is most troublesome at high engine speeds, and sacrifices in performances have had to be made either by retarding the injection timing or increasing the injection period, to keep within the limits. A reduction in the maximum rev/min may come about unless more attention is paid to the engine structure.

Smoke has caused additional difficulties with the fuel-delivery curve shape requirements. A common need for a vehicle diesel engine is a 15 per cent torque back-up from maximum speed to the peak torque speed (about 45 per cent of the full-load speed). Below this speed, it is necessary to reduce the fuel delivery to limit the smoke.

An example of an extreme case is shown in Fig. 9.5. It was a peculiarity of the combustion system of this engine that at constant fuel/stroke, smoke increased rapidly with reduction of speed. For torque back-up, however, fuel must be increased with reduction of speed. Despite this, it was just possible to achieve the required back-up at 1400 rev/min with acceptable smoke. Below this speed, however, it was necessary to decrease the fuelling very rapidly (by 26 per cent down to 800 rev/min) to avoid

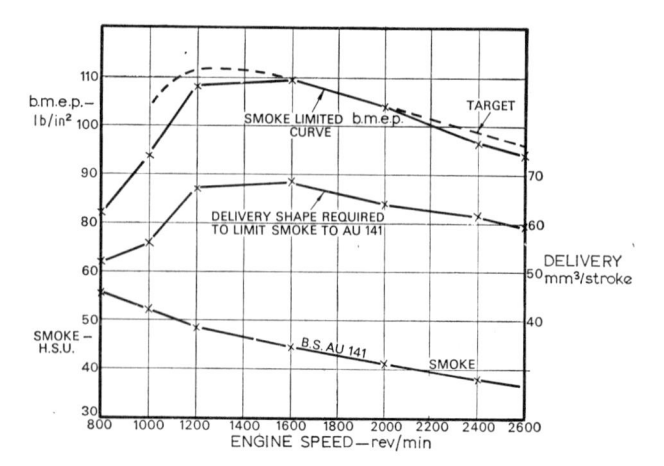

Fig. 9.5. Direct-injection engine

excessive smoke. Normally, this is difficult to achieve and it is clear that a pump having an independent torque-control system is required.

GOVERNING

The majority of current engines use all-speed governors, and to obtain the maximum speed governor performance is not normally difficult.

For tractor operation, the all-speed governor is ideal, because there are a number of agricultural operations that need to be carried out at engine speeds less than the maximum.

For automotive use, all-speed governing is satisfactory, but it is not ideally suited for automatic transmissions. In these cases, direct control of the pump output between the idling and maximum speeds is desirable to prevent excessive torque being applied during gear changing.

The use of automatic gear changing combined with fluid flywheels also creates difficulties at idling. In such cases, low idling speeds of about 380 rev/min in neutral and 350 rev/min in gear are being called for to prevent the vehicle from jerking when the gear is selected, and to prevent the vehicle from creeping when in gear. This is particularly important on public service vehicles. These low idle speeds mean that the governor must have exceptionally good anti-stall characteristics, and must not be sensitive to vehicle attitude. Neither must it be sensitive to vehicle deceleration. Furthermore, care must be exercised to ensure that in providing satisfactory idling governing, excessive pedal loads do not result at maximum speeds.

A good example of the influence of legal requirements on governor design is given by the road speed limitations imposed on tractors by several European countries. It is specified that the tractor must not be able to travel at more than a certain speed on the road. This speed is usually around 25 km/h (not the same in every country), which in top gear usually corresponds to an engine speed well below the maximum engine speed.

Tractor manufacturers were therefore faced with two choices:

(1) to modify the transmission by removing the top gear, or
(2) to ask the F.I.E. manufacturer for a special governor.

For (2) the special governor had to be brought into operation only when those gears were selected that would

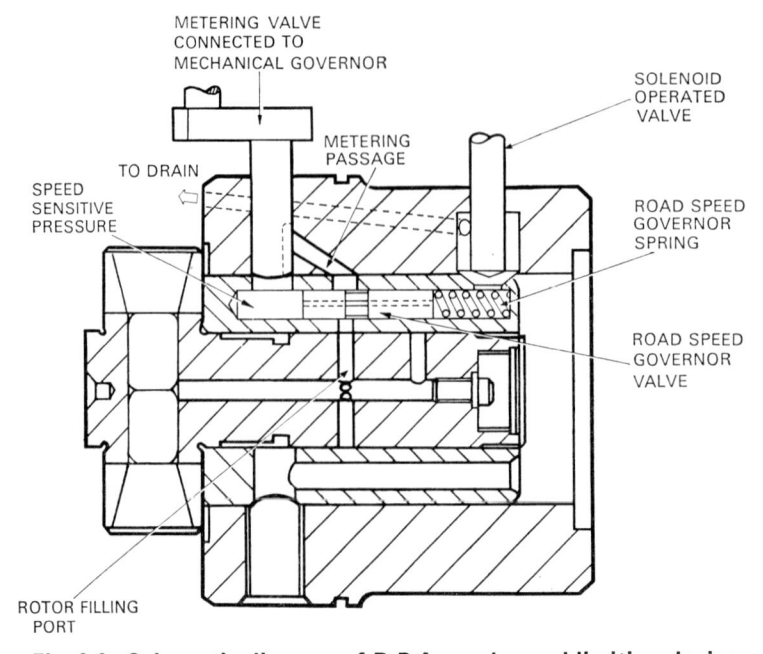

Fig. 9.6. Schematic diagram of D.P.A. road-speed limiting device

otherwise allow the tractor to exceed the legal speed. It must not interfere with the off-highway performance. In addition, it had to be tamper proof and fail safe.

In the D.P.A. pump this was achieved by providing a secondary governor in the hydraulic circuit of the pump downstream of the mechanically governed metering valve. This secondary governor consisted of a valve spring loaded against transfer pressure, which was speed sensitive. It was brought into operation by a solenoid, which kept the governor circuit closed when energized, but when de-energized opened up the circuit, thus allowing the governor valve to operate. The governed speed on the secondary governor was set by adjusting the spring load. A diagram is shown in Fig. 9.6.

Therefore, with the solenoid energized, the engine could operate normally under the control of the standard mechanical governor. With the solenoid de-energized, by a switch on the gear lever, the maximum engine speed could then be limited to that determined by the secondary governor. Control of the engine at lower engine speeds remained unaffected.

STARTING

To obtain the optimum starting, with both I.D.I. and D.I. engines, it is necessary to ensure that the dynamic timing is about 10° b.t.d.c. (3). Therefore, starting tests are initially aimed at establishing the F.I.E. requirements to achieve this, and at determining the unaided starting temperature. Because the starting characteristics of the two combustion systems are so different, from then on tests on the I.D.I. engines are usually aimed at determining which starting aid should be used. However, in the D.I. case the subsequent tests are usually aimed at determining the temperatures at which the aid should be used. Obviously, the choice of a suitable electrical specification also comes into this.

On many D.I. engines it has been found that, providing the optimum timing for starting is chosen, excess fuel is unnecessary. This is illustrated in Fig. 9.7, which compares a fixed-timing in-line pump having excess fuel with a D.P.A. pump having normal fuel, but at the optimum time for starting. However, as the excess fuel in the in-line pump is injected at the end of the injection it is nearer to the optimum timing, so that the addition of a start retard would then only have a marginal benefit.

In some in-line pumps there can be a drastic fall-off in the cranking-speed delivery owing to plunger leakage, which can then result in some engine cylinders receiving insufficient fuel for effective firing. The maximum fuel stop then needs to be repositioned to increase the low-speed delivery. The same is true of some spill-controlled distributor pumps.

It is clear, therefore, that such pumps need to have an excess fuel device, or, more properly, a 'maximum fuel stop override', which the D.P.A., having much better low-speed pumping characteristics, does not require.

Elimination of true excess fuel is beneficial from the point of view of lubricating-oil dilution, as a considerable proportion of the excess fuel finds its way into the sump.

However, excess fuel can help in accelerating the engine once a first fire has been achieved, as it helps to get the engine through the region where the temperature required for combustion is rising faster than the temperature available (3).

Another and perhaps better way to accelerate the engine is to provide external heat. Not only does this improve the time to first-fire, it improves the acceleration, as the compression temperature is then higher than the required temperature. The effect of applying external heat is shown also in Fig. 9.7.

Many modern diesel engines, therefore, are provided with a start retard device, which when combined with speed advance becomes a two-stage advance. A typical advance curve is shown in Fig. 9.8. Such engines will usually start in the U.K. without any other aids. For colder climates, 'Thermostart' starting aids or ether are used. I.D.I. engines usually have electric heater plugs in each cylinder, or 'Thermostart' starting aids fitted as standard equipment.

PRODUCTION PRE-SETTING AND CONTROL

Having illustrated some of the difficulties encountered during development, it is now opportune to discuss pre-setting and production control.

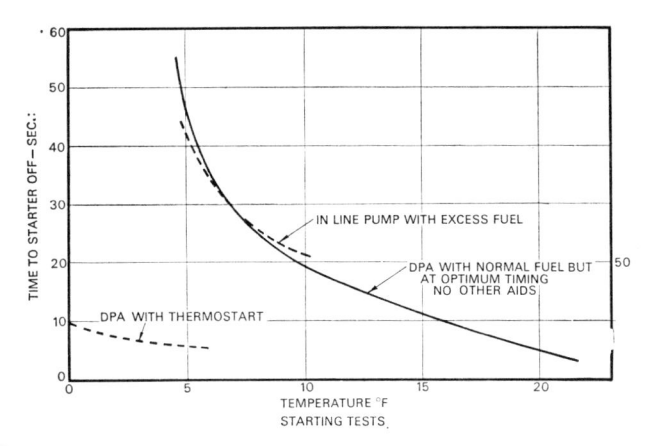

Fig. 9.7. Comparison between in-line pump with excess fuel versus D.P.A. at optimum timing with normal fuel

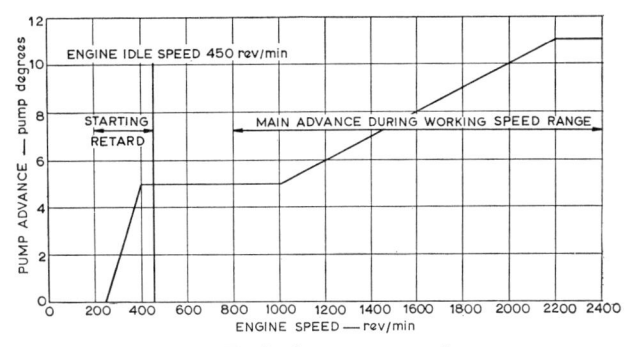

Fig. 9.8. Typical two-stage advance

Pre-setting

Modern production methods call for a minimum of handling, as handling is expensive. Application work must therefore be planned in the beginning to cover all aspects of production tolerances, not only in the fuel pump but in the nozzles, high-pressure pipes, and the engine itself. The final outcome must be a joint effort by the F.I.E. manufacturer and the engine manufacturer.

The D.P.A. distributor pump lends itself to pre-setting because:

(*a*) It has only one fuel adjustment, which controls all outlets.

(*b*) The fuel adjustment is internal, so that once set it is difficult to tamper with.

(*c*) It has 'built-in' delivery curve shape control.

(*d*) It has an integral advance device which can be pre-set.

(*e*) The pump can be easily timed externally, so that mounting to the engine becomes a simple matter.

The production test plan

There are, however, two difficulties. The first is that the inevitable small differences caused by dimensional variations can cause rather large differences of fuelling. This does arise with in-line pumps but is somewhat more frequent with distributor pumps, because the pumping line is cut off from the pump shortly after each delivery. If the cut-off occurs at a time of high-pressure flow, small differences of phasing (i.e. the relationship between cam and rotor positions) cause relatively large differences in the amount of fuel left in the line.

The second is that differences between engine and test machine fuelling may be considerable. This arises partly from the use of standard pipes and test nozzles on test machines. They are therefore different from the engine pipes and nozzles. Even with engine pipes and nozzles on the test machine, however, engine fuelling may be only poorly reproduced. This is only now becoming understood, and is owing to the different torsional oscillation content of the drives on engine and test machine.

For accurate pre-setting, care must therefore be taken during the application work to ensure that:

(*a*) Pump behaviour, in particular the level of maximum fuel, is not at any important part of the speed range abnormally sensitive to dimensional differences within the manufacturing tolerances.

(*b*) Pump-to-pump differences in fuelling to the

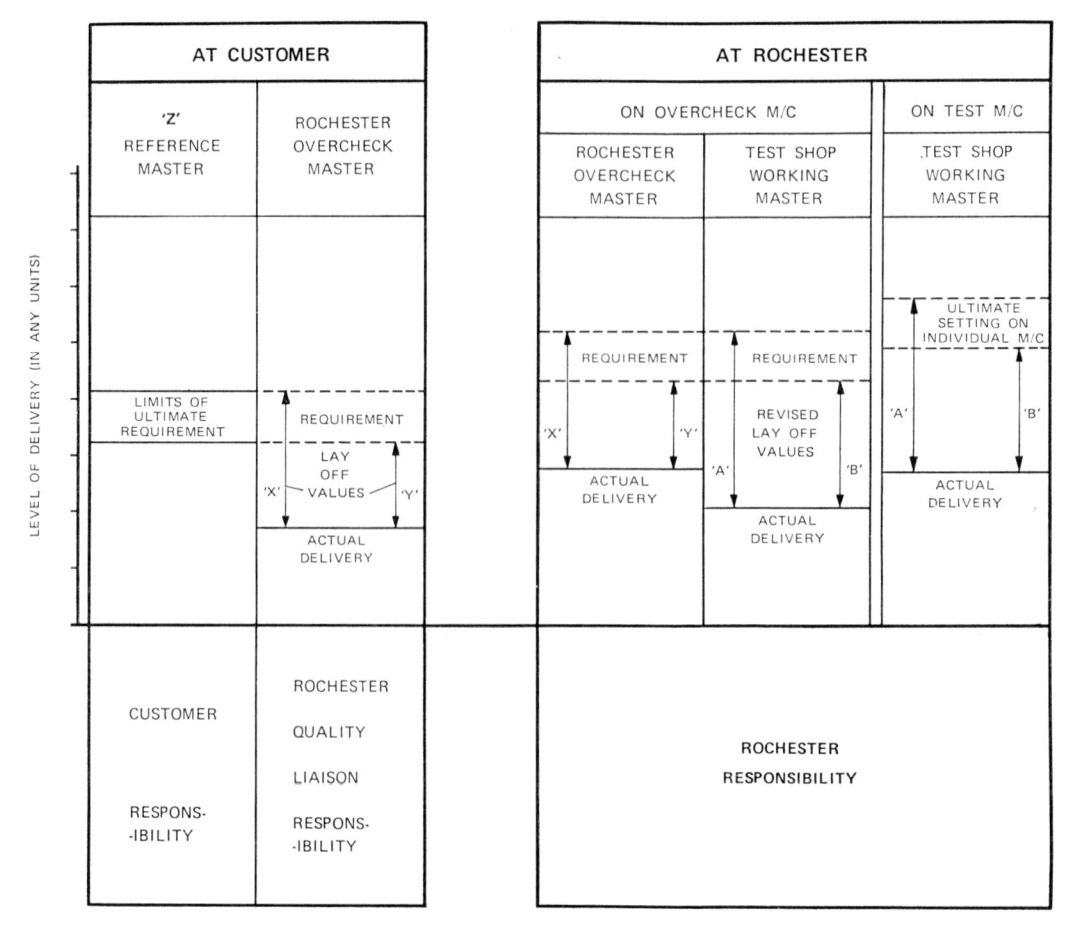

Fig. 9.9. D.P.A. master pump procedure

engine are fairly represented (or preferably over-represented) by corresponding differences in fuel delivery on the test machine.

The master-pump system

Because of the foregoing, C.A.V. have developed a master-pump system for fuel setting. From the application development work, reference pumps are chosen and set on a development engine to represent the top and bottom limit of the production tolerances.

Three master pumps are then set and run in for 200 hours in the manufacturing factory at Rochester; one to become a production setting master, one master to be held by the customer, the third being an overcheck master to correlate the first two.

Fig. 9.9 shows the system. The overcheck pump is tested on the customer's machine and compared with the reference pump limits. Its actual output could be different, so 'lay-off' figures are added to the actual values to bring these to the required level. These are 'X' and 'Y' in the diagram.

This pump is then tested on a production overcheck rig and the working master compared with it. For this, revised lay-off figures ('A' and 'B') are given to the working master such that its delivery on the overcheck machine equals the overcheck master plus the 'X' and 'Y' figures.

The working master is then put on the production test machine and the 'A' and 'B' figures added to its actual delivery. Production pumps are then set to these final figures.

By adopting this system, test machine differences, fuel temperature effects, etc., are eliminated, and small changes in the required fuel level are easily accommodated.

Production results

With the master-pump system, the pumps can be set to within ±1 per cent, but this spreads considerably when applied on production engines, for the following reasons:

(1) Influence of nozzles.
(2) Production engine test-bed measurement inaccuracies.
(3) Influence of engine production variables on fuel delivery.

The overall result can be a *recorded* fuel delivery spread of ±3 per cent on a high-delivery engine, to ±5 per cent on small-delivery engines, where the measurement inaccuracy represents a larger percentage of the fuel delivered (I).

At first glance these figures seem unacceptable, as the spread is more than most engine manufacturers will allow on power spread. Experience shows, however, that this is not so, as the power spread on 'green' engines is not *additive* to the fuel *measurement* spread.

This is clearly shown in Fig. 9.10, in which power points are plotted against fuel delivery for a number of

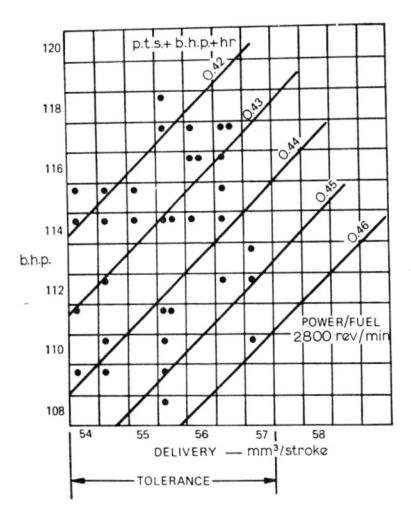

Fig. 9.10. Production engine results, automotive six-cylinder

'green' production engines. Inspection reveals that the maximum power spread, in fact, occurs at one delivery level. Therefore the inference is that even if the fuelling spread were reduced, there would be no reduction in the power spread.

From this it follows that if diesel engines are to be mass produced to a more consistent level, then an investigation into the reasons for engine performance differences must be made.

CONCLUSIONS

This paper has attempted to highlight some of the difficulties facing the manufacturer of fuel-injection equipment.

It is clear that the legal requirements are making the matching of the F.I.E. even more difficult, and closer co-operation and understanding between the engine manufacturer and the F.I.E. manufacturer is required.

It was clearly not possible to discuss all the problems in the time available, but those which have been highlighted lead to the inevitable conclusion that the fuel pump of the future must be more sophisticated, having control of timing over the load and speed, together with means of independent torque control.

The engine manufacturer, for his part, will need to pay more attention to the combustion characteristics and engine structure, to reduce both smoke and noise, and to investigate the reasons for engine differences. Improvements in production fuel and power measurements are also called for.

ACKNOWLEDGEMENTS

I would like to thank the Directors of C.A.V. Ltd for permission to publish this paper. In particular, I would like to thank Dr A. E. W. Austen, Director and Chief Engineer, for his invaluable help and support.

APPENDIX 9.1

REFERENCES

(1) HOWES, P. *A procedure for the specification of distributor pumps for high-volume production* 1969 (September) (Soc. Automot. Engrs, New York).

(2) WALDER, C. J. *Some problems encountered in the design and development of high-speed diesel engines'* 1965 (January) (Soc. Automot. Engrs, New York).

(3) BIDDULPH, T. W. and LYN, W-T. 'Unaided starting of diesel engines', *Proc. Auto. Div. Instn mech. Engrs* 1966–67 **181** (Pt 2A), 17.

Paper 10

FUEL LIMITATIONS ON DIESEL ENGINE DEVELOPMENT AND APPLICATION

H. E. Howells[*] S. T. Walker[†]

The authors discuss the emission characteristics of the high-speed diesel engine, and conclude that fuel quality variations have only a small influence on the emission of black smoke and oxides of nitrogen at full load. However, at idle or light engine loads, good ignition quality is shown to be essential in order to minimize the emission of unburnt hydrocarbons and aldehydes. Emphasis is placed on the need to pay greater attention to fuel system design in order to improve the low-temperature operability of diesel vehicles. The problems of burning residual fuel in medium-speed engines are reviewed and data on the ignition quality of these products are presented.

INTRODUCTION

THE INHERENT RELIABILITY, long life, and fuel economy of the compression ignition engine are major factors influencing its widespread use in road transport and in industrial and marine applications. However, in the face of an ever-increasing demand for more power and the influence of other energy conversion devices such as the gas turbine, the present status of the diesel engine is in jeopardy. In order to maintain its present 'pride of place' there must be advances in engine design such that higher specific outputs, accompanied by significant reductions in noise, vibration, and emissions, are obtained without sacrificing the all-important advantage of fuel economy.

It is reasonable to assume that fuel composition will be able to make some contribution to the future development of diesel engines. The present paper, which discusses the role played by fuel, is divided into two parts: fuels for automotive engines and fuels for industrial and marine applications.

A major consideration in the development of the automotive diesel engine is likely to be the restraints imposed on it by air pollution legislation, and it is therefore pertinent to discuss the emission characteristics of diesel engines in some detail. The other facet of automotive diesel application which we propose to discuss briefly is how fuel system design critically affects the operation of diesel engined vehicles in cold climates. The paper finally discusses some of the problems that can arise with the use of residual fuel oils in medium-speed engines.

EMISSION CHARACTERISTICS OF AUTOMOTIVE DIESEL ENGINES

Legislative control of emissions from diesel engines is at the present time relatively moderate and is primarily aimed at controlling exhaust smoke. However, it seems logical that future standards will continue to become more restrictive and a wider range of exhaust constituents will be controlled by legislation. In that event, the diesel engineer must have information on those engine factors which influence the emission characteristics, as well as a knowledge of the role played by the fuel.

Although there are a great deal of data available on the composition of diesel exhaust gas, surprisingly little is known about the relationship between fuel characteristics and exhaust gas composition. The present paper is intended to stimulate some interest in this subject. However, it is essential that we should place in its right perspective the role of the diesel engine as an emitter of pollutants. Many investigators define the contribution of the diesel engine in terms of the socially unacceptable 'nuisance' value of smoke, exhaust odour, and eye irritation. As a broad generalization it is true to say that the diesel engine produces slightly less total carbon monoxide and unburnt hydrocarbons than its gasoline counterpart. However, the operative word is 'slightly'; indeed at certain operating conditions (notably idle and light load) some diesel engines produce more carbon monoxide and

---

The MS. of this paper was received at the Institution on 2nd February 1970 and accepted for publication on 19th February 1970. 33
[*] Project Leader, The British Petroleum Co. Ltd, BP Research Centre, Sunbury-on-Thames, Middlesex.
[†] Engineer, The British Petroleum Co. Ltd, Britannic House, Moor Lane, London, E.C.2.

Table 10.1. Influence of hydrocarbon composition on the emission of smoke and carbon monoxide from a six-cylinder four-stroke direct injection automotive engine

4·8 in bore × 5·5 in stroke.
Speed, 1200 rev/min. Engine load, 101 lb/in² b.m.e.p. (696 kN/m²).

Fuel	Specific gravity, 60°F/60°F	Aromatics, per cent weight	Smoke, Hartridge units	Carbon monoxide, p.p.m.
A	0·772	3	43	2570
B	0·778	6	46	2620
C	0·788	15	49	2750
D	0·824	21	50	2750

unburnt hydrocarbons, when expressed as a percentage of the fuel supplied, than the spark ignition engine. Pollution levels expressed in this manner are perhaps more realistic than exhaust gas concentration values, which are deceptively low because of the dilution effect of the large quantity of excess air which passes through the diesel engine, particularly at light engine loads. However, in this paper, exhaust gas data are presented primarily in terms of volumetric concentrations, thus allowing a direct comparison with other published work. Some data are also expressed on a basis of fuel supplied in order to give some idea of how the efficiency of the combustion process varies with engine load.

Black smoke

Black smoke is formed when there is a deficiency of oxygen in the vicinity of a burning droplet of fuel. In the diesel engine, smoke emission normally becomes excessive when the overall air/fuel ratio is less than about 20/1. Black smoke can also be formed when there is a local oxygen deficiency, which can occur through overswirling of the air relative to the fuel or insufficient fuel penetration.

Effect of aromatics

Earlier investigators have shown that in diffusion flames, fuels containing a high proportion of aromatics produce relatively more smoke than fuels containing a high proportion of paraffins. However, in the diesel engine the influence of fuel aromaticity is not so pronounced. Indeed, Table 10.1 shows that a significant increase in aromatic content only produces a nominal increase in smoke emission. Thus commercial variations in the aromatic content of marketed fuels are without significant effect on the smoke emission characteristics of typical diesel engines.

Ignition quality

Our earlier work (1)* showed that a change of up to 13 cetane numbers in typical diesel fuels had very little effect on black smoke emission in both direct and indirect chamber engines. This conclusion has been confirmed by more recent work (Table 10.2) carried out on two direct injection four-stroke (automotive type) engines.

Volatility

It is often claimed that the more volatile kerosine type fuels burn with a cleaner exhaust than the conventional gas-oil type fuels. This is a fallacy since kerosines invariably have relatively low viscosity and specific gravity which results in a reduction in the weight of fuel delivered, and consequently an increase in the overall air/fuel ratio. The lower black smoke emission is therefore primarily associated with the engine operating at a lower power output. In general it has been found that on an equal power output basis, kerosine and gas-oil fuels show no difference in exhaust smoke density.

Carbon monoxide

Carbon monoxide is formed during the combustion process if insufficient oxygen is available for complete combustion of the fuel charge. Thus, it is not surprising to find that the emission of carbon monoxide increases as the smoke limited fuelling condition is approached. This is illustrated in Fig. 10.1 which is a typical emission curve for a four-stroke direct injection automotive type diesel

* References are given in Appendix 10.1.

Table 10.2. Effect of ignition quality on black smoke emission from two direct injection four-stroke (automotive type) engines

Fuel	Cetane number	Engine A 1000 rev/min	Engine A 1600 rev/min	Engine B 1000 rev/min	Engine B 1400 rev/min
Gas oil	50	52	39	30	63
Gas oil + 0·1 per cent ignition improver	54	52	44	28	63
Gas oil + 0·2 per cent ignition improver	57·5	48	45	31	64
Gas oil + 0·5 per cent ignition improver	65·5	55	44	29	65
Gas oil + 1·0 per cent ignition improver	>70	—	—	27	61
Kerosine	55	39	28	18	54
Kerosine + 0·1 per cent ignition improver	58	40	29	21	58
Kerosine + 0·2 per cent ignition improver	62	41	30	23	58
Kerosine + 0·5 per cent ignition improver	69	41	29	22	62

The ignition improver was cyclohexyl nitrate.

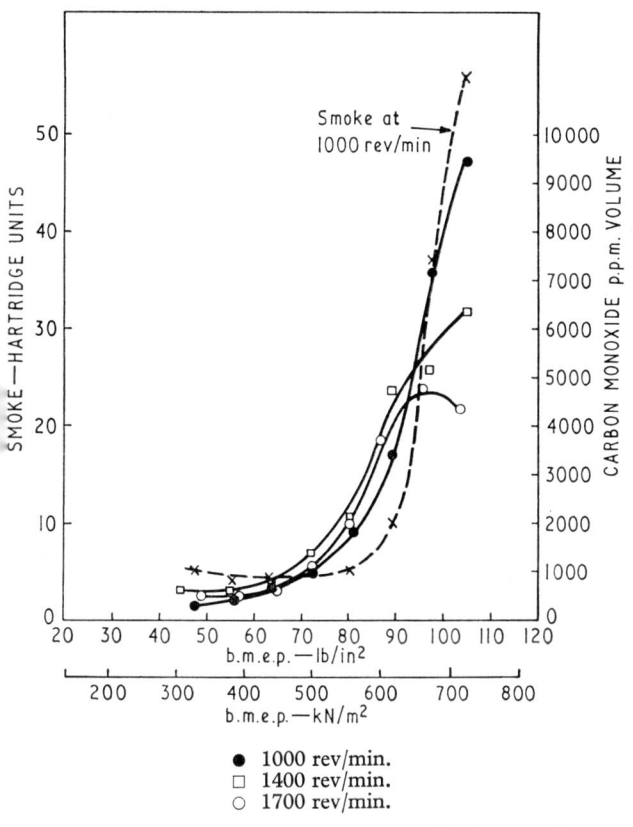

Fig. 10.1. Emission of carbon monoxide from a four-stroke direct injection engine

● 1000 rev/min.
□ 1400 rev/min.
○ 1700 rev/min.

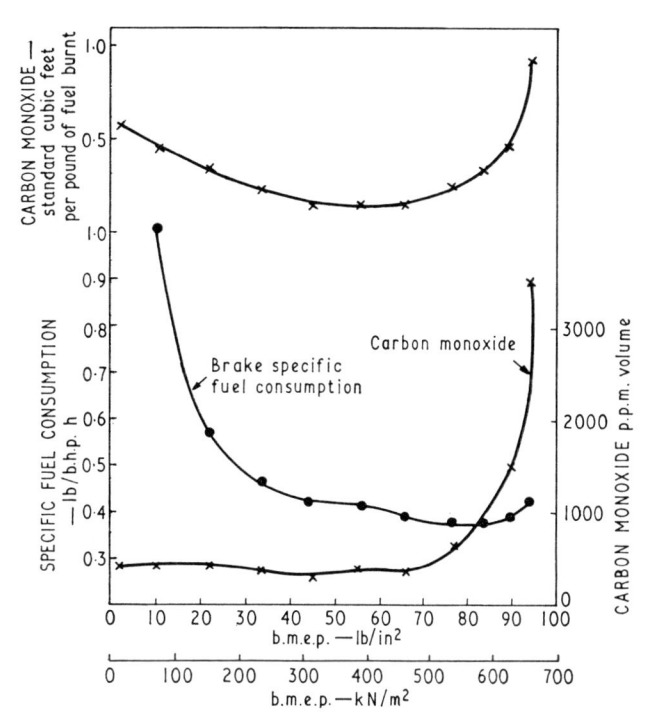

Fig. 10.2. Emission of carbon monoxide from a four-stroke direct injection automotive engine at 1400 rev/min

engine. The figure also serves to illustrate that peak carbon monoxide levels are influenced by engine speed at high engine loads. However, there is little influence at the lower engine loads.

The emission characteristics of a four-stroke direct injection engine over the whole load range (Fig. 10.2) show that carbon monoxide concentrations are low at lighter engine loads. At first sight the concentration of 450 p.p.m. at no load appears insignificant compared with the 3500 p.p.m. emitted at full load. However, if carbon monoxide is expressed in terms of mass or volume per pound of fuel burnt (as also shown in Fig. 10.2), then it becomes obvious that the efficiency of combustion at no load is of the same order as at full load. Light engine load is clearly a critical operating mode in terms of carbon monoxide production.

Since carbon monoxide and smoke emission characteristics appear to go hand in hand at high load, it is reasonable to expect that fuel quality variations would have a common influence. Indeed, the figures in Table 10.1 show that the slight increase in smoke emission which occurs with increase in aromatics content is accompanied by a slight increase in the carbon monoxide emission.

Similarly, the gravity and viscosity changes which significantly reduce the smoke emitted from an engine at a fixed volumetric fuelling rate also substantially reduce the amount of carbon monoxide emitted. Table 10.3 illustrates the changes which occur when the fuel of a four-stroke direct injection engine is changed to kerosine from gas oil.

White smoke

On cold days many diesel engines emit dense and objectionable clouds of white smoke immediately after start-up, and in some cases this emission persists for several minutes.

This white smoke formation is most probably due to a substantial part of the fuel injected into the engine being unable to ignite because it is sprayed on to cold combustion chamber walls. Some of this fuel is vaporized off the walls comparatively late in the cycle by the combustion heat from the small proportion of the fuel that has managed to ignite. However, the bulk of the vapour cannot reach ignition temperature and appears in the exhaust as white smoke.

Tests were conducted on an engine of a type known to give rise to complaints of white smoke emission in service, and a statistical analysis of the data obtained showed that

Table 10.3. Effect of fuel type on emission of carbon monoxide from a four-stroke direct injection (automotive type) engine at 1000 rev/min

Fuel	Specific gravity, 60°F/60°F	Engine load, b.m.e.p. lb/in²	kN/m²	Carbon monoxide, p.p.m
Kerosine .	0·784	110	760	2800
Gas oil .	0·829	115	795	5450

measured cetane number is the fuel parameter which best correlates with white smoke performance. There was also evidence to suggest that, at a given cetane number, increased volatility resulted in decreased white smoke emission.

This work has shown that the ignition quality of currently marketed diesel fuels is sufficiently high to avoid complaints of excessive white smoke emission in service.

Hydrocarbons

At the present time unburnt hydrocarbons as such or partially oxidized hydrocarbons (aldehydes) are not subject to legislative control. However, it is virtually certain that such diesel emissions will be legally restricted in the foreseeable future. Therefore, it is most necessary to have a detailed knowledge of the engine design and fuel quality factors which influence the production of these pollutants. Moreover, these exhaust gas constituents contribute significantly to objectionable exhaust odour, which is already a subject of service complaint.

In our work the total unburnt hydrocarbons are measured by a flame ionization detector method, and in order to avoid the condensation of the heavier hydrocarbons the sample lines are heated to 180°C. The total aldehyde concentrations are measured using the 3-methyl-2-benzothiazolone hydrazone (MBTH) technique (2) and the results are quoted as parts per million by volume as formaldehyde.

In Fig. 10.3 it will be observed that the volumetric concentration of total unburnt hydrocarbon emitted throughout the load range of a four-stroke direct injection engine is relatively constant at all speeds. There is a slight tendency for the concentration to rise at the smoke limited maximum fuelling condition, and again towards light engine loads. The rise in aldehyde concentrations at the extremes of the load range is somewhat more pronounced (Fig. 10.4). However, when these data are expressed in absolute units as a percentage of the total fuel supplied, it can be seen that combustion efficiency, assessed in terms of percentage unburnt fuel, gets worse as the load is decreased. These results are typical of those obtained from a number of four-stroke direct injection automotive type engines. Tests on an indirect chamber engine showed that the overall levels of both aldehydes and total unburnt hydrocarbons were only about half of those observed in the direct injection engines. However, the general form of the emission curves for the two engine types is very similar.

Hydrocarbon emission levels at high engine loads are not greatly influenced by relatively large changes in injection timing or fuel ignition quality. However, factors which influence the timing of combustion at light load are highly significant. Tests have shown that hydrocarbon emission can be increased by (*a*) the injection occurring too early, in which case the delay time increases with the

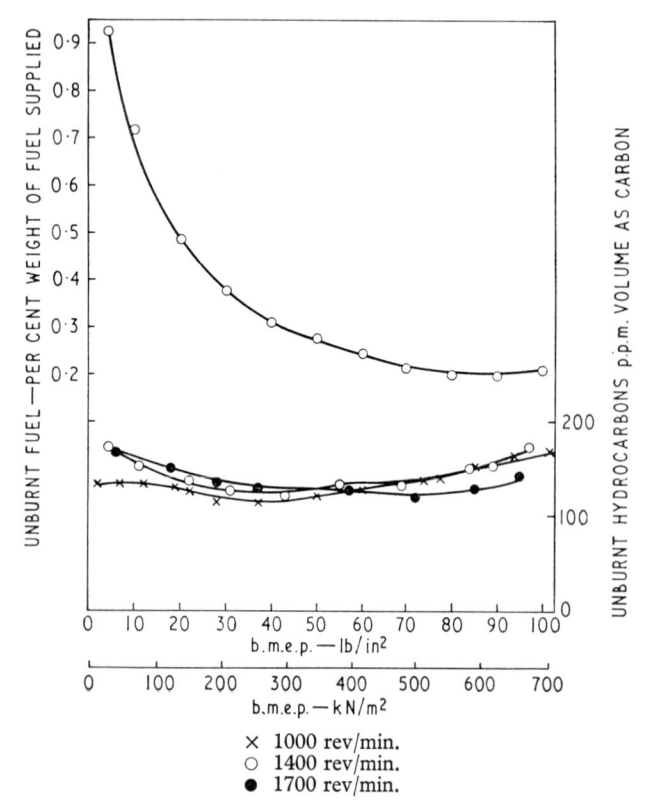

Fig. 10.3. Emission of unburnt hydrocarbons from a four-stroke direct injection engine

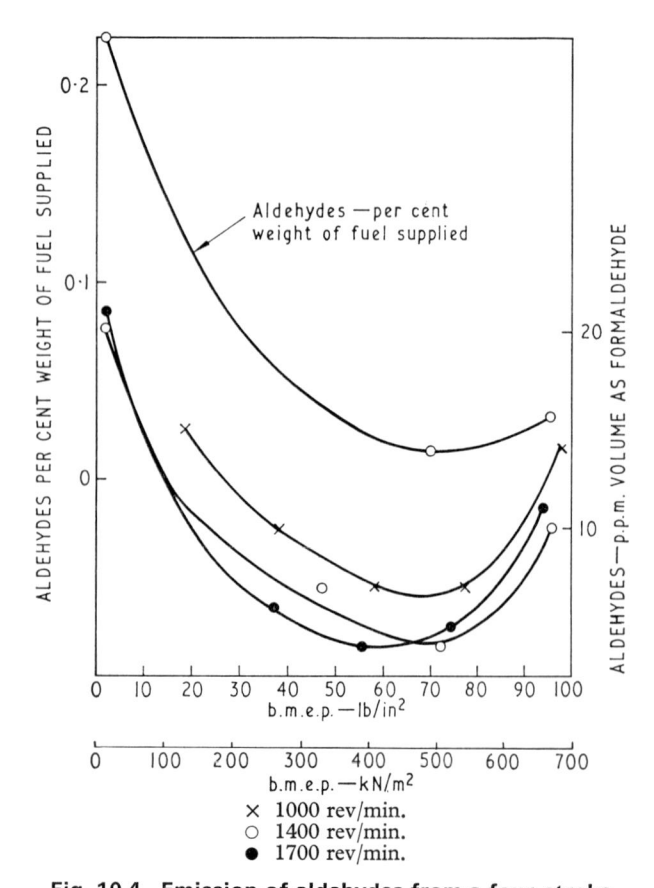

Fig. 10.4. Emission of aldehydes from a four-stroke direct injection engine

result that more fuel can contact the relatively cool cylinder wall, or (*b*) combustion occurring too late, in which case there may be insufficient time for completion of combustion. In both cases the hydrocarbon emission at light load can be reduced by using a fuel of higher ignition quality.

General studies of the relationship between fuel quality parameters and emission characteristics have shown that increased cetane number results in reduced aldehyde emission. However, if we restrict our examination to ignition quality variations likely to be encountered in the market, then the correlation becomes less obvious. Another interesting point which emerged from our studies is that for a given ignition quality, there is a tendency for the more volatile fuels (e.g. kerosines) to produce slightly higher emission levels. This effect is illustrated in Figs 10.5 and 10.6 which show the extremes of aldehyde and total unburnt hydrocarbon concentrations emitted by a number of four-stroke direct injection engines. It can be seen that, when the fuel is changed from gas oil to kerosine, the aldehyde emission is increased mainly at light load, whereas the total unburnt hydrocarbons emitted are increased over the entire load–speed range.

Compared with the direct injection engine, the pre-combustion chamber engine was shown to be relatively insensitive, in respect of exhaust emission of unburnt hydrocarbons and aldehydes, to variations of volatility or cetane number.

Our examination of the fuel quality factors which influence the emission of partially burnt and unburnt hydrocarbons is still at an early stage. However, on the basis of the data available so far, it is becoming obvious that, in terms of minimum hydrocarbon emission, the 'best' fuel for a particular engine is that fuel for which the engine was originally designed. In general, this is a gas oil of moderately high ignition quality. Although this sounds like a 'statement of the obvious', it needs to be re-emphasized in order to counter the fairly widespread belief that more volatile kerosine type fuels are 'cleaner' and produce less odorous exhaust gas than conventional gas oils.

Oxides of nitrogen

Earlier work carried out by McConnell (3) has demonstrated that the maximum concentration of oxides of nitrogen appearing in the exhaust of an indirect chamber engine is significantly less than that found in the exhaust of an equivalent direct injection engine. Two factors are thought to be responsible for this phenomenon. First, the later injection timing (due to the shorter combustion delay time) for indirect chamber engines limits the quantity of fuel that burns before t.d.c. and therefore limits the peak cycle temperature which is a key factor in the rate of formation of oxides of nitrogen. Second, in the indirect chamber engine only a proportion of the total oxygen available is within the combustion chamber during the early critical stages of combustion, and high-temperature combustion is therefore occurring in a relatively oxygen-deficient atmosphere.

Compared with the effects due to engine type, the fuel quality in general and cetane number in particular have been shown to have only a marginal influence on the concentration of oxides of nitrogen appearing in the exhaust.

Recent laboratory work studying exhaust emissions in detail has shown that the conditions which reduce the

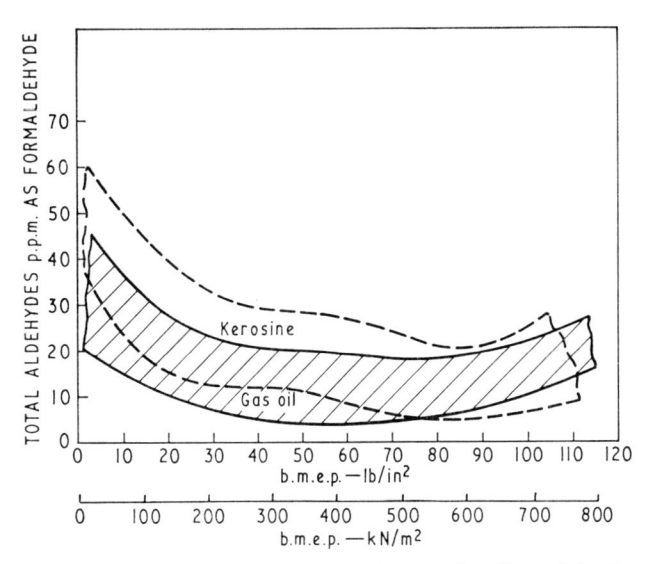

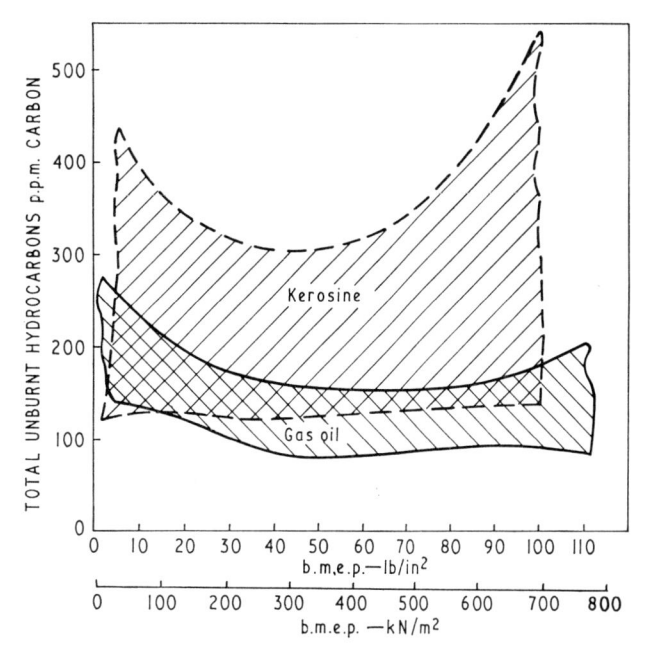

Extreme range of results on three four-stroke direct injection engines. Speed range 1000 rev/min to maximum rated speed.

Fig. 10.5. Emission of aldehydes from a four-stroke direct injection engine

Extremes of results on three four-stroke direct injection engines. Speed range, 1000 rev/min to maximum rated speed.

Fig. 10.6. Emission of total unburnt hydrocarbons from four-stroke direct injection engines

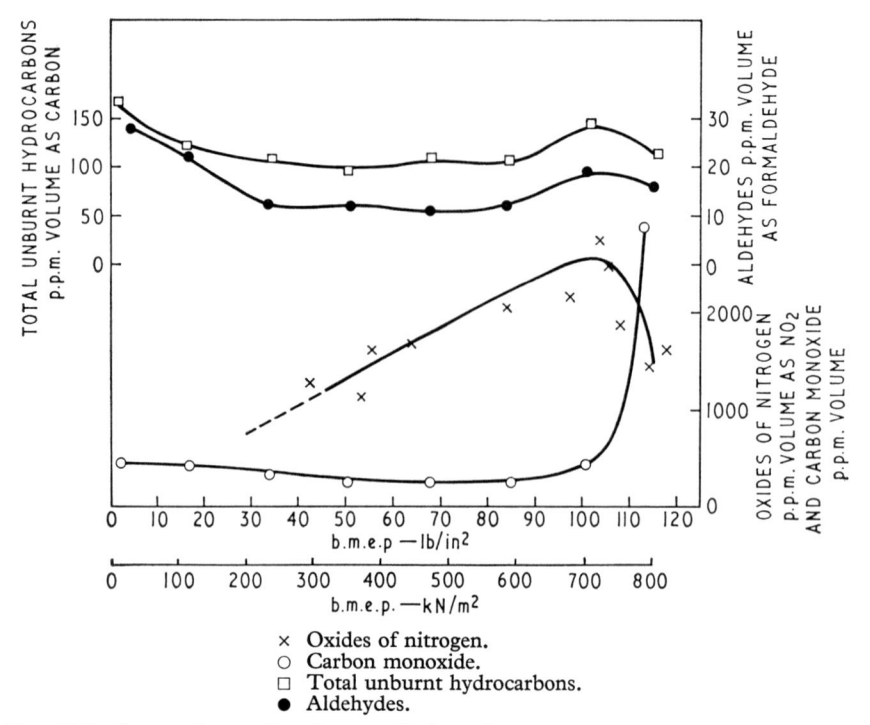

Fig. 10.7. Comparison of various emissions from a four-stroke direct injection automotive type engine at 1500 rev/min

× Oxides of nitrogen.
○ Carbon monoxide.
□ Total unburnt hydrocarbons.
● Aldehydes.

oxides of nitrogen may promote the emission of unburnt or partially burnt fuel. This is illustrated in Fig. 10.7.

Oxides of sulphur

The oxides of sulphur are a natural product of commercial diesel fuel combustion, and the theoretical concentration appearing in the exhaust gas is a direct function of the sulphur content of the fuel and the air/fuel ratio used. In general considerations of atmospheric pollution, the diesel engine has a negligible effect on sulphur dioxide concentration compared with industrial combustion of coal and heavy fuel oil. However, in the specialized context of diesel engines operating in confined spaces (mines and underground engineering projects) other considerations are involved. Normally, safety criteria dictate that air change rate is based on maximum allowable concentrations of carbon monoxide and/or oxides of nitrogen, with oxides of sulphur not really being considered significant. Nevertheless, it is interesting to note that the Federal Office of Accidents in Switzerland has recently introduced legislation whereby the number of air changes is controlled by the sulphur content of the fuel. This is probably related to the fact that, even at concentrations well below the accepted maximum for continuous exposure, the characteristic odour of sulphur dioxide remains objectionable. Therefore, in this particular application in confined areas, the diesel fuel should contain minimum sulphur.

INTERDEPENDENCE OF FUEL PROPERTIES AND THE USE OF ADDITIVES

On the basis of the data presented, it can be concluded that there are two fuel properties which can influence the composition of diesel exhaust gas:

(1) the specific gravity and viscosity of the fuel, which control the amount of fuel injected into the engine and therefore influence the black smoke and carbon monoxide emission, especially at full load;

(2) the cetane number of the fuel, which controls the ignition delay time and therefore influences the emission of aldehydes, total unburnt hydrocarbons, and carbon monoxide, particularly at light engine loads.

In practice, however, there is a high degree of interdependence between many of the diesel fuel properties, and thus it is not realistic to consider these properties independently.

Diesel fuels with good ignition properties, as defined by high cetane number, are prepared by selection of suitable crudes. This quality can be improved by reducing the aromatics content, which in turn decreases the specific gravity and at the same time elevates the cloud point (i.e. worsens the low-temperature filterability). Thus the requirement of high cetane number is in conflict with the demand for maximum fuel economy and the need for good low-temperature handling characteristics.

Similarly, if the volatility of a fuel is increased, then the gain in ignition quality and the improvement in low-temperature properties is offset by lower specific gravity, which gives less power at the same engine fuel pump setting.

Diesel fuels are designed to meet the requirements of the modern high-speed diesel engine in terms of both quality and quantity. Thus, the demand for a change in any particular quality characteristic without accepting a change in other fuel parameters will restrict refinery flexibility and reduce availability. Such a situation must inevitably result in an increase in fuel cost, which is contrary to the basic requirement for cheap power. In this context the use of fuel additives can represent an economic way of overcoming the interdependence of fuel properties and their influence on engine performance, and some of the better known additive types are discussed below.

Ignition improvers

It is well known that organic nitrates, in particular isopropyl or cyclohexyl nitrate, can significantly improve the cetane number of certain types of fuel. The results of the tests described in this paper suggest that ignition improvers of this type may be usefully employed to reduce the emission of aldehydes and total unburnt hydrocarbons at low engine loads, especially at idle. They are also well known for their ability to reduce the white smoke emission from diesel engines operating at low temperatures.

Smoke suppressants

Additives for reducing the emission of black smoke from diesel engines can be classified into two types—detergent additives which help to keep the fuel system clean, and metal-containing additives which modify the combustion process.

The detergent additives do not have a direct effect on black smoke. They work by keeping the fuel system free from deposits of gum and varnish, and this helps to maintain the performance of the injectors. However, as demonstrated by service experience, the effect of such additives is negligible wherever good quality fuels of adequate stability are used.

The metal-containing additives, however, certainly do produce significant initial reductions in black smoke emission. The most successful are based on manganese or the alkaline earth metals—in particular, barium. The mechanism by which these materials reduce the emission of black smoke is not fully understood and work is still in progress to find an explanation. However, despite their undoubted ability to suppress smoke, these additives are not widely used for a number of reasons. Their lack of popularity is mainly associated with their high metal content, which can raise the normal fuel ash content from less than 0·001 per cent weight to about 0·12 per cent weight. This enormous amount of ash must either be retained by the engine, in which case there is a risk that it will build up on critical parts of the engine such as the injectors, or it will be ejected, in which case the solids burden of the exhaust gas, although not visible, will be high. It is also worth bearing in mind that some barium compounds are potentially harmful. Indeed, metal-containing additives in general are likely to be subject to scrutiny because of possible adverse health effects.

Although the additive manufacturers claim only a reduction in smoke for the use of their products, there has been widespread speculation both in the oil industry and in the engine manufacturing industry concerning the use of such additives as a means of increasing the smoke limited power of engines. Such an increase is easily obtainable, but whether or not it is expedient to uprate engines in this manner is open to doubt. The doubt exists in the case of existing engines where use of a fuel additive to increase the smoke limited power means that air/fuel ratios, and hence combustion chamber temperatures, are taken beyond the designed limit for the engine. In the laboratory we have seen that the ability of a six-cylinder direct injection engine to operate for long periods at peak efficiency is seriously impaired if the engine is subjected to periods of modest overloading. The decline in engine performance was traced to a deterioration in injector atomizing efficiency; evidence of needle lacquering and carbon deposits suggests that overheating was the basic cause.

These conclusions apply to existing engines only. It is technically feasible that new engines can be designed specifically to tolerate a higher thermal loading and thus take advantage of the ability of smoke suppressants to increase the smoke limited power. However, the authors feel that the adverse side effects of these additives far outweigh their advantages.

Re-odorants

Diesel exhaust odour can be improved by the use of perfume type additives. They are particularly effective at light engine loads, especially at idle, since these are the engine operating conditions which produce significant quantities of unburnt or partially burnt fuel in the exhaust gas. It must be emphasized that re-odorants do not suppress the formation of the constituents responsible for odour, but simply mask their presence.

LOW-TEMPERATURE OPERABILITY OF AUTOMOTIVE DIESEL ENGINED VEHICLES

All diesel fuels contain wax (solid hydrocarbons) in varying amounts, depending on the nature of the crude oil processed and the distillation range of the particular fuel. However, at normal operating temperatures this wax remains in solution and is no problem.

At low ambient temperatures, wax which has crystallized out from solution can build up in the fuel system of a diesel vehicle, leading to erratic engine operation, loss of power, and ultimately to complete engine stoppage. Thus the low-temperature properties of diesel fuels require to be matched to the prevailing climatic conditions in order to avoid problems due to wax plugging of the fuel systems.

The most widely used laboratory test techniques for assessing the low-temperature characteristics of a fuel are the cloud point and the pour point. In addition, filterability tests such as the Hagemann and Hammerich and Esso Cold Filter Plugging Test are sometimes employed.

The cloud point of a fuel is the temperature at which the first wax crystals appear in the fuel. Therefore, provided that the cloud point of a fuel is lower than the prevailing ambient temperature, or the fuel temperature is kept above the cloud point by heating, there can be no supply failure in the fuel system due to filter blockage or line plugging by wax.

The pour point of a fuel is an indication of the lowest temperature at which the fuel can flow without application of pressure. Thus it is generally considered to be the lowest temperature at which a fuel can be handled in a vehicle.

Since low-temperature problems are not encountered at temperatures above the cloud point, freedom from complaint in the winter can best be guaranteed by specifying a cloud point which corresponds to the expected minimum ambient temperature. It is more risky to rely solely on the filterability tests mentioned above. This ideal use of cloud point does, however, mean that production costs are affected. The use of kerosine to produce fuels of lower cloud point is both costly and restrictive in terms of production, since the requirement for blending kerosine coincides with the peak demand for kerosine as a domestic fuel.

In the long term, the use of additives to improve the low-temperature characteristics of diesel fuels would release kerosine for aviation turbine fuel production, and this is important because the supplier may find it difficult to satisfy both the aviation outlet and the domestic kerosine markets of the future. Since there are no additives known which significantly lower the cloud point of diesel fuels, the solution to the problem of cold weather performance may have to be found in terms of pour point modification. The authors believe that the low-temperature characteristics of fuels can be controlled by additives which modify wax crystal structure, provided that the fuel system is critical with respect to pour point and not to cloud point —i.e. fuel tanks are sited in unexposed positions, fine unheated filters are replaced by coarse ones positioned so as to receive heat from, e.g., the engine or exhaust manifold, and the unnecessary bends and restrictions in fuel lines eliminated.

The desirable features of a well-designed fuel system are enumerated in the report prepared under the auspices of the British Technical Council of the Motor and Petroleum Industries (4). In addition to these recommendations, work that we have carried out has indicated that lift pump delivery pressure can give a significant improvement in vehicle operability. When vehicle manufacturers have adopted the recommendations made by the Council, and fuel systems are improved to take advantage of wax crystal modifiers, pour point can then become the controlling parameter.

The 'well-designed' fuel system is of benefit to both the refiner and the vehicle operator. The refiner would have improved manufacturing flexibility, and the operator would receive fuels of high specific gravity and viscosity. This would be reflected by a higher milage per gallon and less fuel pump wear respectively.

RESIDUAL FUELS FOR INDUSTRIAL AND MARINE APPLICATIONS

Although the burning of residual fuels in large slow-speed diesel engines has been common practice for many years, it is only in recent times, with the increasing demand for cheaper power, that diesel engineers have developed medium-speed engines capable of exploiting the economic advantages given by residual fuels.

Residual fuel oil, as the name implies, consists of the residues from crude oil refining processes diluted to the required viscosity with lighter distillate hydrocarbons. This type of fuel is considered to be a colloidal suspension of high molecular weight asphaltene compounds in an oil base. In addition, it contains the non-volatile ash-forming constituents of the crude oil which have remained unchanged by the distillation process. Thus the main problems which can occur with the burning of residual fuels are due to their poor ignition quality, their slow burning characteristics, and their high ash contents.

Comparatively little information is available concerning the combustion properties of heavy fuel. Most engine builders and users accept the fact that such fuels have low ignition quality and slow burning properties. They also appreciate that if any specifications are applied to the combustion properties of the fuel (for example, a demand for a minimum cetane number), then this would involve the fuel supplier in special blending operations, which would inevitably raise the cost of the fuel and hence invalidate the basic purpose of using it.

Although there is no existing specification for controlling the burning properties of heavy fuel, it is still of interest to both users and suppliers to have some idea of the cetane numbers that are likely to be encountered with such fuels.

Work conducted at the BP Research Centre, using a single-cylinder engine and IP Method 41 (delay angle method), shows that the cetane numbers of residues from naphthenic crudes are in the region of 25–35, whilst residues from paraffinic crudes have cetane numbers of the order 35–45.

The relatively slow rate of combustion of residual fuel presents formidable problems for the engine designer who must ensure that combustion goes to completion in order to minimize the deposition of carbon. The carbon residue content of a fuel was at one time thought to be correlated with the fouling of diesel engine combustion chambers. Although this correlation is now treated with caution, there is no doubt that the significant proportions of carbonaceous residue produced by burning residual fuels under the prescribed conditions of the carbon residue test do reflect their tendency to promote combustion chamber deposition.

Injector life, once a limiting factor in heavy oil burning installations, no longer presents a problem since the requirement for adequate temperature control, especially in the vicinity of the injector tip, has been fully realized. Coolants may be fuel, oil, or distilled water, and the coolant temperature range is limited by:

(a) providing adequate cooling at the injector tip;
(b) ensuring that the temperature of the injected fuel is not significantly influenced (by too low a coolant temperature);
(c) ensuring that the injectors are maintained well above the acid dew-point of the combustion gases.

When medium-speed diesel engines are operated on heavy fuel oil it is frequently found that exhaust valve life is limited by the deposition of incombustible constituents of the fuel on the valve seat. These deposits eventually crack and initiate a gas escape path which subsequently causes high temperature corrosion of the valve seat. Some engine manufacturers who have experienced this problem have achieved satisfactory performance of exhaust valves by the following methods:

(a) suitable selection of corrosion-resistant steels for exhaust valve material.
(b) control of exhaust valve seat temperatures to below the melting point of the fuel ash—in this way adhesion of the ash is discouraged and therefore deposit build-up is minimized;
(c) the use of valve rotators which prolong valve life by preventing the occurrence of localized hot spots.

Modification of the ash fusion characteristics either by water washing the fuel or by the use of additives, whilst technically feasible, does not appear to be economically attractive.

The sulphur content of residual fuels (up to 4 per cent weight) is much higher than that of distillate fuels. This is reflected in a much higher rate of piston ring and liner wear. A proposed mechanism for this higher wear rate has been described elsewhere (5), and lubricating oils have been developed to overcome this problem.

The advantages of fuel cost and disadvantages of increased engine maintenance involved in the use of heavy fuel oil in medium-speed engines must be carefully weighed for each application.

ACKNOWLEDGEMENT

Permission to publish this paper has been given by The British Petroleum Company Limited.

APPENDIX 10.1

REFERENCES

(1) McConnell, G. and Howells, H. E. 'Diesel fuel properties and exhaust gas—distant relations?', S.A.E. Paper 670091, presented at Society of Automotive Engineers' Congress at Detroit, Michigan, 1967 (January).
(2) Hopkin and William Limited *MBTH reagent for aldehydes*, Monograph 73.
(3) McConnell, G. 'Oxides of nitrogen in diesel engine exhaust gas: their formation and control', *Proc. Instn mech. Engrs* 1963–64 **178** (Pt 1), 1001.
(4) 'Low-temperature problems with diesel vehicle fuel systems', Report prepared by Fuels Committee of British Technical Council of the Motor and Petroleum Industries (published by Institute of Petroleum, 1964).
(5) McConnell, G. and Nathan, W. S. 'A wear theory for low-speed diesel engines burning residual fuel', *Wear* 1962 **5**, 43.

Paper 11

DIESEL GENERATION FOR ASCENSION ISLAND

A. K. Mackenzie*

The requirement was to supply 4 MW of electricity and sufficient fresh water by desalination to meet the needs of 500 people on this remote island. The principal factors influencing the selection of the prime movers included proven reliability, ease of maintenance, and compactness. The engines were also required to supply a continually fluctuating electrical load whilst maintaining strict frequency conditions. Maximum utilization of waste heat was required for fresh water production, and two systems working independently from the jacket water and exhaust heat were evolved. The testing of the station as a whole to determine performance and overall efficiencies for the diesel engines is described.

INTRODUCTION

THIS PAPER COVERS some aspects of the design and commissioning in 1966 of the English Bay power station on Ascension Island with which the author was associated. The station is of interest because of the nature of the load it had to supply and the complications of using waste heat for sea water distillation.

Ascension is a dusty volcanic island in the south Atlantic Ocean, 8° south of the equator and 1000 miles off the west coast of Africa. It has no natural water supplies other than a small catchment area constructed by the Navy before the turn of the century. However, the BBC decided to locate a HF radio relay station there and the request to the Ministry of Public Building and Works was to supply and maintain several facilities, including a source of electricity to cope with the maximum demand of 4 MW, and a sea water desalination plant to produce 30 000 gallons of fresh water per day. Installation work for the new project was hindered by the lack of natural resources, extremely difficult communications, and hazardous shipping arrangements. It was in this light that the problems of electricity generation and water production, and in particular of the prime movers involved, were first considered.

PRIME MOVERS

Normal economic calculations for a generating station with an installed capacity of some 6 MW showed a marked preference for using diesel engines, but external factors present in this case justified investigations into other forms of power units. For example, the necessity of producing fresh water cheaply was of prime importance. There were a variety of techniques for desalinating sea water, but at the time the most reliable and reasonably efficient systems were by 'flash distillation' and 'submerged tube evaporation'. In both methods the sea water is boiled at low pressure, and the main form of energy required for this purpose is naturally heat. Initial indications showed that the utilization of waste heat from the power station plant would be an economic proposition, and could provide the anticipated requirements for fresh water.

The amount of heat required for desalination partly depended on the grade of heat available. For example, heat in the form of saturated steam at 50 lbf/in^2, as might be obtained from exhaust boilers on diesels or gas turbines, could be made to produce up to 6 or even 8 lb of fresh water from every pound of steam used. On the other hand, cooling water from diesel engine jackets or steam turbine condensers at a maximum of 180°F would produce less than 2 lb of fresh water per 1000 Btu. Similarly, the other factors listed below were considered in the selection of the power units:

(1) The high cost of site erection and construction work.

(2) The high cost of skilled maintenance once operational.

(3) Frequent large load changes.

(4) High fuel costs.

(5) The presence of abrasive dust, particularly during erection.

(6) Equipment shipping weights limited to 15 tons by the off-loading facilities.

The MS. of this paper was received at the Institution on 20th January 1970 and accepted for publication on 19th February 1970. 33

* *Engineer, Ministry of Public Building and Works, Argyle House (Room L/960), Edinburgh 3EH3 95D.*

(7) The need for readily available sets of proven reliability.

In fact the investigations confirmed that diesel engines would be the most satisfactory form of prime mover, and the field of choice was limited to these.

DIESEL ENGINES FOR GENERATION

For ease of maintenance it was decided to standardize on one size and type of diesel alternator set throughout. An allowance had to be made for two sets being out of commission at any one time, with the full station load being taken by the remaining sets, and a total station installation of six sets was considered to be the best pattern to meet the particular load. As previously indicated, the daily load curve had frequent large step changes, representing the switching on and off of the BBC's transmitters. A variety of other domestic and technical loads had to be supplied, but the expected maximum demand of these was under 1 MW, compared with a maximum of 3 MW consumed by the transmitting station. Fig. 11.1 indicates the anticipated total load curve over 24 h.

Certain loads were designed to be operated remotely from the power station control room, including water heating and pumping and an electrode boiler to help fill in the periods of low load, but this feature could not alleviate the large step load changes which, in conjunction with a requirement for a better than normal voltage and frequency regulation, presented a considerable generation problem.

This was further complicated by the presence of rapid load fluctuations which were in accordance with the syllabic variations of the particular programmes being transmitted. A maximum of four transmitters were in use at any one time, each of which consumed 500 kW when unmodulated and 725 kW at 100 per cent modulation. Thus the syllabic fluctuations of up to 225 kW per transmitter had to be superimposed on the load curve shown in Fig. 11.1 (which was drawn at a mean of 60 per cent modulation) to give a true picture of the load to be supplied. The diesel generators had to produce this load without unstable speed conditions or imbalance of load sharing when in parallel. The specification for the electrical generation stated that during all times of normal operation, voltage swings should not exceed 5 per cent, and the frequency should remain within ± 0.5 Hz, except when the entire transmitter load was switched in or out when a momentary swing of 1·5 Hz (3 per cent speed variation) was permitted. One further criterion was that the combined mechanical and electrical disturbances occurring during modulation should not give rise to an angular deviation of more than $2\frac{1}{2}$ electrical degrees.

To meet the limit of angular deviation, it was first necessary to restrict the mechanical impulses, which encouraged the use of a moderately high-speed, multi-cylinder diesel, with a high rotating inertia, and this high-speed/inertia factor naturally helped to smooth out the electrical disturbances.

The sets finally selected were the English Electric 'V' Range, each having 16 cylinders of 10-in bore by 12-in stroke. They were naturally aspirated, and at full load had a brake mean effective pressure of 89·6 lbf/in^2, and produced 1280 b.h.p. at 750 rev/min. This gave the alternators a full load site rating of 930 kW.

Taking a load change of 285 kW to represent the worst instantaneous electrical disturbance on each engine during periods of normal transmission, it was necessary to have a heavier flywheel (2 tons) as well as a larger alternator frame size than standard, to give the level of inertia required to restrict the angular deviation to ± 2.5 electrical degrees.

In any event, the larger alternator was needed to keep the transient reactance below 15 per cent, which enabled the required voltage stability to be obtained. At this level of reactance a sudden load change of 285 kW per alternator would cause a voltage swing of 4·6 per cent (plus 0·2 per cent due to the change in speed) before the voltage regulator applied the necessary correction. The 5 per cent limit on voltage regulation was thus achieved, but it necessitated using static excitation with thyristor control.

The other supply condition that had to be met was the normal frequency stability to within ± 0.5 Hz. The high rotating mass helped towards this initially, but the diesel governor had to respond quickly, and have a very small droop setting for load sharing. In addition, the diesel engine had to be able to accommodate immediately the large load changes called for by the governor. This last criterion could only be met by using the naturally aspirated version of the engine—unfortunately at the expense of generating efficiency and available waste heat, owing to insufficient excess air being present in the cylinders of the turbocharged version at part load which prevented their immediate acceptance of large load increases. This ability to accept load changes was even more significant when the speed swing limit of 3 per cent, as a result of the larger load change (taken as a 700 kW load increase), had to be

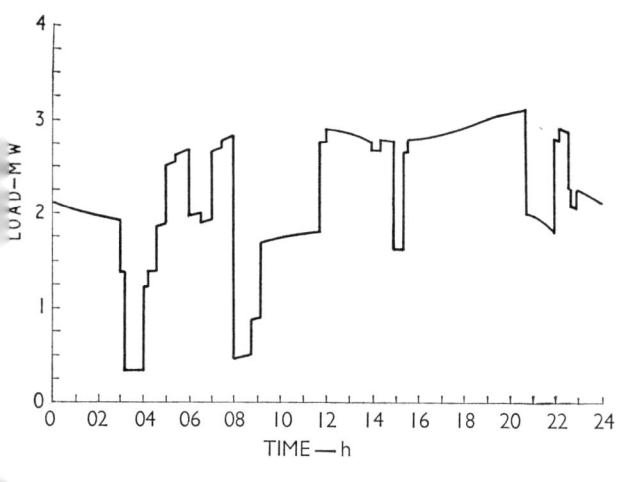

Fig. 11.1. Anticipated mean daily load curve, English Bay power station, Ascension Island

Fig. 11.2. Engine hall, Ascension Island power station

Crown copyright reserved

met. In this case the engine had to be developing the full brake mean effective pressure required for the new load level in less than 1 s after the step change, even with the heavier flywheel.

Fig. 11.2 shows a view of the six sets installed in the engine hall looking from the control room window. The static excitor panels are on the left, together with the larger alternators and flywheels.

In general, the engines have proved to be very reliable over the first few years of operation. Difficulties were experienced initially with a build-up of carbon on the pistons, causing hammering under heavy loads. This problem was alleviated by machining 0·062 in off the piston crowns under the valve heads at the manufacturer's instructions, but a subsequent change of engine oil type and improvements to the intake air filtration appear to have cured the carbon build-up. The only current problem is the ease with which the injector spill pipes can be fractured when removing the injectors; if undetected, this allows the diesel oil to leak into the sump.

Unfortunately the engine reliability has not been matched by that of some of the auxiliary components, but this subject will be dealt with later.

DIESEL ENGINES AS HEAT PRODUCERS

So far this paper has only considered the ways in which the electrical load characteristics affected the choice and adaptation of the driving engines, but the thermal energy output in this case was of equal importance.

Waste heat from diesel engines has been used for many years to provide energy for various processes. Heat at different temperature levels is readily available from the exhaust gases, the engine jackets, and the oil coolers, although the quantity available from the oil coolers is often insufficient to justify the cost and complications of recovery. There are various ways in which the other two sources of heat can be combined, but they all have their own problems of control and of providing alternative jacket water-cooling systems for emergency use. The control of a combined system would have been further complicated in this power station, owing to the continually changing pattern of generators in operation during the day. For simplicity, and to provide reliability and flexibility, two independent heat recovery circuits were designed; the first was a jacket water system pumped through all the engines in parallel, and the second collected steam from the individual exhaust gas boilers into a

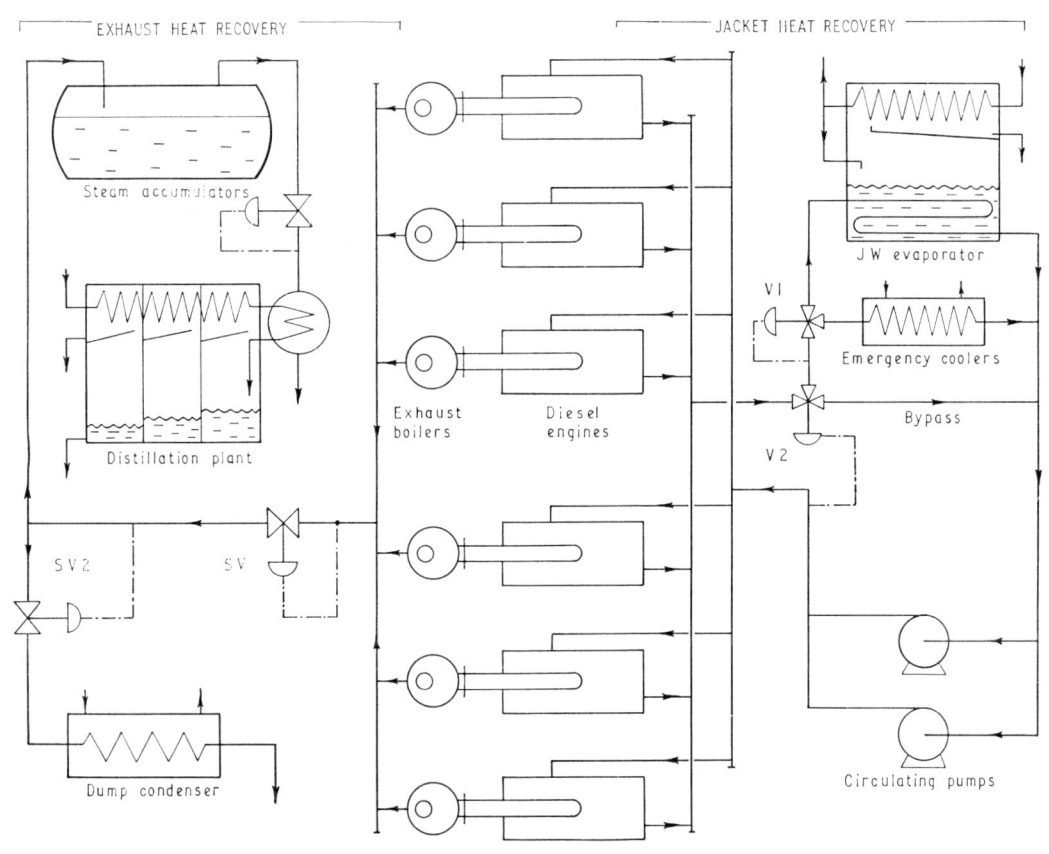

Fig. 11.3. Ascension Island power station. Schematic diagram of heat recovery systems

common circuit. These are shown schematically in Fig. 11.3.

A common jacket water-cooling system naturally presented problems over individual pumped circuits, but it was necessary to provide a constant supply of heat at 180°F from the set of six diesel engines, any one to four of which would be running at intervals throughout the day. The rate of heat absorbed by the engine jacket water was approximately proportional to the shaft horsepower produced, and the available heat thus followed the pattern of the daily load curve. It was considered that the only desalination plant capable of reliable operation under these conditions was a single effect, submerged tube evaporator, similar to those fitted to many marine engines. The principle is simple, with the heated jacket water passing through coils in a tank of sea water which is kept at a reduced pressure of about 3 inHg (abs.). The brine is heated up to its depressed boiling point, and the water vapour is condensed on tubes carrying the incoming sea water. The plant will maintain equilibrium conditions over a wide range of heat input rates, providing there is always sufficient heat to ensure a minimum of boiling action, and also that the rate of change of heat input is not too rapid.

However, the price to be paid for having stable operation from relatively low-grade heat was the poor performance ratio of only 1 lb of fresh water produced for every 1000 Btu consumed, which only gave a predicted 9000 gallons per day on claimed performance rates.

In designing the common engine cooling-water circuit to feed the evaporator, the safety of the diesel engines and the continuity of generation had to be considered of prime importance, and automatic controls were provided to keep the water temperatures within close limits, independent of the fluctuations in generation. Each engine jacket was connected into externally pumped flow and return headers and could be left fully in circuit during running periods, or alternatively fed with hot water through a by-pass across its isolating valve to keep it warm while on standby. This heating of non-running engines was considered justified by reducing wear and giving full load conditions more rapidly on the frequent starts. Water flow and high-temperature alarms were fitted to each engine, which was thus as secure as if it had had its own pump and direct cooling circuit.

The temperature of the water returning to the engines was kept above 160°F by the pneumatically operated control valve V2 (Fig. 11.3), by by-passing some of the hot water coming from the engines across the evaporator and direct to the return. The maximum heat consumption of the evaporator was designed to equal the output of four engines on full load, and in the event of more than four engines being in operation, or of a failure of the evaporator, the flow water would rise above its design maximum of 185°F, causing the control valve V1 to bring into effect the emergency coolers. The flow water would be completely diverted to these coolers if its temperature reached 190°F, and thus a complete failure of the evaporator as a cooler would be automatically allowed for and generation maintained at all station loads.

To keep within the overall limits of 160°F minimum return temperature and 190°F maximum flow, the proportional operating bands on the two control valves had to be kept to 5 degF to allow for the designed 20 degF rise across each engine at full load. This, plus the fairly small thermal inertia in the circuit, necessitated a rapid response from the control valves in the event of a sudden change in heat input. Once commissioned, the operation of the temperature control was quite accurate with the engines running under ideal conditions. However, in maintaining this tight control a certain amount of hunting occurred, particularly in valve V2, resulting in bursts of hot water passing through the evaporator, which caused spells of rapid boiling followed by quiescent periods—the one condition that could upset this plant. The trouble was restricted by precise adjustment of the control valve sensitivity, but it was an indication of the necessity for careful design of diesel engine cooling-water control systems where the heat is required for process purposes, particularly when large load fluctuations are experienced.

With an anticipated 9000 gallons of fresh water per day available from the engine jacket heat, a further 21 000 gallons were needed from the exhaust heat. The type of plant chosen to operate from this source was a multi-stage flash distillation plant. This had a performance ratio of about 5 lb of water from 1000 Btu, and required a supply of steam above 10 lbf/in^2. The steam heated sea water to 190°F in an external heat exchanger, and the water then passed through a series of 10 separate chambers, each maintained at a successively lower pressure. The brine boiled at each stage, with the vapour condensing on tubes in which the fresh sea water was being preheated; the distillate and the brine were drawn steadily through each stage by the pressure differential, and were pumped from the final stage.

Not surprisingly, satisfactory operation of this type of plant was dependent on all external parameters remaining essentially stable—in particular the supply of steam. Steam accumulators were necessary to accept the variable output from the exhaust boilers and to maintain a constant source of steam. A total 15 000 lb of excess steam had to be absorbed during periods of high load and returned to the distiller in the periods of low load. To restrict the physical size of the accumulators it was necessary to have a generated steam pressure of at least 50 lbf/in^2. With the naturally aspirated diesel in particular, running between half and full load, the output of steam from the exhaust gas boiler decreased considerably with increasing pressure, owing to the relatively low temperature of the exhaust gases. The necessity of maintaining an adequate exhaust outlet temperature to avoid condensation and corrosion problems effectively eliminated any advantage in generating steam below 25 lbf/in^2, but even at this pressure about twice the quantity of steam could be produced than at 100 lbf/in^2, at half engine load, with the relationship dropping to about one-and-a-half times the quantity at full load. Apart from making the accumulators more compact, steam generation at 100 lbf/in^2 would have been a great advantage in allowing the use of steam ejectors to maintain the low pressures in the desalination plants, rather than using mechanical vacuum pumps with their inherent high cost and maintenance problems. However, it was determined that the required daily quantity of steam could not be produced from the available waste heat at pressures significantly above 50 lbf/in^2, and this was the operating pressure chosen, giving a full load rating from each exhaust boiler at 1100 lb/h.

The steam circuit is shown in Fig. 11.3. A pressure-maintaining valve SV was included to provide a steady pressure for the boilers to operate against, and the steam then entered the accumulators, from where its pressure was reduced to 10 lbf/in^2 before entering the heater of the distiller. Excess steam could be 'dumped' through a sea water condenser by the automatic operation of valve SV2 when the accumulators reached 55 lbf/in^2. With these two main control valves, the exhaust heat recovery circuit was designed to operate entirely automatically.

PLANT CONTROL

To keep running costs as low as possible, the whole station was planned for operation by only one shift engineer on duty, with the assistance of one locally employed engine driver to keep watch on the engine floor. It was necessary for the engineer to remain near the control room throughout his shift, and he had to carry out all the routine operations of the station from there.

Apart from the normal instrumentation for monitoring and recording the generation side of the station, adequate remote indication, alarms, and controls for the waste heat systems were provided in the control room to cover most eventualities. In addition, as 12 to 15 engine starts each day were not unusual to cope with the fluctuating load and maintenance requirements, the starting and stopping sequences had to be entirely automatic after they had been initiated from the control room. The starting operation in particular had to be quick and reliable, and would normally have been so for the preheated 16-cylinder diesels had it not been for some of the auxiliary control components. Two of these proved to be particularly troublesome when operation first began.

The first item was a centrifugal switch, which was designed to open a pair of contacts above 200 rev/min to shut off the starting air, and to close a second pair of contacts above 525 rev/min enabling the alternator field to be energized. The switch unit was neither robust enough to withstand the vibration and heat of the engine, nor sufficiently accurately made to operate repeatedly at the same speed levels over long periods. This often resulted in the contacts failing to make or break anywhere in the speed range.

The second item was the governor stop–start control, which comprised a pair of solenoids each with its own electrical and mechanical lock-outs. One solenoid returned the governor mechanism to the 'off' position on receipt of the 'stop' signal, and the other released the governor ready for operation when the 'start' button was pushed. In this component the poor design and insufficient strength of the solenoids frequently caused maloperation, sometimes burning the coils out in the process.

These were only small parts but failure of either could, and often did, prevent a diesel alternator being put on load, and in the case of the centrifugal switch an open circuit for a fraction of a second while on load was sufficient to trip the alternator field switch and bring the set off the bars. Three different types of switch were tested over a period of three years before the centrifugal switch problem was solved, but the solenoid units are not yet considered to be 100 per cent reliable. In those three years 90 per cent of the diesel alternator failures were due to these and other components in the auxiliary systems. A need is indicated for more careful design of such minor, but critical, components to give them a reliability to match that of the main machine.

COMMISSIONING TRIALS

The contractual performance specification defined the requirements from the station as a whole, as well as detailing the various limitations of voltage and frequency, and the onus had thus been put on the main contractor—on a design and install basis—to ensure the compatibility of all the plant items. At the time of commissioning, however, the anticipated daily load was not available, nor were there adequate artificial loads to simulate it. The curve in Fig. 11.1 was therefore analysed and approximated into totals of engine running hours at 100 kW levels between 400 and 800 kW. All the engines were first put through the standard load runs, using a 1 MW electrode boiler as the variable test load. Two of the six engines were then selected for further overall efficiency trials at various load levels, and standard test thermometers and flow meters were fitted in the jacket water circuit, the exhaust, and the fuel supply.

Load runs were carried out under steady conditions and the times to consume a measured weight of fuel were recorded. The energy outputs over these periods to the alternator, jacket water, exhaust gas boiler, and exhaust to waste were determined using the manufacturer's figures for only the specific heat and flow rate of the exhaust gases. The average results of the trials for the two engines tested are given in Table 11.1. The mean operating condition of the diesels was then determined by summing the total quantities of energy over the daily engine hours at the specific load levels, and dividing by the total hours (80·5). The result of this is shown in the Sankey diagram (Fig. 11.4) from which the anticipated daily water production was calculated, using the performance ratios of the respective water plants.

These calculations indicated that the output from the exhaust heat recovery system would be close to that predicted, but that the water produced by the jacket water evaporator would be 1000 gallons per day less than quoted. This was because the heat input into the jacket water was measured to be in the range 1200–1400 Btu/b.h.p.-h for engine loads between a half and full load, compared with the manufacturer's figure of 1500 Btu/b.h.p.-h. It was suggested that this latter figure had a margin of safety from the point of view of engine cooling, which had not been deducted when quoting the heat available for process work.

It will be seen from Table 11.1 that the overall thermal efficiency of the diesel alternators (useful energy out over fuel in) came to 74 per cent at full load and the equivalent figure from Fig. 11.4, representing the mean daily operation, was 67 per cent. The same figures for the diesel engine itself (i.e. adding back for the alternator losses and excitation) would come to 77 and 70 per cent respectively. In connection with the limitations on voltage and frequency fluctuations, the static excitation equipment operated precisely after some initial setbacks, and the measurable voltage fluctuations under various types of load changes came within the specification. Control of the frequency was a greater problem. To limit the momentary speed

Table 11.1. Energy output rates from 930 kW diesel alternator at various load levels

Alternator output, kW	400	500	600	700	800	Full load
Engine hours per day	1·5	1·75	23·0	41·25	13·0	0·0
1. Fuel consumption (net CV), Btu/h × 10⁶	1·5	4·8	5·6	6·45	7·3	8·2
2. Net electrical output, Btu/h × 10⁶	1·24	1·58	1·93	2·28	2·65	3·10
3. Jacket water heat, Btu/h × 10⁶	0·75	0·86	1·05	1·28	1·50	1·77
4. Heat to exhaust boiler, Btu/h × 10⁶	0·35	0·50	0·62	0·78	0·94	1·17
5. Exhaust heat to waste, Btu/h × 10⁶	0·90	0·98	1·05	1·10	1·18	1·25
6. Oil cooler, radiation generator losses etc., by subtraction, Btu/h × 10⁶	0·96	0·98	0·95	0·91	1·03	0·91
7. Losses in heat recovery circuits (estimated)	0·21	0·21	0·21	0·21	0·21	0·21
8. Diesel alternator set total efficiency, (2+3+4)/1, per cent	56	61	64	67	70	74
9. Total station efficiency, (2+3+4−7)/1, per cent	51	57	60	64	67	71

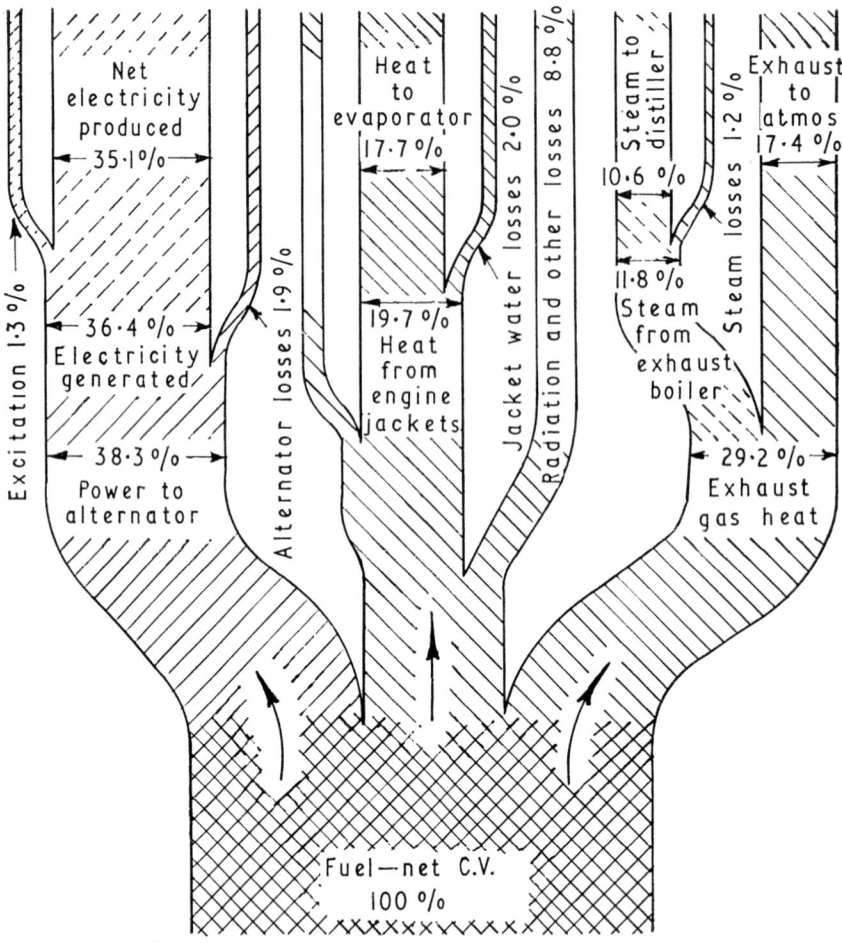

Note: fourth output from the left should be entitled 'Oil cooler losses 4·0%'.

Fig. 11.4. Ascension Island power station. Sankey diagram for generation/ waste heat system under mean daily condition

swing to 3 per cent when 700 kW were applied to, or rejected from, a set, the manufacturers had originally quoted that a 1 per cent droop should be set on the hydraulic-servo governors. These were Europa Regulateurs Curtis-Wright type, and in themselves were found to be very sensitive and precise in operation. However, they worked the normal engine fuel rack systems with their inherent friction and backlash, and were not fitted with any electrical load-sharing equipment. It was found that with a 1 per cent droop a single engine would operate satisfactorily, but that it was unacceptably unstable when running in parallel with other sets. No amount of matching of the governors would produce steady load-sharing conditions much below a speed droop of 3 per cent, and this was the figure eventually set. Fortunately the extent and the effect of the frequency variations were not so significant as originally expected, due partly to an increase in the proportion of domestic loads, and the supply to the transmitters was found to be acceptable.

CONCLUSIONS

The Ascension Island requirement was for a small but complex power station, complicated by (*a*) the nature of the load it had to supply, (*b*) its location, and (*c*) the use of waste heat for sea water desalination. In describing these and other engineering features, it was intended to give some idea of the design policy in evolving the station, and from the user's point of view to show the ways in which the diesel engines were adapted to meet the thermal and electrical loads, and incorporated to form an integrated energy system.

The necessity of carrying out full performance trials on site enabled figures for overall efficiencies under operating conditions to be determined, which showed that even for a station with far from ideal loading, the utilization of waste heat can nearly double the overall thermal efficiency of a diesel generator. Since the original design and installation, the plant at the English Bay power station has been extended to meet increasing demands for electricity and water, and the pattern of operation has changed considerably. The performance figures given in Table 11.1 are therefore no longer directly applicable, although a recent check showed that the overall operating efficiency of the diesel engines was still about 70 per cent.

ACKNOWLEDGEMENTS

The author wishes to acknowledge the design work carried out by the main contractors, Messrs Balfour Beatty & Co. Ltd, and the advice given by many of the equipment manufacturers, all of which contributed to the success of this project. The Ministry of Public Building and Works have kindly given permission for this paper to be published, but it should be noted that all the views expressed are those of the author, and do not necessarily represent the opinion, or the policy, of the Ministry.

Paper 12

IMPROVEMENTS TO CONVENTIONAL DIESEL ENGINES TO REDUCE NOISE

M. F. Russell*

The research reported in this paper has been directed towards making conventional diesel engines quieter by modifying the engine structure. The work has been concentrated on reducing the vibration levels of the external surfaces of the engine, particularly the thinly panelled areas such as the crankcase panels, water jacket panels, timing cover, sump, rocker cover, and tappet covers. Various treatments have been devised for these areas, and they have been applied to a six-cylinder in-line automotive engine and a three-cylinder tractor engine to give overall noise reductions of 5 dBA or more at full rated speed and load.

INTRODUCTION

NOISE FROM ROAD TRAFFIC and particularly from heavy goods vehicles emanates from many sources on the vehicle, but one of the most important and intractable is the engine. This source is particularly prominent when a heavy goods vehicle is accelerating from rest through the gears, or when it is climbing a gradient in a low gear with the engine operating near its maximum speed. These are similar operating conditions to the test conditions laid down in the British Standard method for measuring noise of motor vehicles (1)† and in the Ministry of Transport regulations for the control of motor vehicle noise (2). In order to comply with these and foreign regulations, some vehicle manufacturers are faced with the problem of reducing the noise of their products, and after effectively silencing the engine exhaust and air intakes, with perhaps some attention to the cooling fan, they are left with the engine as the prime noise source.

Previous work by Austen and Priede has shown how the noise of a small diesel engine may be reduced by approximately 10 dB over most of the audiofrequency range by completely redesigning the engine structure, although they retained most of the standard running components. This work has been described fully in reference (3).

The aim of the recent investigations reported in this paper has been to devise noise reduction techniques that may be adopted in production immediately, with the minimum upset.

The MS. of this paper was received at the Institution on 2nd February 1970 and accepted for publication on 20th February 1970. 2
* *Section Leader, C.A.V. Ltd, Wardle Way, Acton, W.3.*
† *References are given in Appendix 12.1.*

NOISE-GENERATING MECHANISMS

A diesel engine embodies a number of mechanisms by which repetitive pulses of vibration are applied to the cylinder block and crankcase structure. The major sources of these pulses in the engines studied so far are the rapid rise of pressure in the cylinder upon combustion (of the fuel injected during the delay period), piston slap, and timing gear rattle (due to crankshaft torsional vibration). Fuel injection pump noise, injector noise, and valve gear noise can be heard occasionally, but they have not been found to affect the overall noise level. Combustion is often the major source on small engines. A one-third octave band spectrum of a cylinder pressure pulse is shown in Fig. 12.1 for a direct injection engine running at 1500 rev/min on full load. The scales and filter bandwidths are arranged so that the exterior noise of the engine can be compared directly with the cylinder pressure spectrum—although it should be remembered that the one-third octave band filters used will encompass several Fourier harmonics of the cylinder pressure pulse and that the one-third octave band level will be correspondingly higher.

The pulses of vibration cause the cylinder block and crankcase to vibrate, and these in turn transmit the vibration to the thin external crankcase and water jacket panels. The vibration is also transmitted to the light section engine covers, such as the sump, rocker cover, tappet cover, and timing cover. The covers and panels of all the engines investigated have exhibited mechanical resonances which, when put together, encompass most of the audiofrequency range. The flexural motion of these panels can cause noise to be radiated with a sound pressure level proportional to the velocity normal to the panel, when the flexural

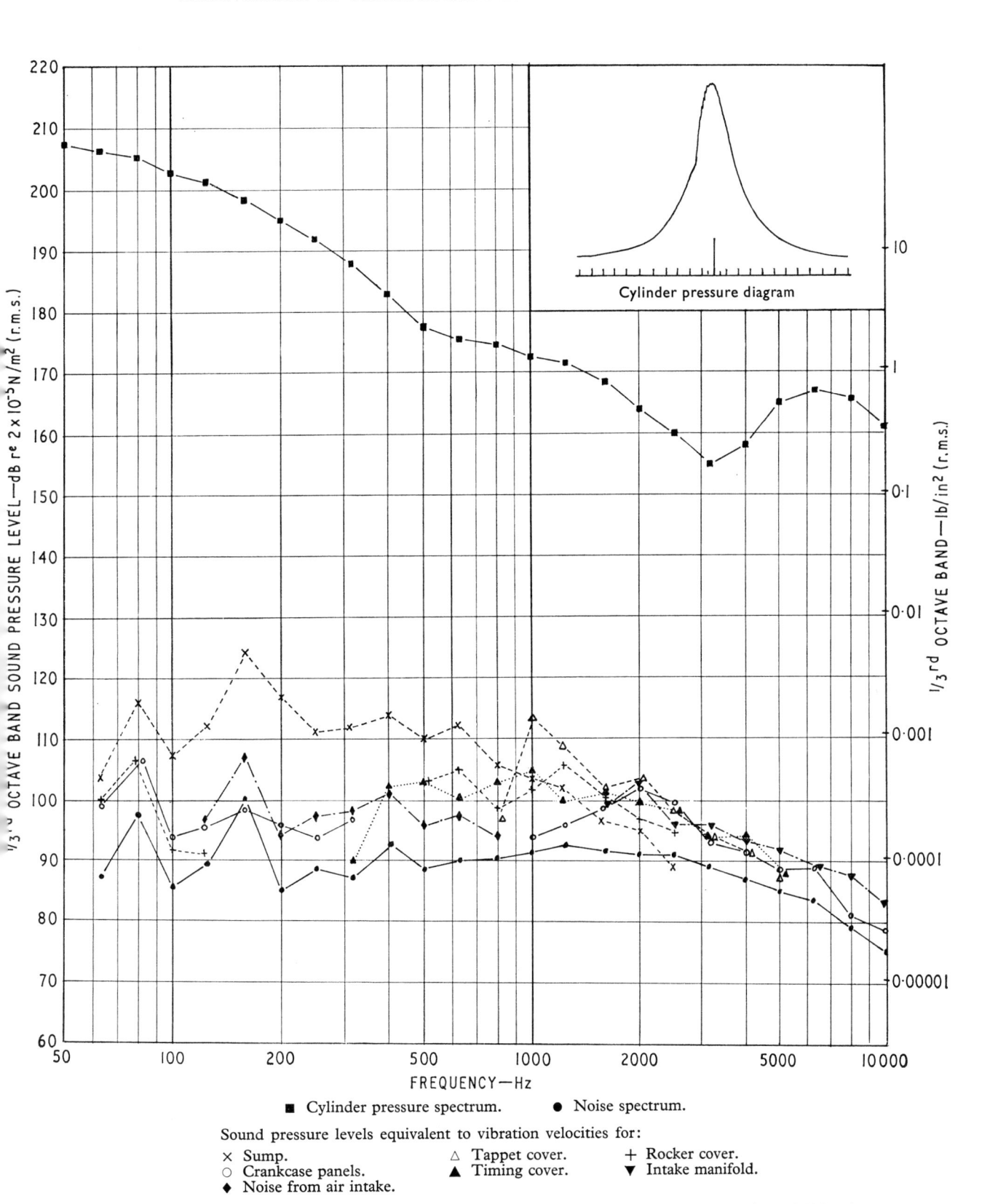

Fig. 12.1. Relationship between cylinder pressure and inlet side noise at 1500 rev/min full load

wavelength of the panel is large by comparison with the wavelength of sound in air at the same frequency. However, the radiation of sound is progressively less efficient at lower frequencies (4). The one-third octave band level vibration of the covers of an engine are plotted as the plane-wave sound pressure levels equivalent to the vibration velocity of the covers to give an idea of the relative importance of the attenuation of the noise by the engine structure, and the attenuation due to the inefficient radiation by the engine panels. As can be seen from Fig. 12.1, the sump (×) is the main potential source of low-frequency noise, giving way to the tappet covers (△) at 1 kHz. At very low frequencies, below 100 Hz, the engine appears to be rocking, giving high levels of vibration on the side of the sump and the side of the rocker cover. However, this mode has a low radiation efficiency at this frequency and the noise of the air intake may be more important. Above 1·6 kHz many of the surfaces of the engine are vibrating with similar amplitudes and no single source can be identified as predominant. The relative importance of these covers depends upon the radiation efficiency associated with their modes of vibration. The wavelength of sound in air is 6·75 in at 2 kHz, which is smaller than flexural wavelength of the crankcase panels (10 in). The crankcase panels can be expected to radiate noise efficiently, and at a frequency which is close to the frequency of maximum sensitivity of an A-weighted sound level meter (2·5 kHz). With some reservations concerning the radiation efficiencies of the various surfaces, Fig. 12.1 shows that all the surfaces of the engine may contribute to the overall noise level, albeit at various frequencies. The resultant noise of the engine plotted in Fig. 12.1 was measured 3 ft from the centre-line of the engine opposite the mid-point of the inlet side of the engine, which was mounted on a test bed. The differences between the surface vibration spectra and the measured noise includes some attenuation due to the distance of the microphone from the noise source.

When the engine is fitted to a vehicle its noise may undergo a number of reflections before it reaches an observer outside the vehicle, and some further attenuation may occur, particularly if the engine is well shielded from the observer as it is in some passenger coach applications.

It is possible to apply noise reduction techniques at any stage, and broadly the possibilities are:

(1) Reduce the combustion noise by (a) controlling the initial rate of injection or by simply injecting for minimum delay period, (b) controlling the piston slap by special pistons or by offsetting the crankshaft or gudgeon pins, and (c) controlling the timing gear rattle by having smaller clearances between gears or by using a chain or toothed belt for the timing drive.

(2) Improve the structure attenuation by reducing the vibration of all external surfaces of the engine.

(3) Place the engine in an acoustic enclosure lined with a sound absorbent to minimize the effects of holes in the enclosure. This will necessitate improved cooling to atone for the added thermal insulation of the enclosure.

The resonances of the thin panels and covers on an engine represent one of the few stages in the path of the noise from source to ear where amplification takes place, and the control of these resonances seems to offer a convenient way of reducing noise by perhaps 5 dBA without incurring any obvious penalty in running cost or reliability. However, a study of the surface vibration (noise) spectra in Fig. 12.1 shows that several covers or panels could contribute to the noise of the engine in each frequency band, and the summation of these bands, with the appropriate weighting, determines the overall noise level in terms of dBA. It is pointless, therefore, to set out to treat one or two areas in such a case; the initial aim should be to treat all the covers and then to look for noise sources of secondary importance that may have been masked by the noise from the original covers and panels.

The outer surfaces of the engine may be divided into categories: cast iron panels, cast iron covers, cast aluminium covers, pressed steel covers, and pulleys. The treatments tend to follow the construction and material rather than the function of the cover on the engine.

NOISE REDUCTION MEASURES

Cast iron panels

Considerable care must be exercised in modifying the cast iron panels covering the engine crankcase and water jacket, as quite small changes may involve considerable alterations to expensive block-machining plant. It is difficult to assess accurately the important modes of vibration of these surfaces by a theoretical analysis, and investigations to date have relied upon experimental determination of the important modes of vibration as a basis for defining noise reducing modification. This has taken the form of a vibration map of both sides of the engine, and the following important modes of vibration have occurred on in-line engines:

(a) Bending torsion of whole crankcase and block, with the 'stiff box' of the cylinder block undergoing torsion about an axis parallel to the crankshaft, and the crankcase walls bending in sympathy. The first (free-free) mode has been found to lie in the frequency range 200–630 Hz for the engines tested (8 litres to 2 litres respectively).

(b) Lateral vibration of the bottom of the skirt on long-skirted crankcases, involving bending of the crankcase panels along a line parallel to the crankshaft. This has occurred around 1 kHz on engines tested.

(c) Crankcase panel flexural vibration, where panels opposite adjacent connecting rods move in antiphase. The vertical rotation of the point of attachment of crankcase wall and the bearing webs may cause some bending within the bearing webs, and consequent axial motion of the main bearings. The frequency of this mode may vary from 800 Hz to 4 kHz.

(d) Water jacket panel flexural vibration where the whole panel moves in the fundamental mode of a plate with built-in edges, all parts on the panel moving with the same phase.

(e) Water jacket flexural modes with nodes forming vertical lines across the panel (mode of vibration is similar to that of crankcase panels in (c)).

(f) At high frequencies complicated modes appear when resonances affect various small areas of the engine structure.

The crankcase flexural modes can be effectively broken up with one or more strong horizontal ribs across the crankcase panels. The extra stiffness of the ribs raises the natural frequency of this mode of vibration to a new frequency where the excitation is lower, and the small panels bounded by the ribs and the bearing webs will have much higher natural frequencies. The increases in frequency may be calculated approximately by considering the crankcase wall as a plate. These ribs have a beneficial effect on the bending torsion mode of the whole crankcase (a).

The skirt-flapping mode (b) may be reduced by stiffening the bottoms of the bearing webs, which are not continuous across the engine at this level. Little information exists as to the effectiveness of such treatment. The water jacket plate modes may be treated by dividing the panel into smaller areas with strong vertical and horizontal ribs.

These measures need to be applied to combat the individual vibration problems of a particular engine, and the importance of the modes mentioned above has varied from engine to engine.

An indication of the effectiveness of stiffening a crankcase panel by using a horizontal rib and increased panel thickness is shown in Fig. 12.2. A reduction in panel vibration of 10 dB has been achieved at frequencies where the noise of this panel was important.

Cast iron covers

On some engines, notably tractor engines, the covers are

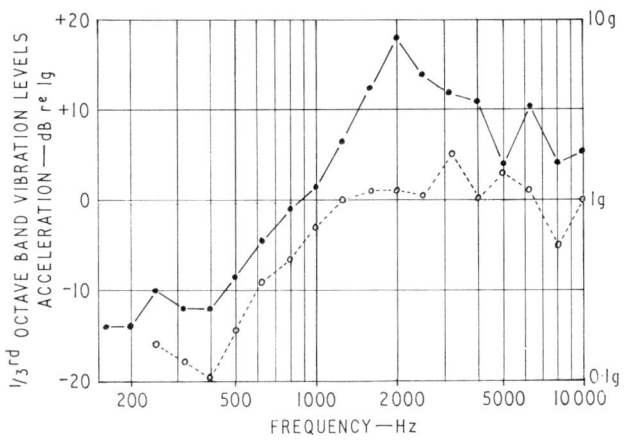

● Thin crankcase panel.
○ Thicker ribbed crankcase panel.

Fig. 12.2. Reduction of vibration of crankcase panel by stiffening

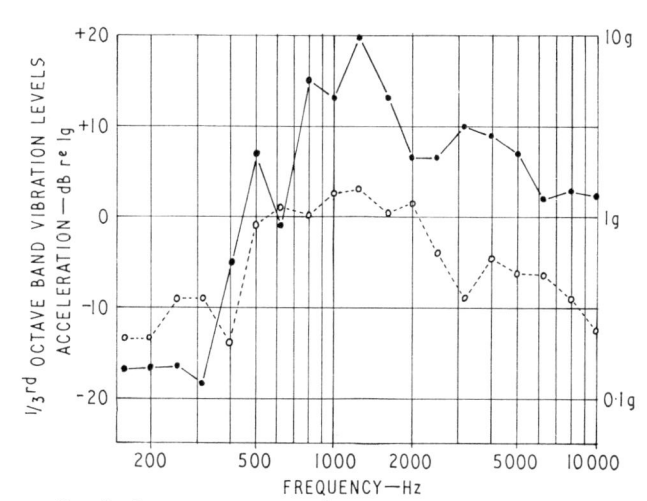

● Standard cover on paper gasket.
○ Standard cover on $\frac{1}{16}$ in thick gasket, with rubber washers under heads of securing bolts.

Fig. 12.3. Reduction of vibration of cast iron timing cover by gasket isolation

made from cast iron, and these are bolted to the crankcase with a thin paper or asbestos gasket. These covers are so heavy that they may be isolated by using a thin rubber gasket combined with rubber washers under the bolt heads, and this would seem to be the easiest way of reducing the noise of such covers. It may be possible to mould a gasket around the edge of the cover and to seat it in a notch to reduce the chance of leakage and to provide lateral support for the cover. The resonant frequency of the cover on its gasket may be quite high, perhaps 200 Hz for a timing cover and 400 Hz for a rocker cover, depending upon the size of the cover. The reduction of vibration of a timing cover isolated with $\frac{1}{16}$ in Nebar cork gasket material is shown in Fig. 12.3. For mass produced engines it may be more satisfactory to replace the cast iron cover with a flexible and damped pressed steel cover, and cast iron sumps used as structural members on tractors may have to be treated in the same way as crankcase panels.

Cast aluminium covers

Cast aluminium covers are usually stiff due to the large number of ribs, bosses, dimples, etc., that have been cast into the cover. These covers also offer scope for isolation at the gasket, but a technique is under development in which slots are left in the cover so that the centre (plate-like) noise-radiating part is held in place by three or four flexible metal strips only. The slots between the metal strips are filled with rubber with the dual purpose of preventing oil leakage and damping the vibration of either the centre part, which is isolated from high-frequency vibration by the flexibility of the metal strips, or the outer part, which is secured to the engine crankcase. Such a 'slit-isolated' technique is believed to be particularly suitable for die-cast covers. The reduction of cover vibration due to gasket isolation and slit-isolation of a die-cast aluminium alloy timing cover is shown in Fig. 12.4. The finished cover is strong enough to withstand

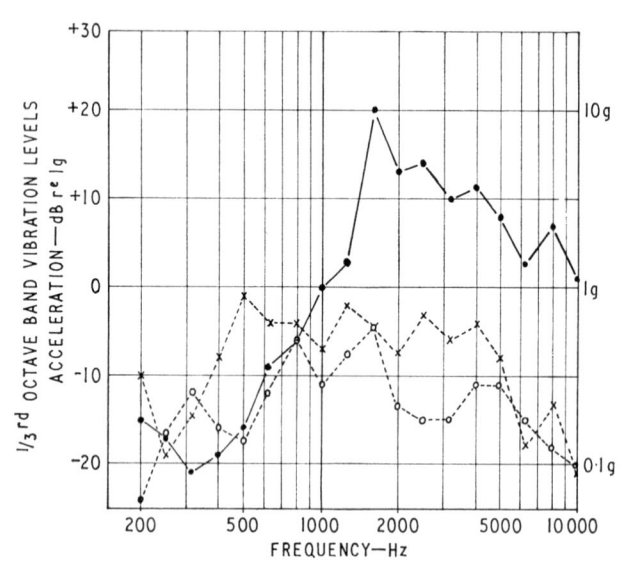

- Untreated die-cast aluminium cover.
× Cover isolated by thick gasket.
○ Cover isolated by slots filled with silicone rubber.

Fig. 12.4. Reduction of vibration of die-cast aluminium timing cover

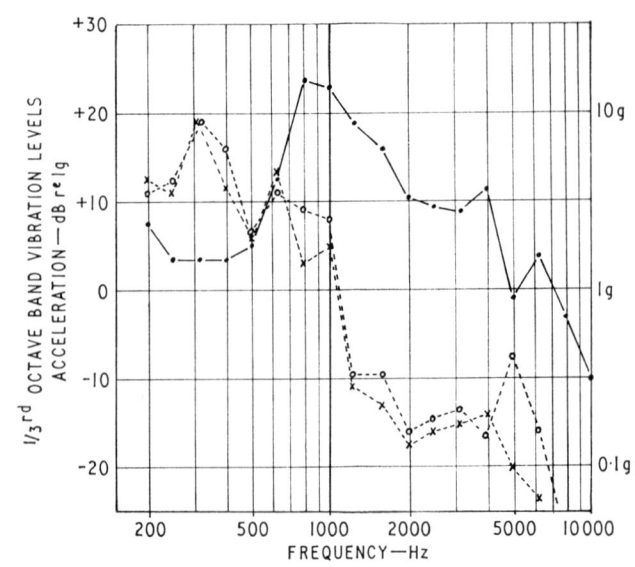

- Untreated sheet steel cover.
○ Cover damped by neoprene bonded cork in sandwich construction.
× Cover damped and isolated on thick gasket.

Fig. 12.5. Reduction of vibration of pressed sheet steel rocker cover

accidental impacts, and the design does not rely upon the rubber in the slots for structural strength.

Pressed steel covers

Covers pressed from sheet steel have so many 'crinkles' pressed into them that they are rigid enough to isolate on a thick gasket with soft washers under the heads of the securing bolts. However, a more practical approach is to damp the vibration of these covers by building upon them a layer of damping material constrained to be sheared, as the cover bends, by being the 'filling' of a sandwich between two sheets of metal, one sheet being the cover. The damping treatment need not cover the whole surface area of the cover, although it may be convenient to press the complete cover from sandwich stock. The aim should be to make such a cover as flexible as possible for the damping to have maximum effect, which means removing all the 'crinkles'. However, the important exception to this is a sheet steel cover that is attached to a massive vibrating member which causes the sheet steel to be bent along its length by the forces which excite the vibration (e.g. a sump fixed to a sump flange which is taking part in skirt-flapping or panel-bending modes). In these circumstances pressed or welded stiffeners should be introduced into the cover to resist the bending, and the damping should be applied beyond the stiffeners.

The reduction of the vibration of a pressed steel rocker cover achieved by covering top and sides with patches of $\frac{1}{16}$ in Nebar and 18 s.w.g. sheet steel to form a sandwich with the original cover is shown in Fig. 12.5.

Other noise sources on the engine

Having stiffened, isolated, or damped all the external surfaces of the engine, a search must now be made for any other noise-radiating surfaces. Any brackets and guards which ring audibly when struck should be abolished; they should be replaced by non-resonant designs or attached to the vehicle chassis rather than the engine. The crankshaft pulley is often a source of noise when its fundamental disc resonance lies in the audiofrequency range. It is usually possible to redesign the pulley so that three or four spokes support the rim from a hub with a small frontal area. The spokes can be made so that they act as isolating springs, allowing the hub to move in the axial direction without causing excessive motion of the rim at high frequencies, but they must be strong enough to transmit torque to the rim and stiff enough radially to prevent the collapse of the rim on to the hub. The three spokes on one low-noise pulley design have each been made from two flat strips with the maximum flexibility in the axial direction. The width of the spokes and the angle of the spokes to a radius of the pulley were chosen so that the pulley would not buckle under load. The thickness of the spokes and the weight of the rim were chosen so that the combination had a natural frequency in the range of poor radiation efficiency of the rim. As the diameter of the pulley was 7·5 in, maximum radiation efficiency was reached at 910 Hz. The pulley rim must be isolated at that frequency so that the maximum frequency for the resonance should be 640 Hz or less. The spoke stiffness was calculated as for a beam built in at both ends, to give a total stiffness such that the rim (mass) and the spokes (spring) would resonate at 300 Hz as a single-degree-of-freedom system. The spokes were made from two sets of leaf springs and they were made integral with the rim (Fig. 12.6). The two springs for each spoke were separated by viscoelastic material to damp their resonance. This simple approach to the design gave a

Fig. 12.6. Reduction of noise 2 ft from front of engine due to crankshaft pulley isolation

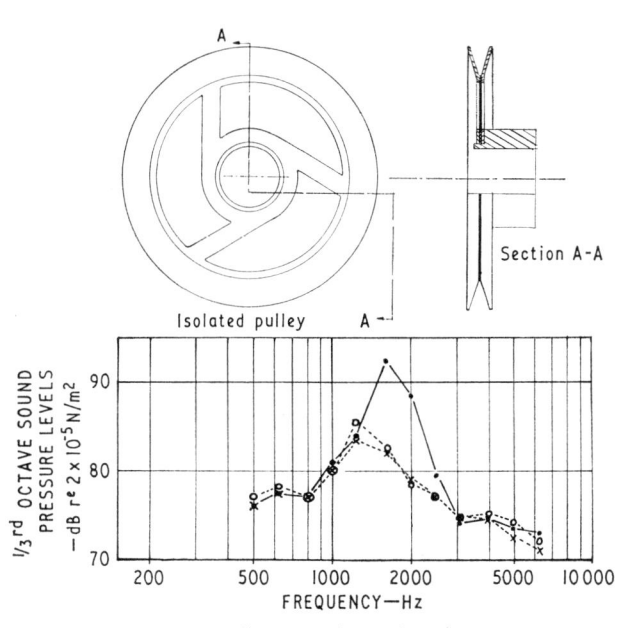

- ● Standard pulley on quietened engine.
- × No pulley fitted to quietened engine.
- ○ Isolated pulley fitted to quietened engine.

pulley that was stiff enough to transmit power but not stiff enough to radiate noise. The effect of such a pulley on the noise of an otherwise quiet engine is compared with the effect of the standard four-spoke cast iron pulley (of the same diameter and on the same quietened engine) in Fig. 12.6. The standard pulley has a resonant frequency of 1·6 kHz, and the full benefit of the quietened timing cover (Fig. 12.5) was not available owing to the noise radiated by the standard pulley. Fitting the quiet pulley resulted in a 5 dBA improvement in the noise 2 ft from the front of the engine, even though all the covers on this engine had been previously treated effectively. The noise at the front of the engine with the quiet pulley was 1·5 dBA greater than the noise of the quiet engine with no pulley at all on the crankshaft. This indicates that the noise of the pulley alone would be approximately 4 dBA less than the noise of the quiet engine without the pulley, and that the noise from the pulley alone has been reduced by approximately 9 dBA.

Other areas of the engine which may become important noise sources include the inlet and exhaust manifolds and pipes and, if the engine is sufficiently quiet, the fuel injection equipment. Isolation of the intake manifold has been necessary on one engine and work is being conducted on quieter fuel injection equipment against the day when engines become quiet enough to warrant it.

Overall noise reduction

The quiet covers were developed on a number of engines, but at the time of writing the complete treatment had been tried on two engines only. The aim in both cases was to reduce noise at the top end of the speed range, and less attention was paid to the noise at low speeds and light loads. The overall noise reductions obtained are shown in Fig. 12.7 in terms of the one-third octave band noise spectra, before and after treatment. These show reductions of 5–10 dB over most of the audio range. The corresponding overall noise reductions were 5 dBA on both sides of the left-hand engine at 2000 rev/min, and 8 dBA inlet side, 5·5 dBA exhaust side, for the right-hand engine at 2800 rev/min. These measurements, however, are not a true indication of the improvement in the noise attenuating properties of the engine structure as changes in injection timing or clearances between mechanical parts may affect the overall noise. The structural improvement may be judged accurately by comparing the structure attenuation of the engine before and after treatment, provided that combustion noise is the predominant source in both conditions; the higher the structure attenuation, the lower the noise of the engine. This may be achieved simply by advancing the injection timing until the external noise of the engine rises at the same rate as the cylinder pressure spectrum; that is, by increasing the combustion noise until other noise sources become insignificant by comparison. In practice, the structure attenuation is plotted for a range of injection timings and engine speeds until the highest attenuation spectrum is repeated at several engine conditions. This spectrum is taken to be the true attenuation of the engine, and attenuations obtained in this way are plotted in the lower part of Fig. 12.7 to show the increased attenuation provided by the complete treatments. These spectra show that there is a real improvement in the noise attenuating properties of the engine structure, although changes in the mechanical noise of one of the engines had taken place over the period of the experimental work.

CONCLUSIONS

The vibration characteristics of the external surfaces of several conventional automotive diesel engines have been examined, and measures to reduce surface vibration amplitudes and, hence, noise have been developed. In the briefest possible terms these consist of stiffening the crankcase and water jacket panels, isolating the stiff covers, and damping the flexible covers.

Provided that *all* the panels which make up the exterior surface of the engine are treated, and that any other major noise-radiating component (e.g. the crankshaft pulley) is located and silenced, then overall noise reductions of the order of 5 dBA may be achieved with a conventional in-line automotive engine. Two of these engines have been treated in the laboratory and a reduction of noise, slightly in excess of 5 dBA, has been achieved at and near the maximum operating speed at full load.

AREAS WHERE FURTHER STUDY IS REQUIRED

Although the treated engines have run successfully in the laboratory, further development and production engineering will be necessary before some of the techniques

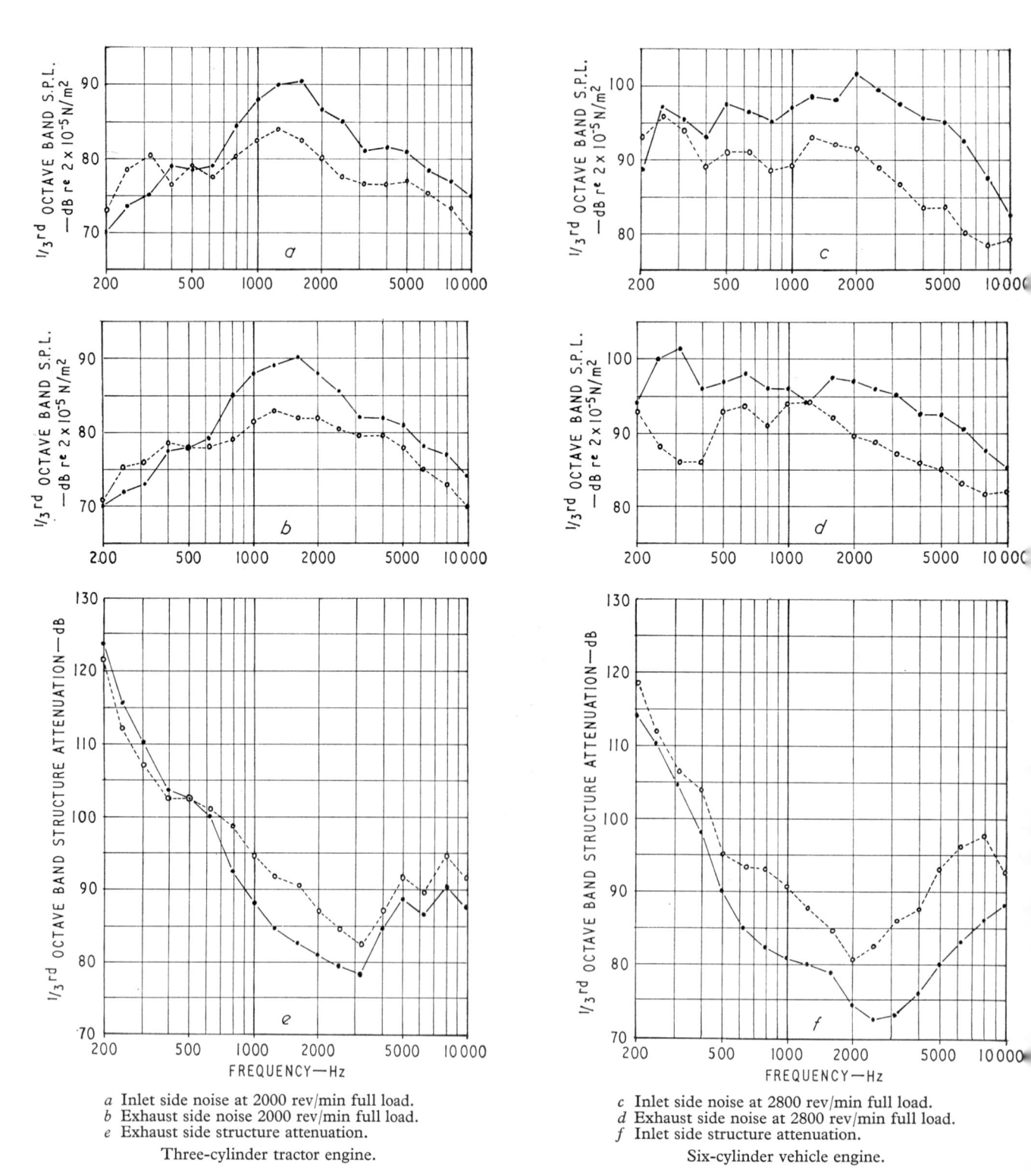

a Inlet side noise at 2000 rev/min full load.
b Exhaust side noise 2000 rev/min full load.
e Exhaust side structure attenuation.

Three-cylinder tractor engine.

c Inlet side noise at 2800 rev/min full load.
d Exhaust side noise at 2800 rev/min full load.
f Inlet side structure attenuation.

Six-cylinder vehicle engine.

Fig. 12.7. Reduction of noise and improvements in structure attenuation obtained with conventional engine designs

described in this paper are fully suitable for mass production. In particular, for the isolating medium for gasket-isolated and slit-isolated covers, a soft rubber or plastic is required which has high viscous losses and will withstand attack by hot lubricant, fuel oil, and salt water, etc., for the life of the engine.

Plastic materials may provide an alternative to pressed steel and die-cast metal covers, in that some of them are inherently flexible and well damped. Problems of sealing and mechanical strength sufficient to withstand abnormal loadings during the handling of the engine (when being fitted to the vehicle or during repairs, etc.) may be overcome by including metal or glass fibre stiffening local to the sealing flange.

ACKNOWLEDGEMENTS

The author would like to thank the Directors of C.A.V. Ltd for permission to publish this paper, and Messrs D. Amos, E. Gardiner, and C. Young, for carrying out the experimental work.

APPENDIX 12.1

REFERENCES

(1) B.S. 3425:1966. *Method for the measurement of noise emitted by motor vehicle.*
(2) *The Motor Vehicles (Construction and Use) (Amendment) Regulations* 1968 (Statutory Instrument 1968, No. 362 Road Traffic).
(3) AUSTEN, A. E. W., PRIEDE, T. and GROVER, E. C. 'Effect of engine structure on noise of diesel engines', *Proc. Auto. Div. Instn mech. Engrs* 1964–65 **179** (Pt 2A, No. 4), 113.
(4) MAIDANIK, G. and LYON, R. H. 'Statistical methods in vibration analysis', *A.I.A.A. Jl* 1964 **2** (No. 6, June), 1015.

Paper 13

SOME ASPECTS OF DIESEL EXHAUST EMISSIONS, ESPECIALLY IN CONFINED SPACES

C. Lunnon*

A study has been made of diesel exhausts, and this paper relates the essential parts of what was found. Exhaust pollutants and their physiological effects are examined, and the steps taken by the National Coal Board for the protection of personnel and against dangers from explosions and fire are considered. The action taken to comply with statutory exhaust gas checks is given, and the necessary minimum ventilation of tunnels to acceptable standards is detailed. Other aspects of exhaust fume control, including that in confined spaces above ground, are dealt with. Finally, methods of treating diesel exhaust emission are suggested.

INTRODUCTION

THE NATIONAL COAL BOARD employs some 700 diesel locomotives and over 100 diesel tractors underground in coal-mines in Great Britain, and we have to satisfy ourselves, and satisfy H.M. Inspectorate, that the exhaust gases from these machines are not going to harm the men who work there. We have therefore made a study of diesel exhausts, and this paper relates the essential parts of what we found. We have only done a very limited amount of experimental work directly ourselves, mainly because the field had already been covered. At the time of the decision to replace London trolley-buses by diesels, a careful study of the physiological effects of diesel exhausts was made, from which certain references are given.

EXHAUST POLLUTANTS AND THEIR PHYSIOLOGICAL EFFECTS

Besides the inevitable oxygen, nitrogen, and carbon dioxide, diesel exhausts contain small traces of sulphur dioxide, several oxides of nitrogen, carbon monoxide, other products of incomplete combustion including aldehydes and polycyclic hydrocarbons, and solid particles, mostly carbon. The effects of these substances on the human body may be briefly summarized as follows:

Sulphur dioxide, SO_2

This gas is formed from sulphur present in the fuel. It is a lung irritant with a pungent smell of brimstone.

The MS. of this paper was received at the Institution on 3rd March 1970 and accepted for publication on 2nd April 1970. 33
* *Traction Engineer, National Coal Board, Headquarters of Production Department, The Lodge, South Parade, Doncaster.*

Oxides of nitrogen

Oxides of nitrogen, including NO, N_2O_3, N_4O_6, NO, and N_2O_4, are formed by the direct combination of nitrogen and oxygen under the conditions of high temperature and pressure found in the engine cylinder. With moisture and further oxygen (both are present in the lungs), they tend to form nitric acid. This is not only damaging in itself but makes the lungs less able to resist bacterial attack (1)†. Whilst the author is not aware of any human death being caused by oxides of nitrogen from a diesel engine, fatalities have occurred outside the mining industry where these gases have arisen from other causes (2). In a colliery, shot-firing can also produce oxides of nitrogen, and tunnels which form the return airways may thus contain fumes from both sources.

There is no doubt from tests on animals (3) that in the exhaust of a well-loaded engine in good condition, the oxides of nitrogen are the most dangerous constituent. However, the body seems able to tolerate and recover from low levels, or intermittent medium levels of these gases, i.e. they are not cumulative poisons. Incidentally, nitrous oxide, N_2O, is not included among the poisons; it is the harmless dentist's or 'laughing' gas.

Carbon monoxide, CO

This turns the haemoglobin in the blood into scarlet carboxy-haemoglobin, and thus suppresses its oxygen carrying function. In a normal diesel in good condition the 'make' of carbon monoxide is negligible, but if the engine is in poor condition, or its breathing is restricted by clogged

† *References are given in Appendix 13.1.*

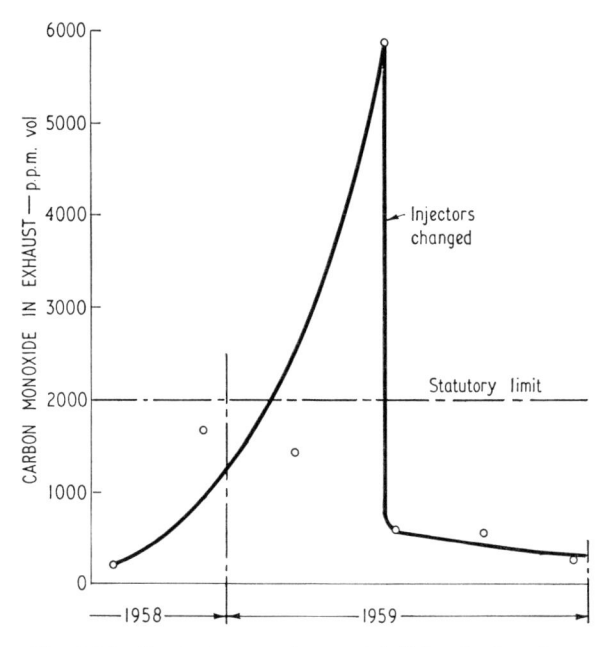

Fig. 13.1. Statutory analyses at full load of carbon monoxide in exhaust

air filters or choked exhaust system, then it can rise alarmingly. Fig. 13.1 shows the quarterly test results on the exhaust of an underground locomotive at full load, indicating the effect of faulty injectors.

Other products of incomplete combustion
These include aldehydes, polycyclic hydrocarbons, and carbon. To aldehydes are attributed most of the smell and eye-watering characteristics of diesel fumes, and these tend to be at their worst under conditions of idling or restricted air supply (3). Polycyclic hydrocarbons such as 3:4 benzpyrene and 1:12 benzperylene are a possible source of lung cancer. However, the amount of these pollutants is minute.

In tests at two London Transport bus garages (4) it was shown that the concentration of these chemicals produced by the buses was about one-tenth of that which existed on one occasion in the air outside the garage. Thus the pollution owing to the buses in the garage, smoky though it appeared to be, was small compared to the pollution from cottage chimneys near-by. Moreover, a study of the incidence of lung cancer among various groups of London Transport staff showed no excess in groups where it might have been expected if diesel exhaust fumes were a serious contributory factor. It should be remembered, however, that the background level of pollution at that time (1956–57) was sometimes a good deal higher than it is at the present time, now that the 1956 Clean Air Act has had time to take effect.

Solid particles
Particles such as soot or carbon can, if they are below a certain size (about 5 μm, but it depends to some extent on shape), accumulate in the lungs and in rare cases, over a large number of years, block up part of them. Larger particles settle in the throat and nose and are expelled by coughing or sneezing. Recently, however (1), it has been demonstrated that in the absence of these small particles, the carcinogenics referred to above are virtually harmless. It was only when carried on solid particles that the benzpyrenes, etc., started bronchial tumours.

If the foregoing makes it seem that diesel exhausts are full of deadly poisons, it should be borne in mind that the concentrations are exceedingly minute. Where animals have been used for tests they have breathed the undiluted exhausts for appreciable periods. In normal circumstances, even underground, the exhaust is quickly diluted by the surrounding atmosphere.

Some of the nastier pollutants are decomposed by sunlight (5) and some decompose of their own accord. For example, when a diesel locomotive is working underground in a colliery's intake airway, one can smell the fumes for some distance down-wind of it; but one does not smell them in the return airway. Solid particles, except perhaps for the very smallest, will settle out.

Compared with other sources of energy, the pollution level produced by well-maintained diesels is quite favourable. Petrol engines furnish most of the same pollutants but with far more carbon monoxide and, in addition, compounds of lead from the anti-knock additive. This has had special attention (5) (6).

President Nixon has recently been persuaded of the need for executive action against dirty jet (i.e. gas turbine) exhausts, justifiably in the author's view. Steam engines cause worse pollution, except where it is possible for power stations, etc., to apply sophisticated purifying treatment (precipitators, etc.) to their exhaust gases and to release them at a great height.

Atomic-energy production involves problems concerning the disposal of radioactive waste. Indeed, hydroelectric energy appears to be the only major source of power which does not involve pollution of one sort or another, and the available amount of this is severely limited.

EXHAUST TREATMENT FOR UNDERGROUND DIESELS

In considering the steps that the National Coal Board takes in treating the exhausts of underground locomotives, tractors, etc., it must be borne in mind that we not only have to protect our men from harm, we also have to guard against explosions of methane seeping out of the coal seams, and against fires arising from the plentiful presence of coal-dust. Methane fortunately requires a high temperature to ignite it, but accumulations of coal-dust can be readily ignited by surprisingly gentle heat. Some of the steps we take are directed against more than one problem.

Briefly, we apply flame-traps at the beginning and end of the engine's breathing canal, we make all parts of that canal more than usually strong and more than usually gas-tight, we cool any surfaces that could exceed 135°C and we bubble the exhaust through water in an exhaust-conditioner box. In addition, we ensure that the gas is suitably diluted with air, artificially or naturally, after it emerges; and we take samples of the effluent, monthly of the roadway air after dilution, and quarterly of the neat exhaust prior to dilution, under specified conditions, for analysis for certain gases.

Exhaust conditioner box

The exhaust-conditioner box serves several purposes. It ensures that the gases are cooled at least to the boiling point of water before emission. It dilutes them with water-vapour, it traps much of the soot and tar, and it partly dissolves out those pollutants which are soluble in water. As these pollutants are mainly acidic, the water in the box becomes progressively more acid, and will tend to attack the walls of the box. For this reason the box is made of stainless steel or thick copper-bearing steel.

Further, the water is made strongly alkaline by the addition of soda (caustic or carbonate) at the beginning of each day's work, the quantity being regulated to ensure as far as possible that the contents will remain alkaline throughout the day. This not only reduces corrosion but also accelerates the neutralization of acidic oxides such as sulphur dioxide. However, in the brief period of contact between water and gas, only part of the oxides of nitrogen are dissolved. Carbon monoxide, being insoluble, is not dissolved.

The design of the box should ensure that in the event of the engine being inadvertently rotated backwards, it is impossible for water to be drawn up into the cylinders; this would of course cause broken pistons and bent connecting rods. A big bell (Fig. 13.2) and a relatively low height of water outside it will produce the required result. The difference between levels inside and outside gives the

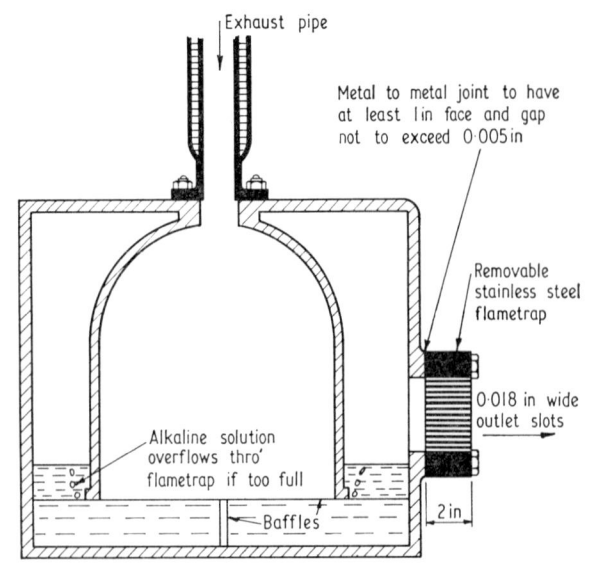

Fig. 13.2. Exhaust conditioner and flame-trap

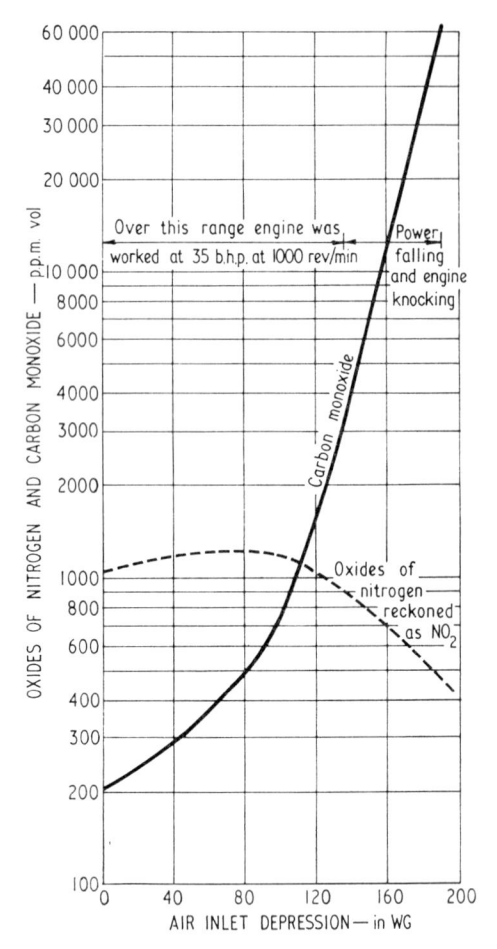

Fig. 13.3. Effect of restricted breathing on exhaust

number of inches water-gauge back pressure on the engine. This must be added to the restrictions caused by inlet and outlet flame-traps, air-filter, etc., in estimating the total restriction, and the amount of derating that is necessary to offset it. This is commonly of the order of 15 per cent. A graph is given in Fig. 13.3 which shows the effect of various air restrictions on the output of carbon monoxide and oxides of nitrogen (measured and reckoned as nitrogen dioxide) in the exhaust.

If an exhaust flame trap, which is basically a series of narrow slits designed to quench a flame by contact with relatively cool metal, should start to get choked, it would impose additional back pressure, reduce the air flow, and cause the production of soot at an increased rate. The sooting-up of the slits would be accelerated and a vicious circle would be set up. The production of carbon monoxide would be increased and eventually the engine would choke itself to a standstill. To avoid this situation, we take great care to remove and clean exhaust flame-traps every day or even, where appropriate, every shift.

STATUTORY EXHAUST GAS CHECKS

Undiluted exhaust gas checks

Every 90 days we are required by law to take samples

the undiluted exhaust gas of every underground diesel engine, both when it is idling and when it is developing full power. Two samples are taken in each condition, and, of these, one is analysed for carbon monoxide and one for total oxides of nitrogen. We are allowed up to 2000 parts/million by volume of carbon monoxide and 1000 parts/million by volume of oxides of nitrogen, reckoned as if they were entirely NO_2. The analysis of the sample for oxides of nitrogen, normally done by the phenoldisulphonic acid method, is a long and complicated procedure culminating in the comparison of the colour of the final solution with samples of various known strengths.

Great care is necessary over the purity of the reagents used, otherwise false results are obtained. All the oxides of nitrogen are oxidized to NO_2 and reckoned as if that is what they had been to start with; but this need not be regarded as an error because they would tend to behave similarly in the lungs. For this reason, the abbreviation NO_2 will be used in the remainder of this paper to represent 'oxides of nitrogen reckoned as nitrogen dioxide'. Because of the difficulty of the analysis, and the difficulty of establishing the 'full power' condition of the engine, a high degree of repeatability is not attainable in the estimation of this poison.

Roadway air gas checks

Every 30 days we are required by law to sample the roadway air down-wind of the diesel under prescribed conditions and to analyse it for carbon monoxide. If this exceeds 50 parts/million CO by volume, we must take steps to improve it. If it exceeds 100 parts/million, we must stop the diesels running.

It is of interest to note, with reference to Fig. 13.1, that on the day of the highest carbon monoxide reading, the air in the underground locomotive garage concerned contained 7 parts/million of CO. This compares with a figure of 300 parts/million, which was measured in the Blackwall Tunnel before the recent improvements.

NECESSARY MINIMUM VENTILATION OF TUNNELS

From 2000 parts/million to 50 parts/million is a 40-fold dilution, and since diesels generally produce about $2\frac{1}{4}$ ft³ (0·064 m³) per min hp of exhaust, some 90 ft³ (2·55 m³) per min hp of ventilating air are needed. To be on the safe side, our practice is to provide at least 150 ft³ (4·25 m³) per min of ventilating air for every hp of installed diesel power likely to be using the roadway concerned at any given time.

There is no test in the roadway air for oxides of nitrogen. This is because we are not yet entirely satisfied that there is a test sufficiently sensitive and accurate for the extremely low concentrations involved. If an exhaust which is just on the limit of 1000 parts/million of oxides of nitrogen is diluted in the available ratio of 150/2·25, it arrives at a strength of 15 parts/million. This is just acceptable, bearing in mind that an engine employing the usual mechanical transmission does not actually work at full power, except for brief moments.

It sometimes happens in an underground roadway that a diesel locomotive or tractor moves at about the same speed and in the same direction as the current of ventilating air. One would expect that excessive concentrations of fumes would then arise, and certainly the smell becomes somewhat strong. However, practical investigations made in this country (7), and similar investigations elsewhere (8), have shown that dangerous conditions are most unlikely to arise.

OTHER ASPECTS OF EXHAUST-FUME CONTROL

If we get a high carbon monoxide analysis, we find that attention to maintenance, to improve the condition of the engine, almost always puts matters right. A high NO_2 analysis is less easy to deal with, because improving the condition of the engine may well make it worse. To keep the NO_2 figure down, we take the following steps:

(a) We use low-sulphur fuel oils. Besides reducing sulphur dioxide, this reduces NO_2 because apparently the sulphur acts as a catalyst in NO_2 formation.

(b) Where possible, we use indirect-injection engines. These inherently give less NO_2 than do direct-injection engines. In a special experiment on NO_2 emissions, an

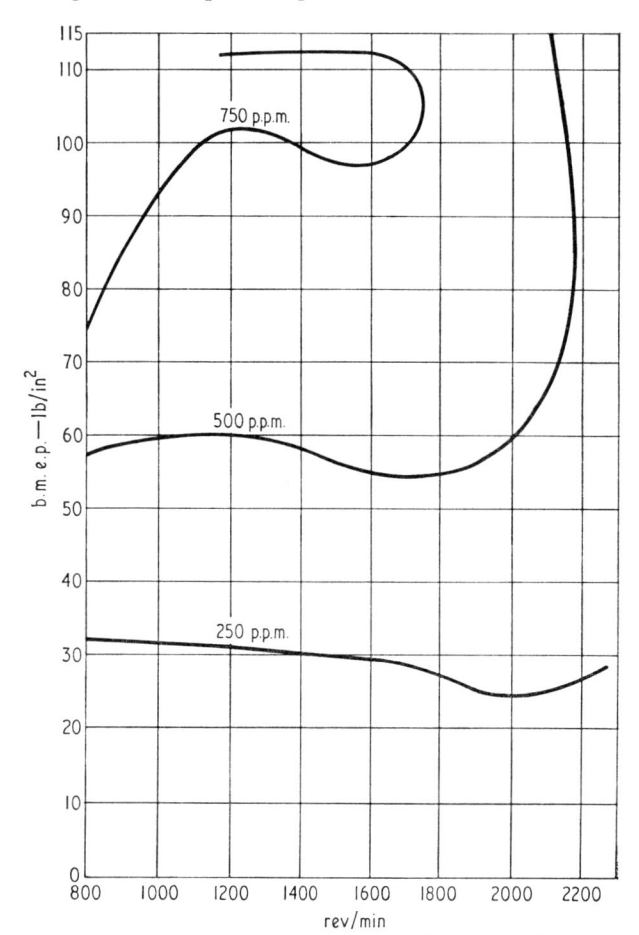

Fig. 13.4. Contours of constant NO_2 proportion in exhaust, precombustion-chamber engine

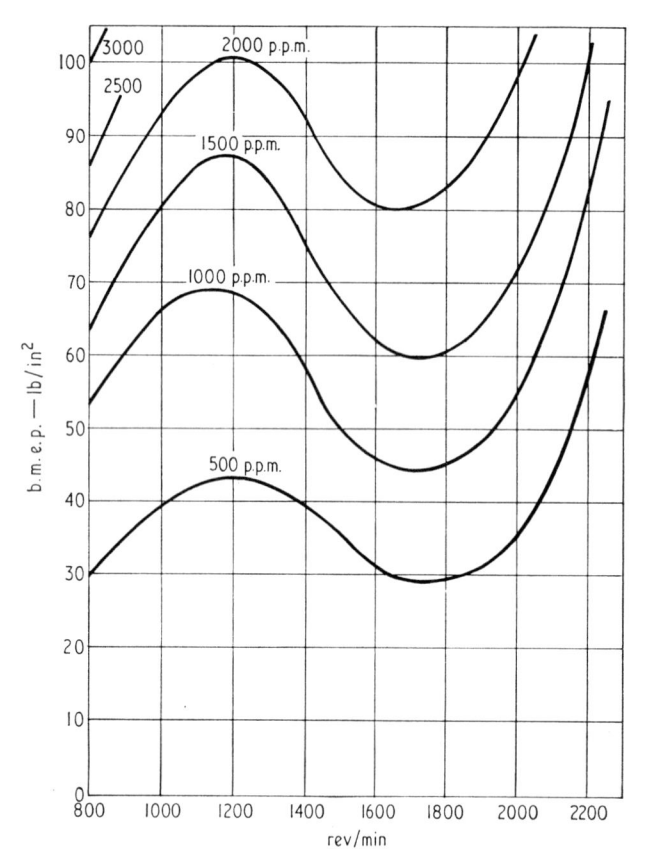

Fig. 13.5. Contours of constant NO$_2$ proportion in exhaust, direct-injection engine

indirect-injection engine gave the contour diagram shown in Fig. 13.4. It was converted overnight to a direct-injection engine by the fitting of new pistons, connecting rods, cylinder head, fuel-injection pump, etc. The following day it gave the contour diagram shown in Fig. 13.5, i.e. two to three times as much NO$_2$.

(*c*) Where we have to use a direct-injection engine we commonly retard the injection by 5°. This is simply a means of reducing cylinder temperatures and pressures; it slightly reduces efficiency in addition to reducing the output of oxides of nitrogen.

We have not adopted a suggestion of reducing the air intake to reduce NO$_2$, for fear of producing an excess of carbon monoxide.

A number of fuel additives have been examined with a view to the possibility that they might reduce the output of NO$_2$. None was markedly effective. A 'smell suppressant' was also tried and found to be of doubtful value. We do *not* use an additive claimed to reduce black smoke. We found that this, in fact, reduced the size of the carbon particles from the visible range to the invisible range, which, as may be inferred from what has gone before, would actually make them more harmful to the lungs rather than less harmful.

Smoke is clearly objectionable, but we consider that it should be avoided by attention to design, rating, and engine condition, not by additives of this sort. If smoke occurs, it should be taken as a warning that maintenance attention is needed. To suppress it by other means is to suppress a valuable and much needed warning.

CONFINED SPACES ABOVE GROUND

Reference has already been made to diesel bus garages, and brief attention may now be directed to what may be done to improve conditions in such places as bus stations, railway stations, narrow streets, etc. It would not appear necessary in these environments to fit exhaust conditioners, nor to avoid the ubiquitous direct-injection engine. Carbon monoxide will not be a hazard unless air filters are clogged or engines are in a very poor condition. But smoke is to be avoided most carefully, and steps to this end should include careful design to give good combustion at all speeds and all loads.

The author considers that design and power rating can and should be such that no visible smoke is produced when the engine is loaded till the fuel rack abuts the full-load stop, at any speed from maximum to idling, and at any state of the barometer likely to be encountered in the locality. Some mass-produced, automotive diesel engines appear to smoke even when new, and one wonders whether this is owing to a lack of care, or to unduly wide tolerances during assembly. The part that good maintenance plays in clean running is well known.

Last but not least, careful driving, and particularly the avoidance of long periods of idling, can play a substantial part not only in avoiding the production of fumes at the time but also in preventing the onset of fouled injectors etc., which will increase the amount of smoke when the engine is put back on to load. Smell is not a good criterion by which to judge a diesel engine, but exhaust smoke is— it should not be visible. It is probably true that background and lighting affect the appearance of exhaust smoke. A reading of 25 per cent on the Hartridge meter is therefore suggested as the target. If it is objected that the author's proposals represent a loss in output horsepower, the reply is that they represent a gain in thermal efficiency.

CONCLUSIONS

There are minute quantities of a number of objectionable substances in diesel exhausts, and where engines are used in enclosed spaces it is good practice to wash the exhaust in an alkaline solution, to take and to analyse certain air and exhaust samples at regular intervals, and to ensure current of air in the roadway sufficient to effect a 67-fold dilution. Indirect-injection engines are to be preferred to those with direct injection, but even the latter can be used with a degree of injection-retardation and a degree of derating. Over a period of 30 years these steps appear to have been effective in keeping underground workers free from harm.

In spaces that are confined but not actually enclosed, the washing process may be dispensed with and the direct injection engine's timing may be standard; but it is most desirable that rating, design, manufacture, maintenance, and usage should be such that the engine produces no visible smoke. Except in very restricted spaces, smoke is the most objectionable of the pollutants because of its long-term effects. Although the engine normally produces negligible amounts of carbon monoxide, certain faulty conditions can greatly increase the output of this very poisonous gas.

ACKNOWLEDGEMENTS

The writer acknowledges with gratitude the assistance he has had from colleagues in the National Coal Board, especially in Scientific Control, Medical, and Mechanical Engineering branches, in the work described. Messrs F. Perkins Ltd provided valuable assistance in the comparison of direct- and indirect-injection engines. The author also thanks Mr N. Higginson, N.C.B. Chief Mechanical Engineer, for his encouragement and permission to give this paper. Any opinions expressed or implied are the author's and not necessarily related to the official policy of the National Coal Board.

APPENDIX 13.1

REFERENCES

(1) WOLKONSKY, P. M. 'Pulmonary effects of air pollution', *Archs envir. Hlth* 1969 (October), **19**, 586.
(2) CHIEF INSPECTOR OF FACTORIES (GT BRITAIN) *Annual Rept* 1947.
(3) PATTLE, R. E., STRETCH, H., BURGESS, F., SINCLAIR, K. and EDGINTON, J. A. G. 'The toxicity of fumes from a diesel engine under four different running conditions', *Br. J. ind. Med.* 1957 **14**, 47.
(4) COMMINS, B. T., WALLER, R. E. and LAWTHER, P. J. 'Air pollution in diesel bus garages', *Br. J. ind. Med.* 1957 **14**, 232.
(5) WALLER, R. E., COMMINS, B. T. and LAWTHER, P. J. 'Air pollution in a city street', *Br. J. ind. Med.* 1965 **22**, 128.
(6) DRINKWATER and EGERTON. Combustion processes in the compression ignition engine', *Proc. Instn mech. Engrs* 1938 **138**, 415.
(7) DURHAM DIVISION OF N.C.B. Private communication.
(8) HOLTZ, J. C. and DALZELL, R. W. 'Diesel exhaust contamination of tunnel air', *Rept of Investigations 7074* 1968 (U.S. Dept of the Interior, Bureau of Mines).

Paper 14

DIESELS AND THE GENERATION OF ELECTRICITY

F. Ratcliffe*

Diesels used by the Central Electricity Generating Board as prime movers for supplying stand-by or emergency auxiliary supplies of electricity require to have a high level of reliability and availability. These are the important features rather than the production of cheap electricity. When used exclusively for generating electricity for the grid system, their use is restricted because of limitations in MW output potential. Those in use must (1) produce cheap electricity, (2) assist in the continuity of supply, and (3) have peak-lopping facilities in certain applications. The major operational cost is fuel, and four fuels are considered; these range from 35 s to 3500 s fuel oils. Diesel maintenance costs can be substantial and methods of reducing these are reviewed in depth. Continuity of supplies of electricity into the National Grid and for local loads is critical. The electrical, mechanical, and chemical aids used to maintain diesel reliability are discussed.

AUXILIARY MACHINERY

WHILST THIS PAPER discusses diesels used exclusively for generating electricity, it would be an incomplete picture if some mention was not made of diesels used as auxiliary plant in nuclear, conventional, or gas-turbine power stations. Such an example is, therefore, given.

Electricity generated at modern power stations is fed to the National Grid system from substations operating at 132 kV, 275 kV, and 400 kV. The distribution therefrom must be arranged so that the chance of supplies being interrupted is eliminated as far as practicable should a fault develop in one section of the system. One established method is to interconnect the substations in such a manner that an alternative section of line can replace a faulty section. Transmitting electricity from a power station has a similar security of a choice of lines or outlets. A power station isolated from grid supplies normally loses the use of its electrically driven auxiliaries, without which the prime movers cannot operate and reconnection to the grid rendered impracticable.

This situation would be critical, and one method of safeguarding power stations is to have a stand-by emergency diesel engine directly driving a generator. The generator in turn will give sufficient supplies of electricity to energize an adequate quantity of auxiliaries to get a prime mover

The MS. of this paper was received at the Institution on 7th April 1970 and accepted for publication on 7th May 1970. 33
* *Station Superintendent, Central Electricity Generating Board, South-Eastern Region, Ashford 'B' power station, Victoria Road, Ashford, Kent.*

fully operational. It is important that the diesel can be started without electrical supplies, and it is usually started from an air bottle, which can be replenished by a petrol-driven air compressor.

Fig. 14.1 shows such a lay-out in a 110 MW gas-turbine power station. Two 55 MW A.P.4 Rolls-Royce/A.E.I. gas turbines generate at 11·8 kV, which is stepped up to 132 kV. Auxiliary supplies at 415 V are normally obtained from two auxiliary transformers (6·6 kV/415 V) or from the unit transformers. With both gas turbines stationary, loss of grid supplies would also make dead the 6·6 kV supplies to the auxiliary transformers. The unit transformers cannot be used in this situation. By running the emergency diesel, auxiliary supplies sufficient to start up both gas turbines consecutively can be generated.

Assuming loss of the 415 V auxiliary supply, the total time required to obtain reconnection with the 132 kV bars, and generating 110 MW, would be about 30 min using the emergency stand-by diesel. In this situation the critical factors relating to the diesel are:

(1) Constant availability with minimum support from ancillary plant.
(2) Absolute reliability in starting.
(3) Speed in attaining full-load conditions.

ECONOMY OF DIESELS USED FOR GENERATING ELECTRICITY

The Central Electricity Generating Board (C.E.G.B.) does not expand its use of diesels intended exclusively for

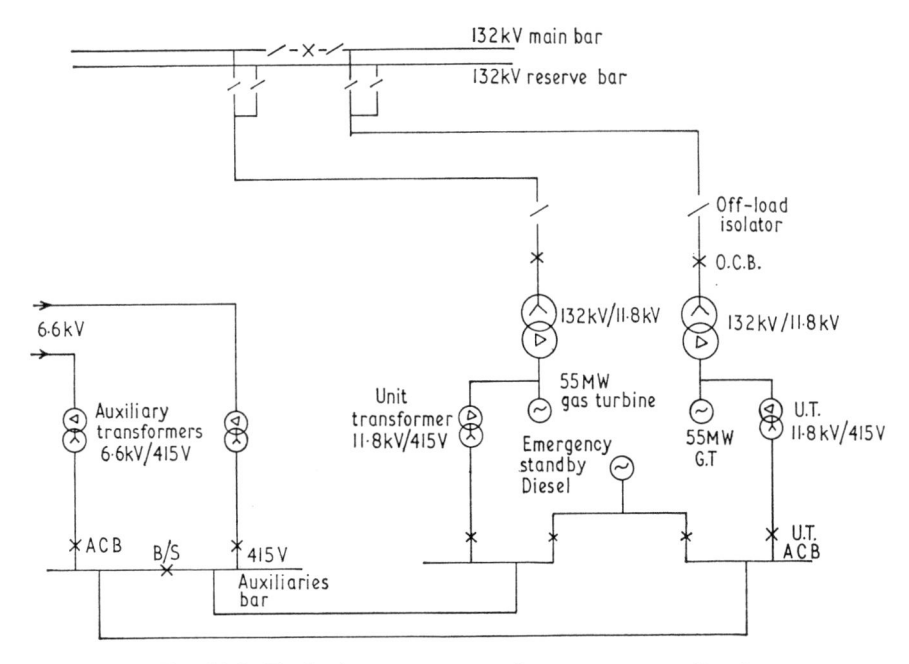

Fig. 14.1. Typical arrangement of an emergency diesel

generating electricity because of the large increments in plant capacity occurring and the limitations diesels have in this direction. This limitation also influences capital and maintenance costs. Such diesels that have been installed, however, have a specific and critical function.

The prime movers at the C.E.G.B. Ashford 'B' power station are five Mirrlees KVSS.12 engines, producing a maximum output of 10 MW running an annual 3500 h plus. The Board requires two essentials from this plant: economy and reliability, and the former is discussed first.

This plant was conceived originally as an alternative to more expensive transmission reinforcement and because of its facility for peak lopping. Throughout its 15 years' life its role has changed slightly, and running periods are consistently extended to a regular 14 h/day, 7 days/week, except for outage periods. The sole reason for this change is an economic one, for Ashford 'B' is high in the order of merit table.

Experience has shown that the 15 in bore Mirrlees K range of engines has had sufficiently low operational costs to offset the substantial maintenance expenses usually attributable to diesel engines. Table 14.1 shows a previously published (1)* list in descending order of the 10 most economical diesel power stations in the United Kingdom for 1968. The majority of these sets are the 15 in bore type under discussion, and Ashford 'B' lies fifth in this list with a total running cost of 0.977d./unit.

Of this 0.977d./unit total running cost, approximately 60 per cent can be allocated to fuel costs and the remainder to salaries, overheads, and maintenance costs. Except for maintenance costs, which fluctuate with the amount of running, the remaining 40 per cent of the total cost is met largely whether the station runs or not. Running costs

* References are given in Appendix 14.1.

Table 14.1. The 10 most economical diesel power stations in the U.K.

Name of undertaking	Total running cost, old pence/unit
1. States of Guernsey	0.735
2. Jersey	0.794
3. N.S.H.E.B., Gremista	0.862
4. Station 'C'	0.898
5. C.E.G.B., Ashford 'B'	0.977
6. Douglas Corporation	0.990
7. C.E.G.B., Macclesfield Group	0.998
8. N.S.H.E.B., Kirkwall	1.024
9. Isle of Man, Peel	1.030
10. C.E.G.B., Haverfordwest	1.032

alone, therefore, compare favourably with those of major conventional power stations and represent overall thermal efficiencies within the range 33–35 per cent.

FUEL ECONOMY

Since fuel is the major running cost, fuel is the chief consideration and there can be but one policy. The cheapest fuel alone is burned, without the aid of additives. Whilst the theory of certain fuel additives is correctly founded, an operator should have an understanding of the manufacturer's design philosophy of producing exhaust temperatures below any critical range. There is a lot of published material on the characteristics of a variety of eutectic mixtures but it may be accepted that much work is required as yet in this direction.

At temperatures of about 1652°F (900°C) and above, it is known that mixtures of the sodium sulphate order (normally high up in the eutectic temperature tables) are aggressively corrosive in the combustion spaces of, say, gas turbines. The resulting problems of metallurgy or surface coatings do not appear to be relieved by fuel additives. The solution is exclusively using a fuel where the total alkaline metal does not exceed 0·6 p.p.m. in the fuel or 0·02 p.p.m. in the air, on the understanding that it is more difficult to restrict the sulphur content. This specification usually means using 34–48 s gas oil having a sulphur content of about 0·5 per cent and keeping the water content to below 50 p.p.m.

Towards the bottom of these melting point tables appear the eutectic mixtures combining sodium and vanadium, of which the most critical is probably sodium vanadyl-vanadate (5–1–11) $5Na_2O$, V_2O_4, $11V_2O_5$, which may solidify at about $1000°F$ ($538°C$). This means that exhaust-valve seats should not exceed this temperature or, symptomatically, exhaust temperatures after the valves should not be much in excess of $820°F$ ($438°C$) with uncooled valve seats (2). The designed maximum exhaust temperature at cylinders of the KVSS.12 is $800°F$ ($427°C$), and at Ashford this temperature is only reached or exceeded during abnormal running periods.

This subject has been discussed in detail elsewhere, so it is sufficient to state that the manufacturer has provided the means of burning residual fuels. The co-operation of the fuel supplier is required to keep the fuel as consistent as possible. Disturbing stories of the proportion of low residuals in 220 s blended fuel can be true, and its effect on injectors particularly can be critical. In this range of blended fuel a high proportion of ash can be present, with its effects on exhaust-valve seats. The operator must co-operate by adequately cleaning and preheating the fuel.

The five diesel sets at Ashford have engine cooling of the closed-circuit type, with the primary water passing through Serck heat exchangers. Cooling secondary water is drawn from a three-section, induced draught, cooling water cooling tower. Make-up water is drawn from the adjacent River Stour.

Engine-jacket water is maintained at about $165°F$ ($74°C$) because temperatures in excess of this figure will affect the temporary hardness of the water and increase the failure rate of water joints. A tapping from the primary water system serves two Serck oil-heaters and thence through heating coils in the $4 \times 24\,000$-gal main oil-storage tanks. The two Serck heaters can maintain a fuel temperature for all sets of $140°F$ ($60°C$), and the storage-tank coils can maintain a free flow up to 950 s oil.

Ashford power station ran on ordinary diesel fuel for an initial period, and the heating arrangements described were more than adequate. Then 220 s blended fuel was burned successfully for many years. To increase the fuel temperature to $160°F$ ($71°C$) it was necessary to install 2×12 kW electric line-heaters, which ran with a normal load of 18 kW when the diesels were on full load. Over the last few years, the station has burned 950 s fuel, for which an additional 40 kW electric line-heater was installed. With sets on full load, the total electric heating load averages 45 kW.

Currently a scheme has been developed to change to 3500 s fuel, for which extra heating will be obtained from steam generated from the exhaust gases. By allowing a proportion of the exhaust gas flow from one engine only to by-pass the silencer, and circulate through a thimble-tube waste-heat boiler, steam can be generated. From one diesel engine exhaust there is sufficient heat to generate 1000 lb steam/h at 100 lb/in^2. This is sufficient steam heating to raise the 3500 s fuel temperature at the rail to $250°F$ ($121°C$) by using suitable heaters, and provide sufficient heat to the main oil-storage tank coils to stock the heavier fuel.

All fuel is centrifuged prior to use and great care is exercised in achieving the correct fuel temperature at the injector. Care is also taken to ensure that the correct fuel/air ratio is achieved at all times and that cylinders are correctly balanced out. These simple precautions, together with normal routine maintenance of components, ensure that temperatures throughout do not become outside the manufacturer's specification.

During the years of burning residual or blended fuels, no serious increases in maintenance costs have been attributable solely to the fuels used.

ECONOMY FROM CORRECT FUEL/AIR RATIO

Correct fuel/air ratio is the most critical aspect of an economic diesel prime mover. The fuel monitoring is relatively easy, usually depending upon the efficiently serviced fuel pumps and injectors and ensuring that the fuel temperature at the injector is precisely correct. Monitoring the correct air flow requires more thought and attention, particularly if outage periods are inconvenient and expensive.

The original installation at Ashford 'B' provided the following path for the air flow for one bank of one diesel engine. Air was drawn from the environment about the control panel area inside the engine room. In summer this ambient temperature could exceed $100°F$ ($39°C$) substantially, and in winter remain quite warm. It was drawn through an oil-wetted, basket-type filter filled with ferrules. The air was consistently warm and impregnated with oil-vapour and carbonaceous debris. To eliminate completely the contaminants in the engine-room air would require more maintenance staff and time than was economic. This type of filter did not cope with such aerosol forms of contaminant. Much of the contaminants, therefore, condensed on the turbo-charger compressor blades that followed the filter. From the turbo-charger the air passes through a gilled tube intercooler, which condensed a further proportion of contaminants. On its way through the air manifold to the air valves, a proportion of the remaining contaminants condensed out.

This situation was unacceptable. The original air was dirty and not as dense as it should have been and passed through a filter not specifically designed to separate the main contaminant. It passed through an inefficient turbo-charger compressor, owing to the blades being oil-wetted and rough, and then through an intercooler, which was

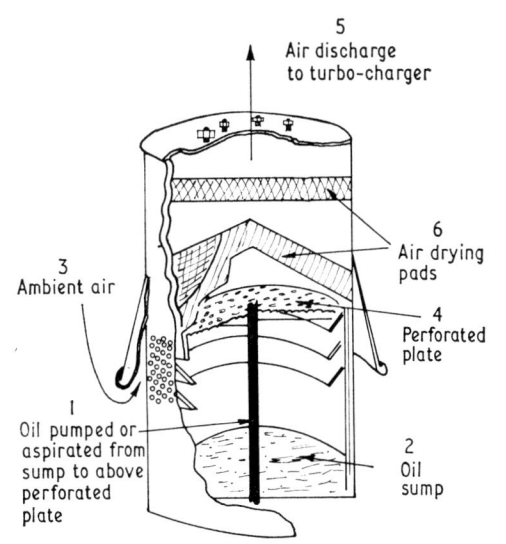

Fig. 14.2. Sketch showing arrangement of air filter

partially blocked by oil-wetted carbonaceous material. The cooler restricted the air flow and only partially cooled it. Finally, since the air manifold sides were dirty, smooth air flow must have been affected.

Clearly, it followed that, except for a few hours following a complete service, the air flow of 13·3 lb/b.h.p. h could not be sustained. The tendency was for one bank to become dirtier than its neighbour and a progressive state of imbalance throughout the engine would develop, with all its attendant consequences: loss of full load, high exhaust temperatures with valve failures, build-up of deposits on turbo-charger blades, increased injector maintenance and constant removal of intercoolers for cleaning.

Preventing all traces of oil-vapour contaminating engine-room air in a continuously running plant is expensive. It was cheaper and more positive to transfer the air intakes to outside the engine room and provide external Cycoil air filters. In a single modification the ambient temperature of the air is reduced by as much as 50°F (28°C), providing more dense air before cooling, whilst silica and other particulates are removed by the filter to an acceptable level and the air is oil-dried also by the filter (Fig. 14.2). There were no further worries on the air flow, although it is advisable to fit U-tube manometers to indicate accurately the air pressures. Turbo-charger compressor blades and intercoolers do not now require attention from one annual overhaul to another. The air filters are remarkable in that little maintenance throughout the year is required.

ECONOMY OF LUBRICATION

Much has been written about oil-lubrication and its technological and economic influence. Without duplicating what has been explained in detail many times elsewhere, its economic significance applied to Ashford 'B' is mentioned.

Lubricating oil in an engine must be regarded as much a doctor's thermometer to the health of an engine as, say, exhaust temperature of individual cylinders. As such, regular chemical monitoring is critical where prolonged running without outages is essential.

In addition to performing the single function of lubricating moving parts, the oil must reduce corrosion and piston-fouling caused by acids formed primarily in the combustion spaces. Generally speaking, the choice of base oils and additives up to a Supplement I specification is, by experience, sufficient for the needs of the KVSS.12 engine. Provided there is oxidation stability and adequate dispersive power, chemical monitoring will indicate the reserve of alkalinity necessary for maintenance of good engine performance. This simple rule reduced the complex subject to a single managerial yardstick. In addition, it is necessary to maintain to an acceptable level the quantity of insolubles and fuel dilution by regular centrifuging—at Ashford 'B' once every 100 h.

Industrial engine-oil formulations vary widely in the permutations of base oils and additives used. Industrial engines themselves vary in design and performance and, indeed, in how they are managed. Successful lubrication is the right combination of oil to engine. A high-cost, brand-name lubricating oil may be the ideal choice for a particular engine and unnecessarily expensive for another engine. With the KVSS.12 engine running within the design pressures and temperatures, a lubricating oil made from acceptable materials to Supplement I level of detergency appeared to be suitable. Like fuel oil, the cheapest acceptable oil is obtained because of the quantities used.

At Ashford 'B' a 15-month test was carried out using a high-cost, brand-name oil in one engine, a moderately priced, brand-name oil in three engines, and a C.E.G.B. specification oil to Supplement I level obtained by open tender. The specification oil was relatively low priced. Throughout the 15-month test period, chemical tests revealed little difference between the three oils. At the conclusion of the test, the engines were stripped and all engines were equally free from deposits, ring sticking, or wear.

The station has run on specification oil successfully for a number of years and the sump, which has a capacity of 350 gal, is changed each year (3500 h plus), for reasons other than technical ones. The combination of regular centrifuging, make-up of about 20 gal/engine per day, and suitability of the oil makes its life indefinite.

RELIABILITY OF DIESELS USED FOR GENERATING ELECTRICITY

Optimum reliability

Optimum reliability is critical to any prime mover generating electricity. Diesels will perform their function reliably if worked to specified conditions over a specified period. Optimum reliability is the maximum realistic figure nearest to 100 per cent. The probability of not reaching 100 per cent can be expressed as the equation:

$$a\% = 100\% - (b\% + c\% + d\%)$$

where a is the percentage reliability, b the probability that the diesel will be available, c the probability that the diesel will start when required, and d the probability that the diesel will remain on load until no longer required.

It is a feature of diesel plant that with normal servicing routines, incidence of faults developing during stationary periods are rare enough to be unusual. When reviewing loss of availability records, therefore, remedial action to improve b and c covers a smaller field but is usually a very significant area. Accidental presence of residual fuel in injector pipelines at shutdown, or an overhaul programme not completed on time, would be two examples. Isolated cases of this kind can be quickly identified and remedial action injected into any system. The parameter of defects or breakdowns covered by d requires by far the biggest proportion of any study.

Invariably such a study will emphasize the type and frequency of servicing of components. Economics usually dictate the frequency and type of servicing and eventual renewal policy. Once the policy of servicing, say, a set of exhaust valves is implemented, tests over a long period on the breakdown rate can be taken. At Ashford 'B' these have shown that an isolated exhaust valve can fail some time after 700 h. To increase this period requires a more expensive quality control, so exhaust valves are serviced, economically, at 600 h for pure reliability. It is known that other similar installations have service periods extended to well over 1000 h and, whilst it may be tempting to try the extended period, statistics on reliability firmly suggest the 600 h. For the same reason, injection equipment is serviced every 250 h.

CUMULATIVE RECORDS

Analysing cumulative records is one useful tool in improving reliability. An extract from such a record for one engine covering a 10-year period is shown in Table 14.2. The total outage time is that period when both planned or unplanned outages occurred during periods when the engine would otherwise be required for load. The causes of the individual outages are expressed as a percentage of the whole *outage* time for the 10-year period and are discussed below.

If the outages are expressed as a percentage of the total period the diesel should have been running, the reliability equation for this engine can be completed. For the 10-year period, the items in Table 14.2 affecting category b had a value of 1.9 per cent, and those affecting category d a value of 2.3 per cent. None of the items recorded affected directly the starting of the diesel when required and, therefore, category c is 0 per cent. Hence, it can be shown that the overall reliability of the diesel over a 10-year period was 95.8 per cent, if the annual overhauls and two items affecting the alternator are deleted:

$$95.8\% = 100\% - (1.9\% + 0\% + 2.3\%)$$

It should be recorded that the reliability percentage has progressively improved over the 10 years, as a direct result of the factors brought out in this paper.

This is a critical factor in the application of any plant to the generation of electricity. The 100 per cent starting

Table 14.2. Table of planned or unplanned outages at Ashford 'B' power station over 10 years on one KVSS.12 engine—1st April 1959 to 31st March 1969

Reason for outage	Proportion of total outage time expressed as a percentage
Annual overhaul	78.6
Bed-plate	0.2
Column	1.4
Cylinder blocks	1.7
Crankshaft	1.2
L.E. bearings	
Gears	0.7
Liners	0.4
Pistons	0.9
Cylinder heads	0.9
Camshaft and cams	0.6
Fuel pumps, injectors, and pipes	0.4
Exhaust valves	0.2
Inlet valves	
Exhaust system	
Cooling water	0.2
Intercoolers	0.3
Air manifold	0.4
Turbo-charger	3.6
Governor system	
Lubricating oil	
Push rods	
Barring gear	
Alternator	7.3
Slip rings	1.0
Oil circuit breaker	
Pedestal bearing	
Total	100

reliability cannot be improved upon, and the overall reliability percentage of 95.8 per cent over a 10 year period is a very good figure to improve on in future years.

ANNUAL OVERHAUL

At 78.6 per cent the annual overhaul period occupies the major proportion of the total outage period and requires an explanation. For almost the whole of the period under review, the diesels at Ashford 'B' power station are taken out of service consecutively for a period of weeks in the summer. They are serviced by apprentices drawn from the South-Eastern Region of the Board under a planned and supervised series of apprentice training courses. Under careful supervision, second-year apprentices strip, clean, inspect, repair as necessary, reassemble, and finally recommission a diesel–alternator set within a specified time-scale.

The Education and Training Department have regarded the system as valuable experience for craft apprentices. Overall skills required for diesel maintenance stretch the apprentices to a satisfactory degree. The challenge of being a member of a team responsible for completing high-merit plant within a given time-scale has always been met responsibly.

Without the annual apprentice training courses, the overhaul periods would be substantially reduced. On the other hand, such training of apprentices has been successful enough to continue this unusual factor in the application of diesel engines.

VIBRATION

Table 14.2 shows specific outages for defects connected with bed-plate, columns, cylinder blocks, and crankshaft. The majority of these defects were assisted by engine vibration, which occurred early on in the 10-year period. The bed-plate defects were exclusively the loosening or fracturing of holding-down bolts; the columns had cracks occurring between the crankcase and cam-inspection doors, and cylinder blocks developed movement between joint interfaces, which leaked oil. Crankshaft alignments required more frequent attention than was economic or convenient. The rate at which cylinder-liner, water-seal rings failed also lessened when the vibration was eliminated.

Two remedial steps cleared the majority of these defects. The dynamic balance of the crankshaft was improved by fitting counter-weights; the changing of the air filters and resiting them in the open air assisted the balance of the engines, as previously described.

GEAR TRAIN

The rate of failure of gear wheels in the crankshaft–camshaft train dropped dramatically when the teeth of each wheel were strengthened by decreasing their number.

PISTONS

Following early modifications to pistons, Ashford 'B' has been running for the majority of its life with the same design of piston widely used in the KVSS.12 engine. An incident in which the piston crown became detached from the body started an investigation. It was found that this piston had fractured from the corner of the oil chamber, horizontally outwards to the third compression-ring groove to the point of detachment.

A system of ultrasonic non-destructive testing was speedily developed at Ashford, in which an 'Ultrasonoscope' was used at a frequency of 1·25 MHz. The low frequency is used because of the coarse-grained material of the piston. A single cylindrical probe is applied to the periphery of the crown, the waves being directed downwards through the suspect area. This system has been infallible and several pistons were found to have similar cracking, but before the detachment stage had been reached. It was deduced that the cracking started at the small end or gudgeon boss and extended circumferentially clockwise and anticlockwise until the full circumference had been completed. It is also believed that the cracking starts in the cooling chamber and travels outwards.

Pistons are tested annually in this manner and renewed if cracking has commenced. Cracking would appear to start at about 13 000 h of life. Fig. 14.3 gives a brief indication of the position of the cracks.

It should be mentioned here that all large end-bearing

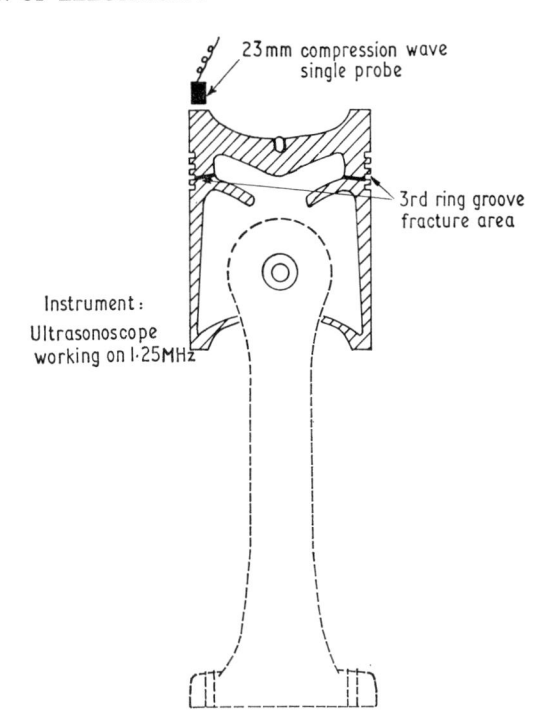

Fig. 14.3. Ultrasonic testing of suspect pistons

shells are tested by ultrasonics annually for white metal adhesion.

CAMSHAFT AND CAMS

The problems related to fuel-cam failures only, the frequency of which dropped to negligible limits when the thickness of the hardened cam surface was increased.

CYLINDER HEADS, FUEL PUMPS, INJECTORS AND THEIR PIPES, EXHAUST VALVES

Breakdowns of these components are invariably predictable and are eliminated by correct servicing at the optimum frequency.

INTERCOOLERS

There is no record of intercoolers failing prematurely. Outage periods recorded here relate solely to unplanned cleaning periods prior to the air filters being changed.

AIR MANIFOLD

The length of the air-manifold accounts for cracking and loss of air. Correct welding repairs and timely replacements should make enforced outages unnecessary. The elimination of the excessive engine vibration referred to above contributed also to a reduction of cracking.

TURBO-CHARGERS

Outage periods are long when failure of a turbo-charger is involved because it invariably means the removal of at least a part of the assembly from the engine. Since they are of continental manufacture, strict maintenance routines are necessary, covered by adequate spares to prevent waiting periods whilst spares are dispatched from abroad.

The chief concern and main reason for long outage periods is the corrosion of the jacket walls of the gas inlet and outlet casings. Despite the use of zinc sacrificial strips in the cooling-water spaces, corrosion is fairly rapid and not precisely predictable.

CONCLUSION

Critical factors in the application of diesel engines to the generation of electricity must relate to economy and reliability. Where use is made of them for auxiliaries, starting and availability reliability must be high. When used solely for generating electricity, economy must be added to this reliability.

This paper tends to emphasize these points and indicates the methods used to keep these factors controlled. Without duplicating too many technical details previously published, the paper confirms the relative economy of the prime movers at Ashford 'B' and the reasons for it. It also gives a factual account of the reliability of the plant over a 10-year period and indicates the critical factors that influenced it.

ACKNOWLEDGEMENTS

The author wishes to thank the Regional Director of the South-Eastern Region of the Central Electricity Generating Board for permission to publish the paper. He also wishes to thank Mr C. R. Garnett, Group Manager of Group 'F', South-Eastern Region, and Mr F. Lennon, Education and Training Officer, South-Eastern Region, for their advice and assistance in its preparation.

In acknowledging the work of his colleagues at Ashford power station in collating the relevant records, the author also wishes to emphasize that the views and opinions expressed are his own.

APPENDIX 14.1

REFERENCES

(1) 'Working cost and operational report', *D.E.U.A. Pubn 324* 1968.
(2) PEARSON, B. and WALLACE, W. B. 'Medium-speed diesel engines using residual fuels', *D.E.U.A. Pubn 330* 1970.

Paper 15

DIESEL ENGINE PISTON RING DESIGN FACTORS AND APPLICATION

D. A. Law*

> The range of piston ring materials currently available to the designer is given, showing their relative importance in durability and fatigue. Ring wall pressures for given duties illustrate the wide variations that exist. The piston ring free shape and its association with the ratings of diesel power units highlight the need for a better understanding of thermal gradients in ring section. Surface coatings and their relative merits to ring wear, scuff, and liner wear are illustrated. Oil control and blow-by characteristics from break-in to stable operation are discussed. Piston ring groove temperatures coupled with cylinder liner wall temperatures adjacent to the top ring location need early appraisal in design and development. Control of surface finish for both cylinder liner and piston ring is important for the critical break-in period. Control of quality is outlined.

INTRODUCTION

THE CONSTANT QUEST of the diesel engine designer to extract more power from a given cubic capacity has resulted in many interesting thermal problems in the area of the combustion chamber. Coupled with these are, of course, the normal physical issues resulting from ever-increasing firing pressures. These firing pressures are very often accompanied by extremely high rates of pressure rise. Factors such as these give the piston ring manufacturer many challenging problems to overcome. In many instances only by close co-operation between the engine designer and the piston ring designer can satisfactory conditions be achieved within a given diesel engine to ensure adequate operation of the piston ring in its difficult environment.

The piston ring is therefore a critical component in that it must seal under conditions of high temperature and pressure; it must work with minimal lubrication; it must be extremely compatible with the bore surface of the cylinder; it must have sufficient strength to withstand high shock loads owing to firing impulses. It must be of such a free shape that when fitted and at operating temperature it will conform readily to the cylinder bore with minimal high pointing around its periphery.

The MS. of this paper was received at the Institution on 1st April 1970 and accepted for publication on 21st May 1970. 22
* Assistant Managing Director, Wellworthy Ltd, Lymington, Hampshire.

PROPERTIES OF BASIC PISTON RING MATERIALS

Physical properties

For many decades the conventional grey iron, low in alloy content but rich in graphite, has been found to be extremely acceptable as a piston ring material. Its natural properties, i.e. extremely good wear resistance owing to its structure, its low modulus giving rise to relatively low wall pressures, and its favourable modulus over tensile ratio, whereby ring radial sections can be proportioned to give a wide seating face width, have deemed it an extremely useful material.

It has been produced for many years by single-cast or static pot-cast methods which have given it the attractive 'A-type' graphite structure. Latterly, intensive work has yielded a comparable iron in terms of structure, using centrifugally cast techniques. As diesel engine ratings have increased, firing pressures and rate of pressure rises have forced ring materials to become more sophisticated.

A general growth in this area has taken us through low-alloy materials into the carbide-containing materials, to the malleable materials through the heat-treated carbide-containing materials, to the spheroidal graphite irons, and ultimately to the steels. Appendix 15.1 lists the salient properties for a number of typical piston ring materials.

Each step has given the piston ring designer a new range of stress in which he can work with safety. It will be readily appreciated that a piston ring must be so stressed that an acceptable level of stress is induced when fitting a ring to its piston, thus avoiding breakage. At the same time an adequate margin under the maximum stress level must be maintained when the ring is fitted in the cylinder in its operational condition.

A graph for a typical piston ring iron is shown in Fig. 15.1. It gives the fitting and fitted stress levels for each given radial width of piston ring. The graph shows the maximum stress that can be tolerated for the material in question. It is assumed for this graph that the piston ring radial section stays constant around the full circumference of the ring. A number of detail changes can be made to fit rings of wider radial section to avoid overstressing. A typical way is to scallop out at the inner surface of the ring gap so that the ring is opened less than would be the case if the radial was maintained.

Fatigue properties

Piston rings can readily fatigue in diesel engines if certain conditions of pressure rise, clearance between ring and groove, and piston movement are satisfied. A typical engine test to establish fatigue levels on an accelerated basis can be that of increasing the top ring to groove side clearance and advancing injection so that a steeper rate of pressure rise and maximum pressure within the cylinder are achieved. This will readily bring about a broken piston ring should the material be inadequate to withstand these conditions.

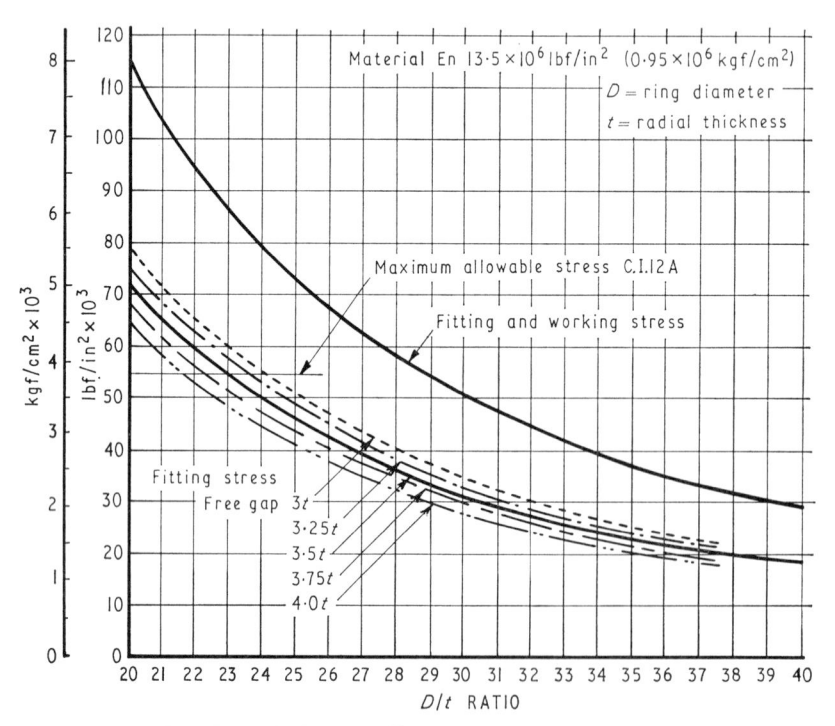

Fig. 15.1. Fitting and fitted stress for piston ring

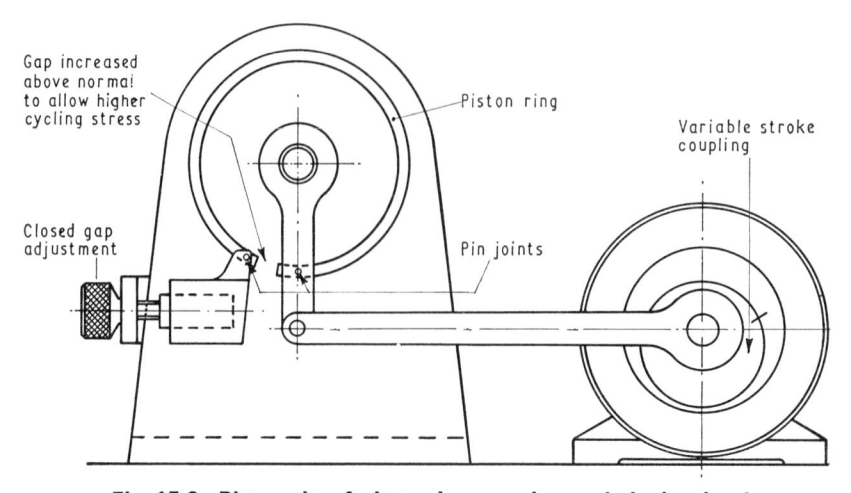

Fig. 15.2. Piston ring fatigue rig—opening and closing load

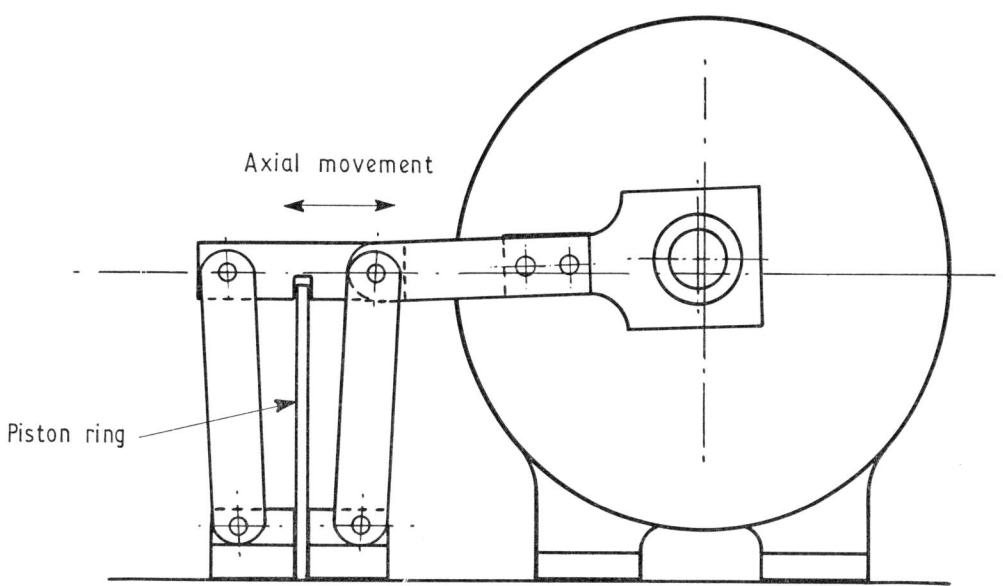

Fig. 15.3. Piston ring fatigue rig—axial load

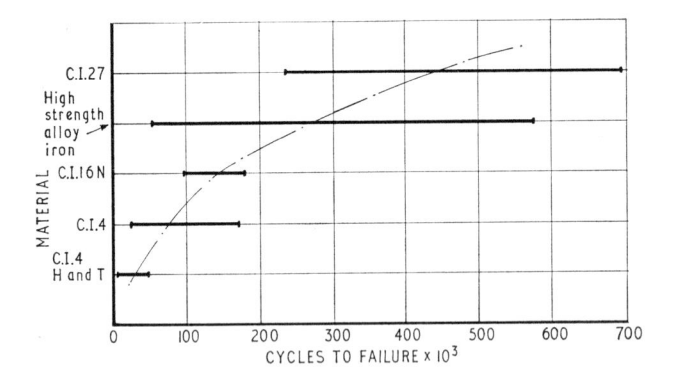

Fig. 15.4. Fatigue grouping of piston ring materials, axial loading

Engine testing can be quite costly, and therefore a number of fatigue rigs have been used to establish the fatigue limit of piston ring materials in two modes of vibration. The first mode is that of opening and closing the ring, thus simulating radial movement of the ring within the cylinder. The second mode is to either vibrate or bend the ring in a plane parallel to its running face.

Fig. 15.2 shows the former rig, illustrating the method of load application. Fig. 15.3 shows the ring-bending rig in both diagrammatic form and as a finished rig. This rig operates on the basis of constant strain, the ring being clamped across its joint faces whilst a grooved loading-bar moves the ring through a given strain angle; the speed of the bar is 1500 cycles/min, and the strain angle is adjusted for a given group of materials.

We found this to be necessary, as the high-strength irons need a relatively large strain angle to produce a meaningful result. Using the same angle for grey irons produces an extremely short cycle-to-failure figure, and the scatter over a small number of cycles, as one would expect from fatigue testing, is unacceptable at the high strain.

The rig is extremely simple, robust, and has produced some useful results covering a wide range of materials. The grouping of these is shown in Fig. 15.4 and follows a pattern that would perhaps have been guessed without any results, but indicates the scatter band associated with different types of material.

Wear properties

The wear properties of piston rings cannot be discussed in isolation as they are associated with the matching component, the cylinder liner. If we assume a normal centrifugally cast cylinder liner having an 'A-type' graphite structure in the bore region with a ferrite control of less than 5 per cent, then the relative wear rate increases as materials move from grey iron to spheroidal graphite. Invariably the grey-iron component will function more readily with minimal lubrication, owing to its inherent self-lubricating properties, than will any other cast iron.

Adjustment in overall wear properties can be improved by surface-finish treatment through the base iron, such as phosphating or steam oxide. These coatings have inherent wetability, which helps to minimize wear and induce lubrication over the rubbing surfaces. Ferrite is again a feature which must be controlled in the manufacture of piston rings, preferably to below 5 per cent, so as to achieve optimum wear rates. Fig. 15.5 shows the order of wear relative to the basic materials used for piston rings. Improved rates of wear can be achieved with ring coatings, as discussed later in this paper.

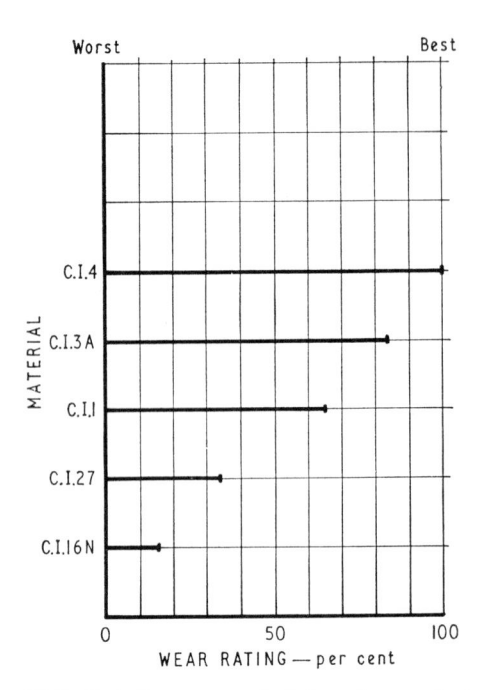

Fig. 15.5. Relative wear of piston ring materials, uncoated

Compatibility properties

Poor compatibility between materials results in excessive wear of either one or both of the metals. Sometimes the wear is so severe that the phenomenon of scuff occurs. Materials in the untreated condition again follow the typical wear rating. Grey iron has excellent compatibility with a number of surfaces, whilst the malleable or spheroidal graphite irons are poor unless coated with some form of plated or sprayed surface. Grey irons appear to operate extremely well, associated with equivalent grey-iron cylinder bores in either the as-cast or hardened and tempered condition. They operate moderately well in chromium-plated bores, depending on the surface-finish condition of the bore.

The carbide-containing irons again appear to operate extremely well against similar bore surfaces, and in the hardened and tempered condition can achieve extremely good wear rates in chromium-plated cylinders.

Similar success can be achieved with hardening and tempering spheroidal graphite irons for use against chromium. The malleable or spheroidal graphite irons invariably need to be plated on their periphery to operate satisfactorily against both as-cast or hardened and

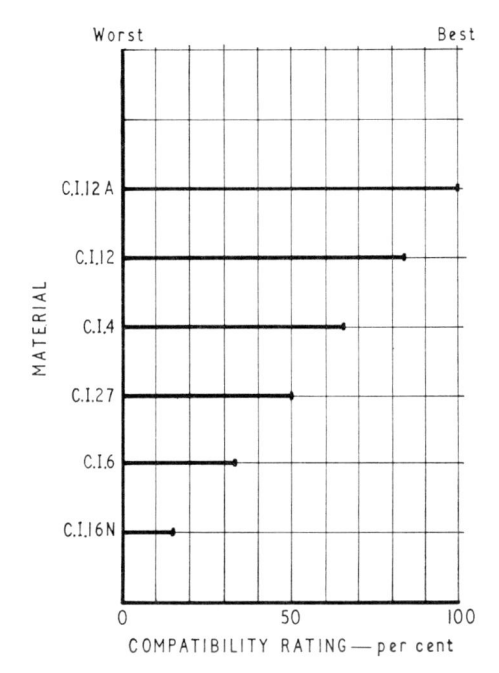

Fig. 15.6. Relative compatibility rating of piston ring materials, uncoated

tempered cast iron. Fig. 15.6 shows a compatibility rating for various materials.

HEAT COLLAPSE

Piston ring materials must be able to withstand soaking at relatively high temperature with the minimum of collapse. It is obvious that if a ring is susceptible to collapse after operating in an environment of 350°C, the gas-sealing effectiveness of the ring will be impaired. This will result in heavy blow-by past the ring periphery, causing overheating of both the ring and the piston, with disastrous results.

Fig. 15.7 shows the collapse of piston rings of various materials after soaking at varying temperatures for 36 h. It will be seen from the curves that a figure of around 10 per cent is that which has been found to be acceptable in service, assuming ring temperatures of 300–400°C.

The method of test in assessing the collapse pattern is to close the ring to its fitted diameter and constrain it in a pot similar to the engine cylinder. The assembly is then soaked at the appropriate temperature for 36 h. The ring is then removed and the free gap measured. This measurement is then compared with the original figure. Fig. 15.8 shows the collapse effect for differing radial dimensions at 400°C.

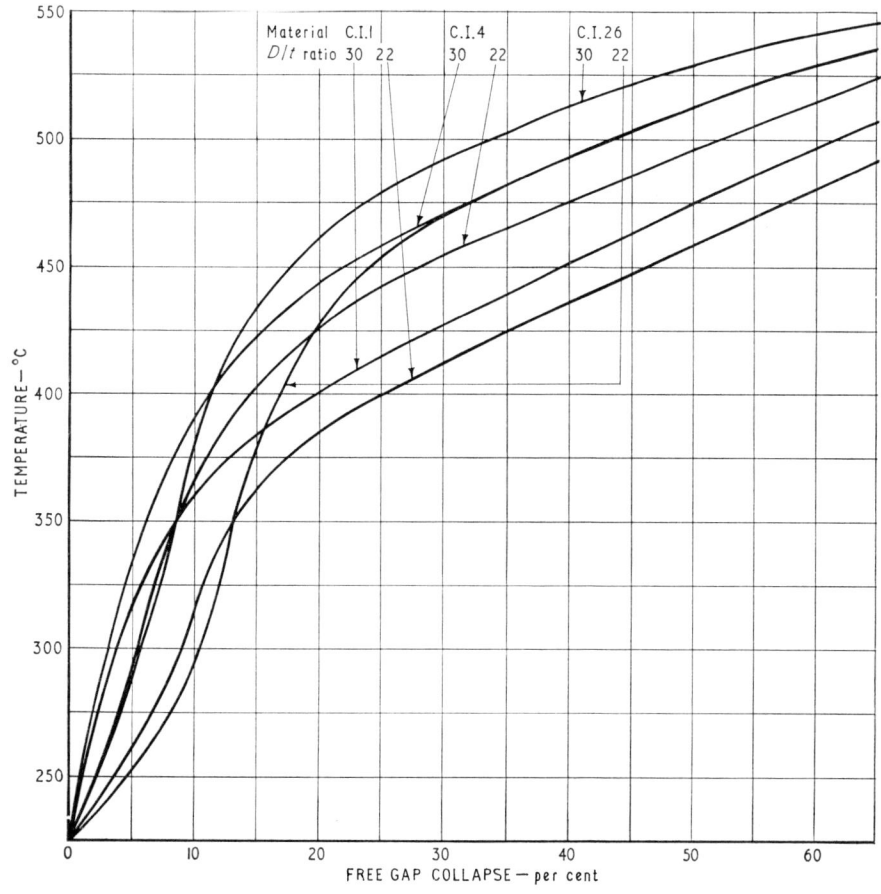

Fig. 15.7. Collapse of various piston ring materials after 36-h soak at temperature

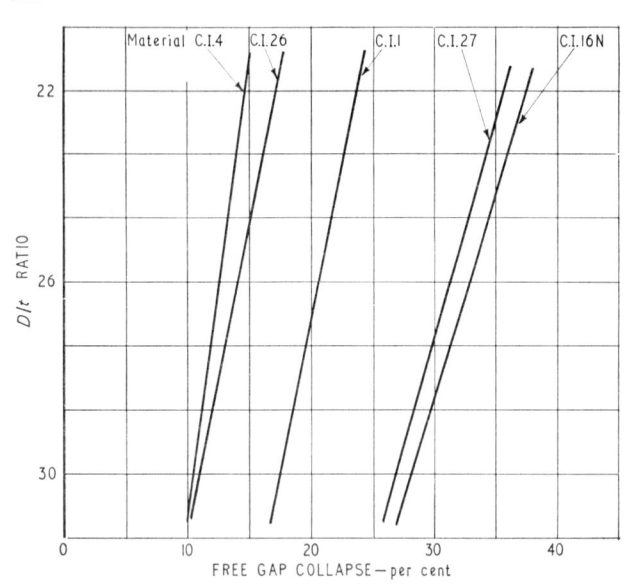

Fig. 15.8. Effect of differing ring radial dimension on heat collapse at 400°C

PISTON RING FREE SHAPE

Ring-fitted pressure pattern

Much work has been done and much still is being done to optimize the free shape of a piston ring for a given duty and thus establish a satisfactory pressure pattern when operating in a diesel engine. The computer is an invaluable tool in aiding the piston ring designer to formulate shapes that will give satisfactory piston ring life.

For many years piston ring shape assessment has been based on ovality, i.e. the difference between measurement, with the ring constrained in a flexible tape, across the gap axis and at right angles to it. The difference is designated either positive, if the ring is larger across the gap axis than the 90-degree axis, or negative, should it be the opposite.

This method of assessment, unless backed by more sophisticated pressure-measuring equipment, can be misleading in terms of the ultimate pressure pattern of the ring when constrained against a cylinder bore. Thus it is important to establish the true free shape of the ring to give the requisite pressure pattern, and this must be

Fig. 15.9. Piston ring pressure-pattern measuring rig—load cell method

closely adhered to in the manufacturing cycle. A number of sophisticated rigs have been made for measuring pressure pattern; one developed by Associated Engineering Developments Ltd is shown in Fig. 15.9. Another, which is used in a number of piston ring manufacturers' plants throughout the world, is that where strain gauges are used to estimate pressure as the ring is rotated in the fixture. Fig. 15.10 illustrates this type of fixture, which is readily manufactured for a large number of ring types at reasonable cost.

The modern four-stroke engine in its turbo-charged form would appear today to need a pressure pattern approaching that of constant pressure around the ring periphery when fitted. If anything, the pressure adjacent to the ring gap could be lower than the remainder of the ring. This appears to be compensated for in operation by temperature gradients across the ring, which modify the ring form so that the joint ends fill out to suit the bore size. Conversely, if the pressure adjacent to the points of the ring was high, this would result in an even higher pressure, with considerable distress in this area. Fig. 15.11 shows a number of rings produced with differing pressure patterns to illustrate the effect it has adjacent to the ring gap.

The ported two-stroke engine has its own problems, especially at the ring gap. Invariably this area is heavily relieved and washed away to enable the ring to traverse the port areas without clipping. Again, this relieving

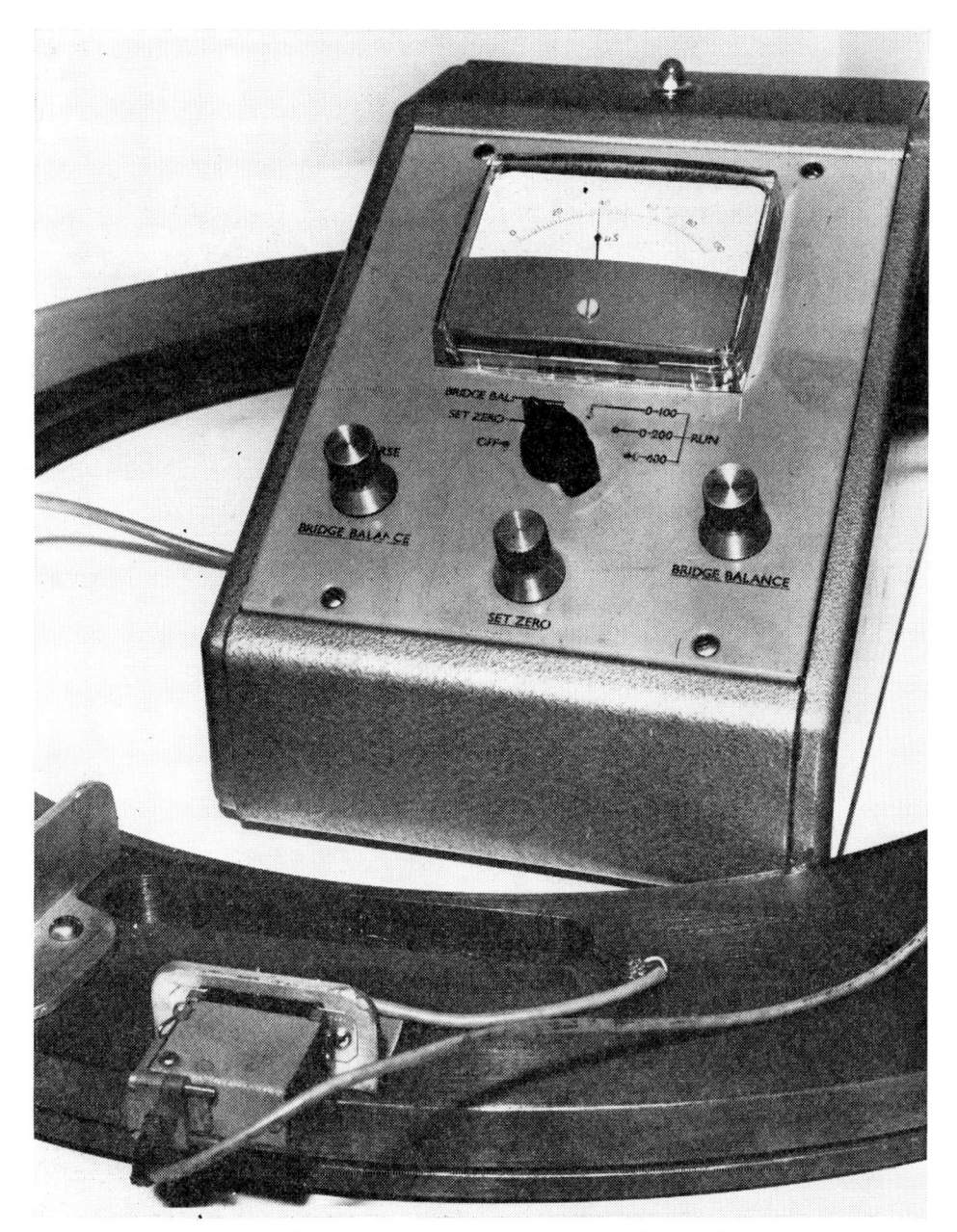

Fig. 15.10. Piston ring pressure-pattern measuring rig—strain gauge method

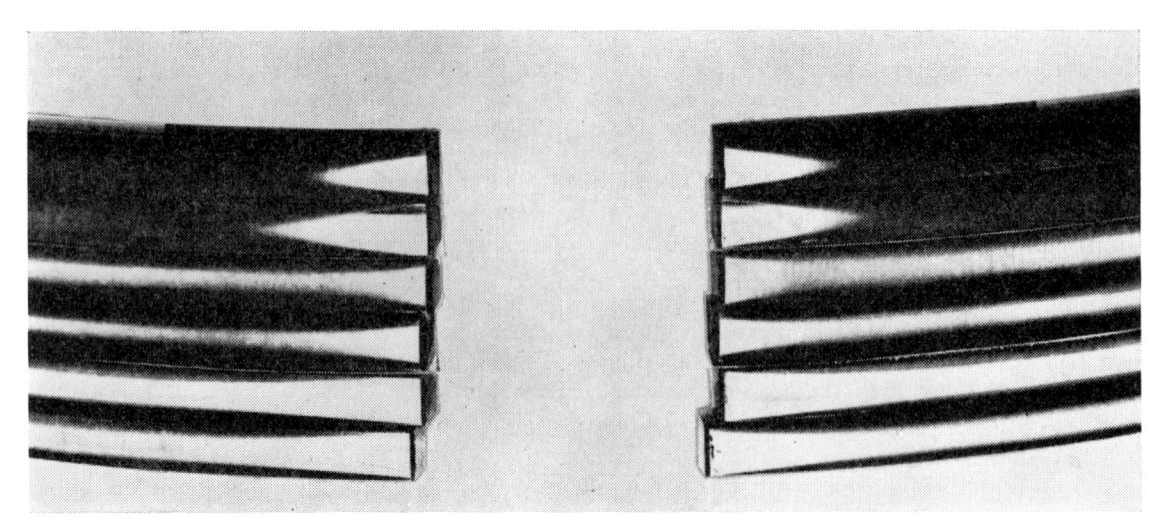

Fig. 15.11. Pressure-pattern variation effect on ring-tip pressure

compensates for ring distortion owing to temperature gradients across its section. These detailed ring shapes can be produced in many ways, either by heat forming (Fig. 15.12 shows the work head of a modern induction heat-forming unit), or by cam turning [Fig. 15.13 shows the modern vertical cam-turning unit for rings up to $47\frac{1}{2}$ in diameter (1200 mm)].

Fig. 15.12. Induction heating coil for induction heat forming

Fig. 15.13. Cam-turning machine for rings up to 1200 mm diameter

Appendix 15.2 shows the formula used to establish piston ring free shapes, the work being based on Professor Swift's determination of piston ring free shape. The computer is a ready tool for determining radial ordinates.

PISTON RING MEAN WALL PRESSURES FOR GIVEN SIZE AND DUTY

Considerable variation in piston ring mean wall pressure exists between rings on a given piston assembly, and between engines. For plain-faced rings the wall pressure of a given ring is independent of its width, provided the radial dimension remains constant. Today, however, only in the very large engines do we see continuity of plain-faced rings. The majority of smaller engines in bore sizes, say, 20 in downwards have adopted over the years more and more sophistication in the piston ring forms, which have given rise to wall pressures considerably higher than hitherto. For engine sizes up to, say, 8 in diameter a range of wall pressure factors is shown in Fig. 15.14 covering differing forms of ring. The basic factor can be converted to wall pressure by multiplying the first three digits from the En modules of the material chosen. The radial depth-to-diameter ratio is assumed constant.

For piston rings up to, say, 20 in diameter, a new set of rules applies, as the radial dimension of the ring alters. A further change occurs for rings above this size, i.e. into the marine propulsion range.

RING SURFACE COATINGS

Innumerable coatings exist for piston rings. These coatings are increasing daily with the help of spray techniques, either by conventional oxyacetylene or by using plasma equipment. If we first consider the use of coatings when operating against as-cast or hardened and tempered cylinder liners, it has yet to be shown that, provided scuff conditions do not exist, chromium plating has a substitute. It is being used more and more in many engines in all ring positions to give good compatibility, good scuff resistance, and excellent life. It has also been found that a copper flash of approximately 0·001 in thick over the chromium can give rise to acceptable break-in conditions where hitherto marginal acceptance has been the case. This technique is not new, having been used by a number of engine builders for many years.

For high-speed automotive and truck engines it is becoming almost universal to embody a ring pack having all ring peripheries chromium plated when used in conjunction with cast-iron surfaces for the cylinder liner.

Many sprayed coatings have been developed in an attempt to minimize scuff conditions in an engine where

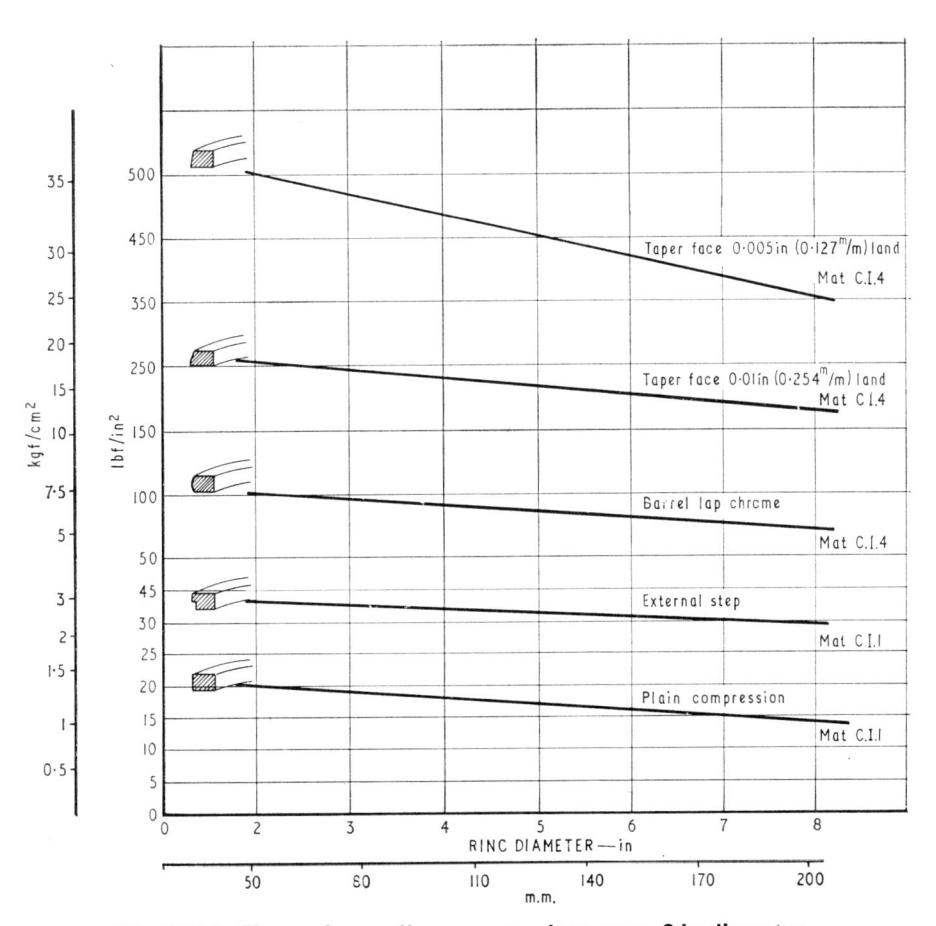

Fig. 15.14. Piston ring wall pressures, rings up to 8 in diameter

it is known that temperatures are such that boundary lubrication is virtually non-existent. These coatings have had limited success; but invariably within the diesel engine, temperatures are such that coatings begin to break down after relatively short periods of operation.

It has been found also that sprayed coatings appear to give rise to high cylinder-liner wear; for instance, molybdenum is known in certain engines to have given wear rates of between two and three times that of chromium as deposited by electrochemical deposition, provided the chromium, of course, did not scuff. Fig. 15.15 shows wear rate for a series of sprayed deposits as evaluated in an unlubricated reciprocating wear rig. Electrochemically deposited chromium is shown as a base-line in the same rig. It is obvious that many engine tests will be necessary to show the correlation between the rig and actual operating conditions; but already indications are that the order of wear is comparable with the rig assessment. It is therefore important when assessing any ring coating to pay close attention to the relevant liner bore wear.

Alternative coatings that have been used over many years are shown in Fig. 15.16. They still perform extremely effectively in quite highly rated engines, but undoubtedly as our knowledge of the sprayed coatings improves and the make-up of the sprayed deposition is optimized, we will be able to perhaps achieve comparable cylinder-bore wear rates with the earlier coatings. The sprayed coatings are normally simpler to apply and easier to control than many of the earlier techniques, hence their attractiveness to the production engineer.

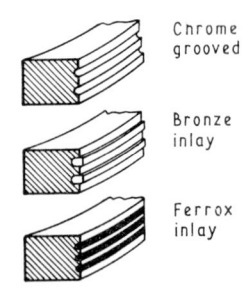

Fig. 15.16. Established top ring surface coatings

RING SURFACE TEMPERATURES

It is virtually impossible to measure the surface temperature of a piston ring whilst it is operating in a diesel engine cylinder. In spite of this, comparative assessments may be made using known instrumentation techniques which are improving continuously, thus enabling the development engineer to assess whether optimum operating conditions exist.

A typical method for assessment of piston ring temperature under laboratory test conditions is shown in Fig. 15.17, the method depending very much on the size of ring involved. It is possible sometimes, especially with chromium-plated rings, to assess the surface temperature of the ring from the fall in hardness of the plated deposit. This can only be reasonably accurate if a ring from the same plating batch is used as a marker, and is calibrated in terms of temperature–hardness. An average curve for chromium hardness reduction at temperature is shown in Fig. 15.18.

LINER SURFACE TEMPERATURE

The temperature at the inner face between piston ring and cylinder liner at the topmost position of the ring travel is

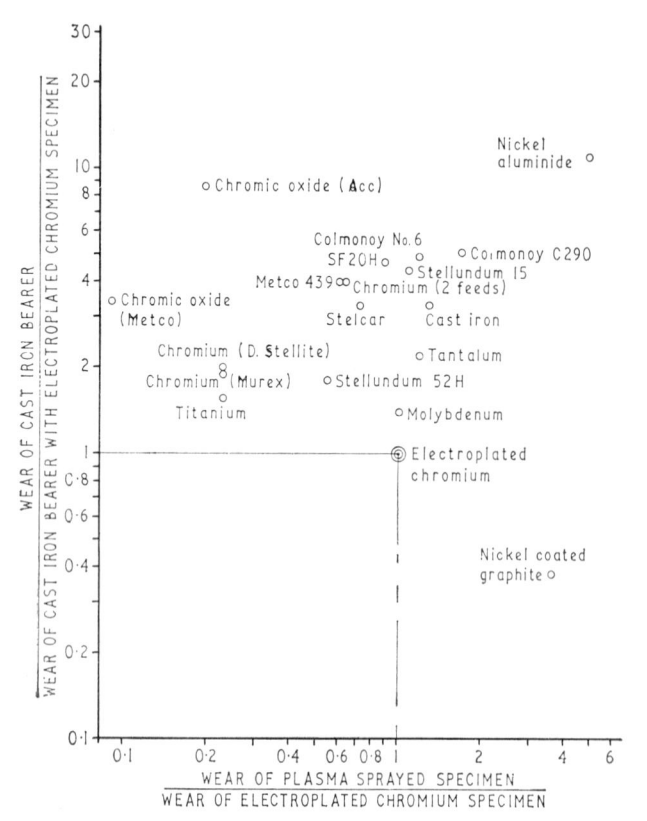

Fig. 15.15. Piston ring coatings, unlubricated reciprocating wear tests

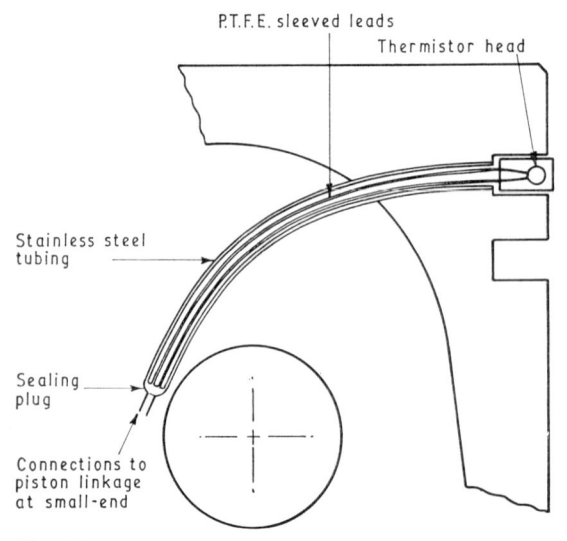

Fig. 15.17. Piston ring temperature measurement

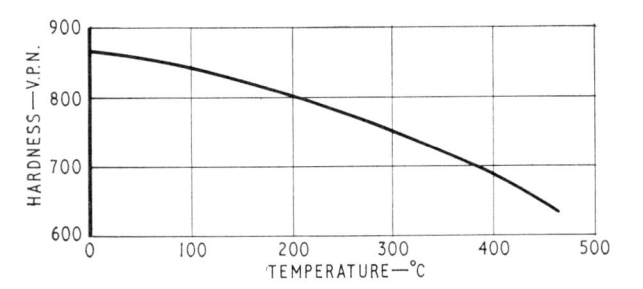

Fig. 15.18. Reduction of hardness at temperature, electroplated chromium

of vital importance. If this interface temperature is too high, then lubrication will break down completely and local welding in the form of scuff will commence. The typical method of assessment of liner temperature in this zone is by the insertion of either thermal couples or thermistors through the cylinder liner wall to within about 0·040 in of the bore.

A typical installation is shown in Fig. 15.19. In this instance, hypodermic tubes were pushed through the cylinder block into the cylinder area, with thermal couples inserted. The maximum temperature around the block should be between 160° and 180°C at this point. The temperature difference between maximum and minimum is preferably kept to within 20–30 degC. This will minimize distortion, and also limit the thermal gradient between liner bore surface and ring mating surface.

PISTON RING GROOVE TEMPERATURES

To ensure adequate freedom within the piston ring groove, the piston ring must operate at a temperature of between 220° and 240°C. When temperatures exceed these levels, then, progressively, problems arise, which result in inadequate performance of the piston ring. Higher temperatures invariably cause carbon build-up in the ring groove zone resulting in packing of the ring. This can eventually cause either scuffing of the ring surface or premature wear. Alternatively, ring stick can occur, which will result in either ring breakage or blow-by. Both these faults will almost certainly result in overheating of the piston assembly. Again, temperature surveys can be made by a number of methods to give comparable temperatures in the top ring-zone area to ensure during the development stage that temperature levels are satisfactory.

Many design features for piston rings have been developed in an effort to overcome ring sticking in an engine

Fig. 15.19. Thermocouple installation for liner temperature measurement

that is essentially over the limiting temperature. These features are essentially palliatives and in some cases have given passable results in service, but in others have never really solved the problem. In many cases such palliatives, i.e. wedge-type rings, either single or double, shaker rings behind the conventional ring to ensure movement of the carbon as it begins to build, are still in use today. However, as ratings rise, invariably more positive steps need to be taken to cool the piston assembly, thus reducing temperature levels down to an acceptable limit, which in many instances enables more conventional ring equipment to be used.

ENGINE DESIGN FEATURES TO ENSURE ADEQUATE COOLING OF THE PISTON RING

To ensure adequate operation of a piston ring within a diesel engine brings one into a discussion which ends with a circular equation. Should the ring be capable of performing satisfactorily, irrespective of the environmental conditions, or should these conditions be such that the ring is bound to operate satisfactorily? The major feature associated with the solution to this equation is that of the lubricant. If the lubricant breaks down, then the chance of satisfactory operation is virtually nil.

We thus tend to be limited with current lubricants irrespective of their additives to the operational temperatures mentioned earlier. Many engines in production, especially in turbo-charged form, live and sometimes die with higher than acceptable limits. Invariably owing to the load factor on the engine, it gives satisfaction in service.

Levels of brake mean effective pressure are, however, constantly lifting, which result in more and more problems around the combustion area. We are thus faced with a need to appraise the cooling of the cylinder head, the cylinder, and the piston more carefully than ever before. Certain of the design features which can be seen in many high-volume-produced engines of today are shown in Fig. 15.20. They indicate without doubt that attention to detail in terms of heat flow would give greater service life than is possibly the case.

It is realized that the goal of any piston ring manufacturer must be to produce a piston ring that will operate without lubricant and at a temperature of 1000°C for a life of 20 000 h. This responsibility is accepted, and little by little we are working towards it!

PISTON RING SURFACE-FINISH CONTROL

Owing to its very operation the piston ring must have at its periphery surface-finish conditions that will enable it to operate satisfactorily under marginal lubrication. It should be capable of operating at a very low coefficient of friction factor, which can be induced either by suitable coating or be inherent in the very material of which the ring is made.

The edge condition of the piston ring must be continous so as to avoid any irregularities that may pierce the thin oil film between the ring and the cylinder wall. Uncoated as-cast rings are invariably turned on their periphery to give oil-holding qualities during break-in. Most plated rings, other than the inlaid type, invariably embody a lapped periphery.

CYLINDER LINER SURFACE-FINISH CONTROL

Much work has been done over the last few years relative to the nature of a cylinder bore to ensure adequate control during manufacture and an acceptable surface for quick break-in under production test conditions. It has been determined that cylinder bores do not operate satisfactorily during break-in with finishes produced by synthetic diamond honing stones.

These stones are extremely good as a manufacturing

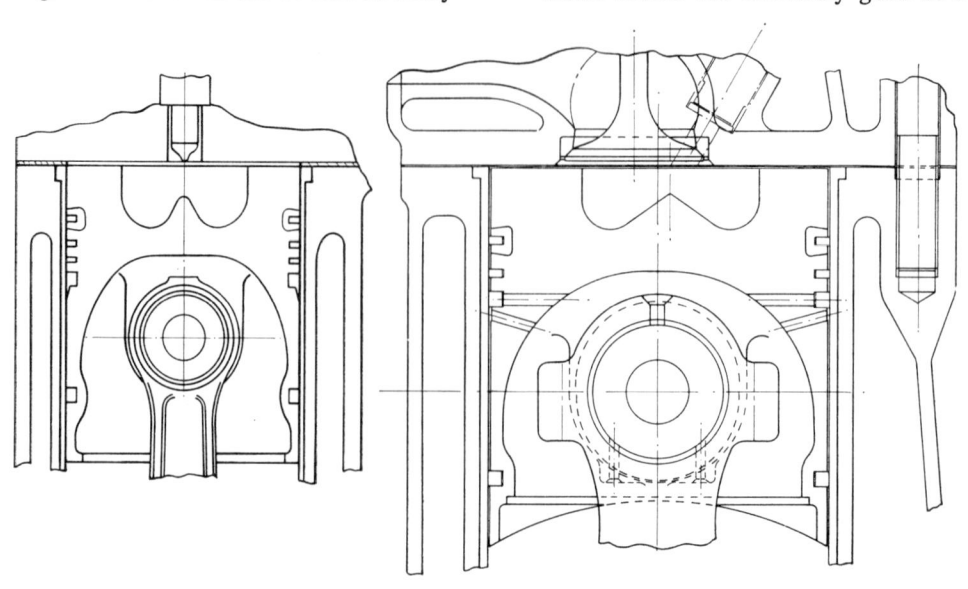

Fig. 15.20. Simple solution to poor heat flow from piston rings

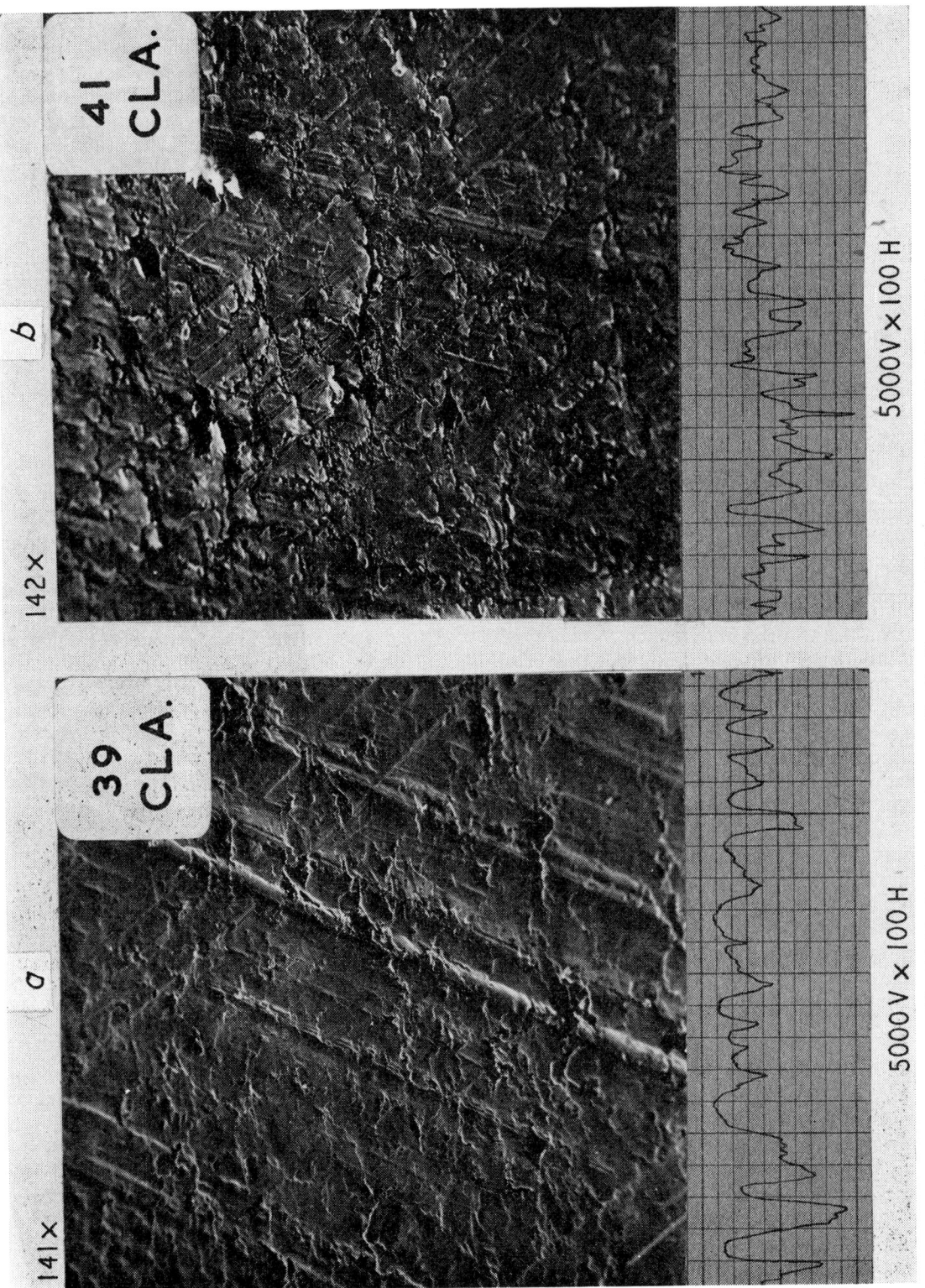

Fig. 15.21. Surface texture synthetic diamond finish and Wellworthy CK finish

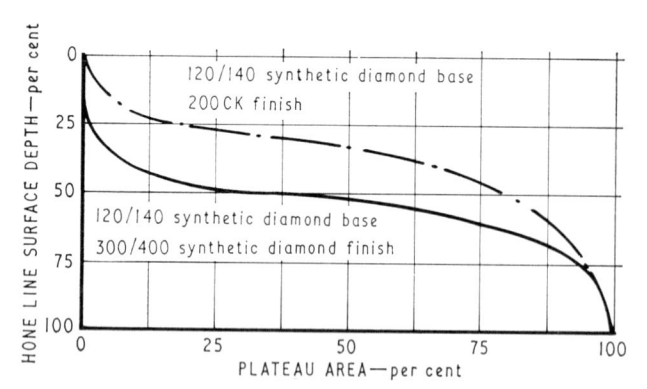

Fig. 15.22. Liner surface assessment for plateau with differing honing finishes

tool, and give extreme consistency of both size and surface finish, as measured in centre-line average μin. The very nature of the surface produced by this technique is such that it gives rise to considerable debris during initial operation of the engine. At the same time, sharp asperities on the crests of the honing lines can break through the thin oil film between cylinder liner bore and ring and give rise to conditions of scuff.

The trend has therefore been away from this stone type as a finishing tool to either silicone carbide or decomposition stone, which will give what is commonly called a plateau finish to the bore before the engine has run. It could be argued that this is a free-running surface which may take a little longer to settle down in terms of oil control, but which will give a satisfactory combination in terms of mating the ring to the cylinder. Fig. 15.21 shows the surfaces generated when using diamond honing stones as against silicone carbide.

An alternative technique used by Wellworthy is to have a soft matrix stone using cork with silicone carbide embedded. This provides a very flexible and insensitive technique giving both a plateau finish and an extremely clean bore finish after manufacture. Some interesting work has been carried out by Campbell of Perkins (see bibliography, Appendix 15.3), interpreting this for diamond-honed bores against the Wellworthy cork honing designated CK. It will be apparent from the graphs shown in Fig. 15.22 that the desired plateau has been achieved.

LUBRICATING-OIL CONSUMPTION

Lubricating-oil consumption in the U.K. has, for many years, been related to fuel-oil consumption. The figure is usually shown as a percentage. Present-day diesel engines return values between 0·25 and 0·5 per cent. Values lower than 0·25 per cent can give rise to excessive ring–liner wear. This will occur without scuff in the normal sense being evident. Experience has shown, however, lubricating-oil consumption values 0·5–1·0 per cent with high incidence of scuff. Therefore scuff can, in certain circumstances, be independent of lubricating-oil consumption.

The basic factors affecting lubricating-oil consumption can be summarized as follows:

(a) Ring wall pressure.
(b) Ring side flatness.
(c) Ring peripheral contact.
(d) Piston ring outer edge form.
(e) Ring groove flatness.
(f) Ventilation in scraper ring location.
(g) Ring–ring groove side clearance.
(h) Cylinder bore circularity.
(i) Cylinder bore finish.

The combination–permutation of these factors prevents specific recommendation in this paper.

Fig. 15.23 is a nomograph giving ready relationship of lubricating-oil consumption in a variety of ways.

BLOW-BY

The efficiency of the piston-ring seals can be assessed by measurement of blow-by. Figures based on a large number of engines show that an acceptable level is in the order of 12 ft^3/litre of engine capacity/h (0·34 m^3/litre/h). This figure applies to normally aspirated and turbo-charged power units.

Care is needed in measurement when dealing with blown units owing to blower-seal leakage. This can give a false figure for the piston assembly. Ideally, the blower should be isolated from the crankcase and vented to atmosphere for test purposes. Basic factors affecting blow-by are as follows:

(a) Ring side flatness.
(b) Ring peripheral contact.
(c) Ring-fitted gap.
(d) Piston ring groove flatness.
(e) Cylinder bore circularity.

Specific recommendations cannot be given here, as each condition must be satisfied for the engine in question.

Fig. 15.24 gives blow-by values for a range of engine capacities.

QUALITY-CONTROL FEATURES FOR THE PISTON RING

Detail control in the manufacturing cycle is imperative for technical performance.

Close control at the material casting stage to ensure homogeneity across the ring section is vital for good fatigue life. The many physical features of the finished ring are checked by known and proved methods. Greater use of automatic inspection techniques will enable closer control of vital features.

The fundamentals are: diametral load, light tightness, width, surface-finish periphery, flatness, gap as fitted, and burr-free edges.

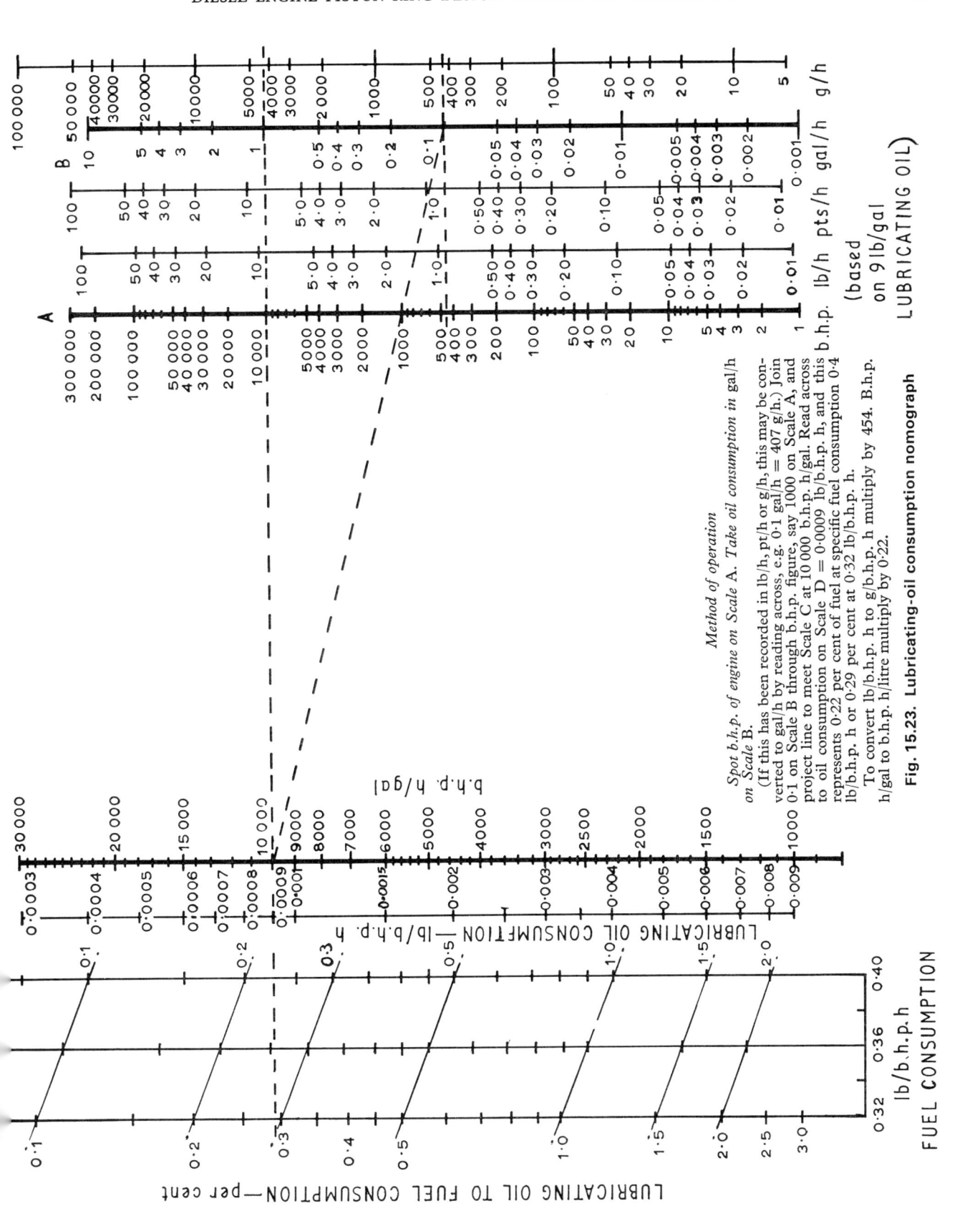

DIESEL ENGINE PISTON RING DESIGN FACTORS AND APPLICATION

Fig. 15.23. Lubricating-oil consumption nomograph

Method of operation

Spot b.h.p. of engine on Scale A. Take oil consumption in gal/h on Scale B.

(If this has been recorded in lb/h, pt/h or g/h, this may be converted to gal/h by reading across, e.g. 0·1 gal/h = 407 g/h.) Join 0·1 on Scale B through b.h.p. figure, say 1000 on Scale A, and project line to meet Scale C at 10 000 b.h.p. h/gal. Read across to oil consumption on Scale D = 0·0009 lb/b.h.p. h, and this represents 0·22 per cent of fuel at specific fuel consumption 0·4 lb/b.h.p. h or 0·29 per cent at 0·32 lb/b.h.p. h.

To convert lb/b.h.p. h to g/b.h.p. h multiply by 454. B.h.p. h/gal to b.h.p. h/litre multiply by 0·22.

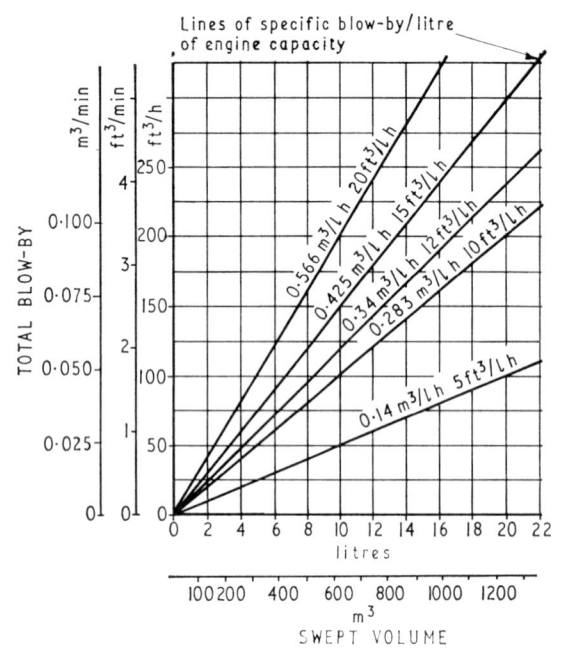

Fig. 15.24. Blow-by—swept volume

CONCLUSIONS

Refinement of the diesel engine piston ring continues in order to keep pace with the ever-increasing needs of the industry. Improved techniques of measurement to assess performance of the ring in its operational environment are constantly emerging from test laboratories.

Greater attention is needed in the cylinder area relative to bore finish and cooling. Piston cooling is all too often left as a last resort when ring problems exist. Charge-air cooling for turbo-charged units can provide relief to ring-belt areas of the piston.

The need to instrument engines more fully in order to understand the basic problems emerges from many ring pack development programmes.

ACKNOWLEDGEMENTS

The author wishes to express his appreciation to the Directors of Wellworthy Ltd and the Directors of Associated Engineering Ltd for permission to present this paper.

APPENDIX 15.1

MATERIALS

Wellworthy material ref.	CI.1	CI.3A, H and T	CI.4
Microstructure, ×200	See Fig. 15.25	See Fig. 15.26	See Fig. 15.27
Material description	Unalloyed grey cast iron Centrifugally cast Matrix, pearlite	Low alloy cast iron Centrifugally cast Matrix {tempered martensite}	Alloy cast iron Centrifugally cast Matrix, fine pearlite
Typical chemistry			
Total carbon	3·2	3·2	3·0
Silicon	2·1	2·2	2·1
Manganese	0·8	0·8	0·8
Sulphur	0·1 max.	0·1 max.	0·1 max.
Phosphorus	0·5	0·5	0·35
Nickel	—	—	—
Chromium	0·4	0·4	0·95
Molybdenum	—	0·6	0·9
Mechanical properties			
Ring tensile strength			
Minimum . . tonf/in^2	17·5	21·0	26
kgf/cm^2	2750	3300	4100
Nominal modulus of elasticity			
Minimum . . En lbf/in^2	15·5 × 10^6	16·5 × 10^6	20 × 10^6
kgf/cm^2	1·09 × 10^6	1·16 × 10^6	1·41 × 10^6
Hardness BHN	230/295	230/305	269/302
Mean coefficient of linear expansion, in/in °C	12·2 × 10^{-6}	12·2 × 10^{-6}	12·4 × 10^{-6}

Wellworthy Material ref.	CI.6	CI.7	CI.12A
Microstructure, ×200	See Fig. 15.28	See Fig. 15.29	See Fig. 15.30
Material description	High nickel austenitic cast iron Centrifugally cast Matrix, austenitic	Unalloyed grey cast iron Static cast pots Matrix, pearlite	Unalloyed grey cast iron Centrifugally cast Matrix, pearlite
Typical chemistry			
Total carbon	2·6	3·2	3·5
Silicon	2·0	1·5	2·25
Manganese	1·0	0·8	0·9
Sulphur	0·1 max.	0·1 max.	0·1 max.
Phosphorus	0·45	0·4	0·45
Nickel	13·5	—	—
Chromium	2·0	0·3	0·4
Molybdenum	—	—	—
Copper	6·2		
Mechanical properties			
Ring tensile strength			
Minimum . . tonf/in^2	16·0	16·0	18·0
kgf/cm^2	2520	2520	2830
Nominal modulus of elasticity			
Minimum . . En lbf/in^2	14 × 10^6	14 × 10^6	12 × 10^6
kgf/cm^2	0·985 × 10^6	0·985 × 10^6	0·845 × 10^6
Hardness BHN	160/240	200/245	228/311
Mean coefficient of linear expansion, in/in °C	18·0 × 10^{-6}	12·2 × 10^{-6}	12·2 × 10^{-6}

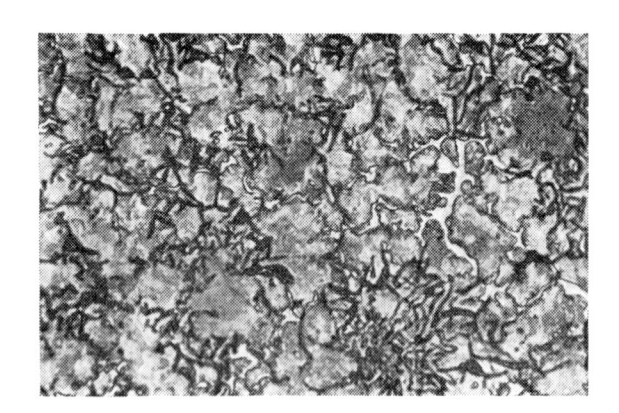

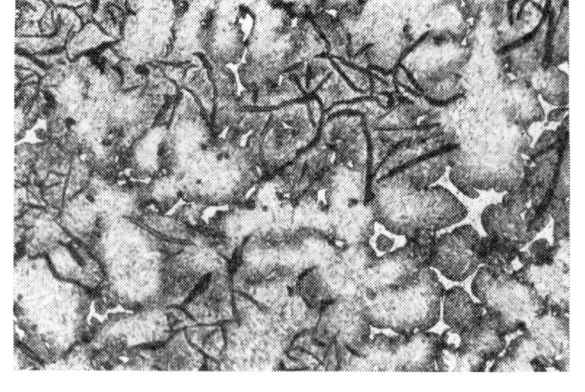

Fig. 15.25

Fig. 15.26

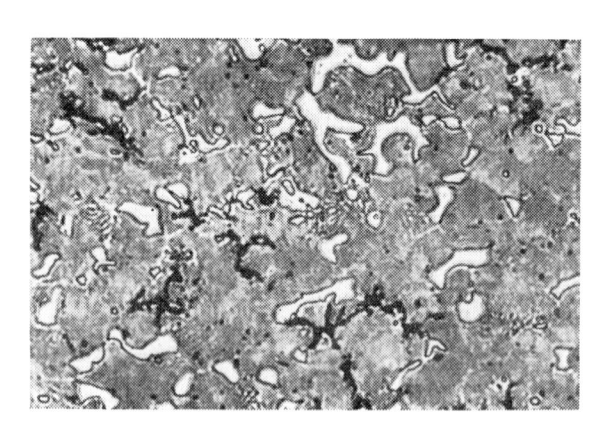

Fig. 15.27

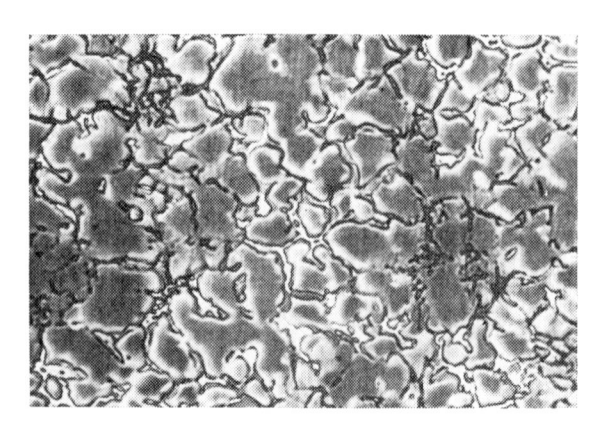

Fig. 15.28

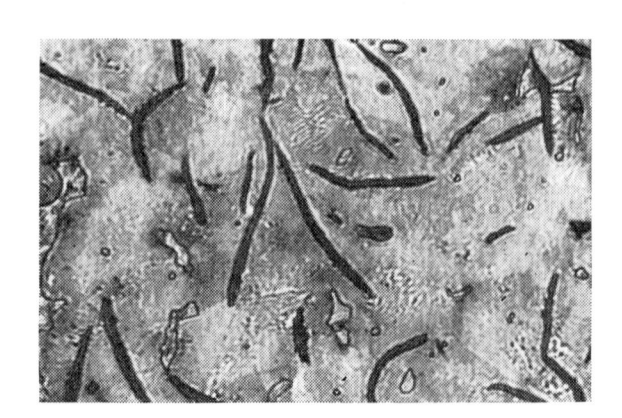

Fig. 15.29

Fig. 15.30

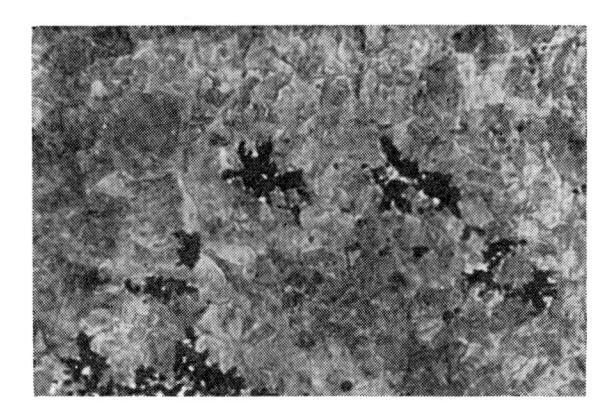

Fig. 15.31

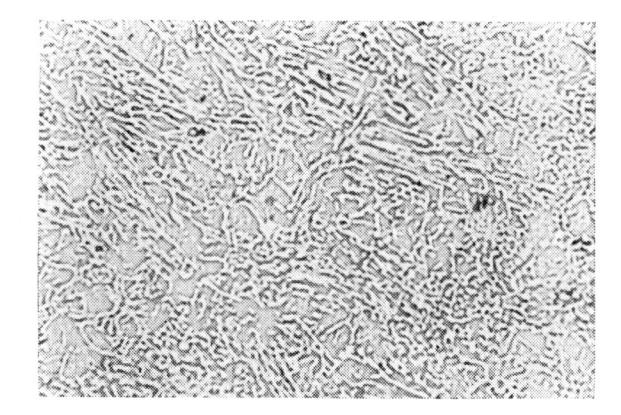

Fig. 15.32

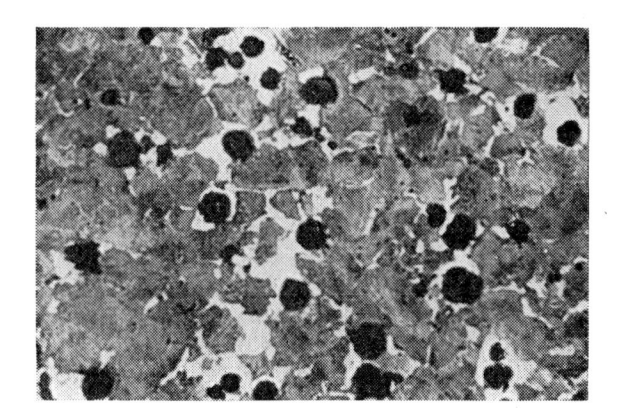

Fig. 15.33

Wellworthy material ref.	CI.16N	CI.26	CI.27
Microstructure, ×200	See Fig. 15.31	See Fig. 15.32	See Fig. 15.33
Material description	Unalloyed malleable cast iron Matrix, pearlite	High chromium-alloy cast iron Matrix, chrome-rich ferrite	Unalloyed spheroidal graphite cast iron Matrix, pearlite
Typical chemistry Total carbon Silicon Manganese Sulphur Phosphorus Nickel Chromium Molybdenum Magnesium	3·1 1·1 0·7 0·1 max. 0·1 max. — 0·15 — —	1·6 1·9 0·8 0·1 max. 0·1 — 32·0 — —	3·3 2·2 0·5 0·02 max. 0·08 max. — — — 0·05

Mechanical properties
Ring tensile strength
  Minimum . tonf/in^2
           kgf/cm^2
Nominal modulus of elasticity
  Minimum . En lbf/in^2
           kgf/cm^2
Hardness BHN
Mean coefficient of linear expansion, in/in °C
	35
	5500
	
	20 × 10^6
	1·41 × 10^6
	230/300
	14·4 × 10^{-6}
		35
		5500
		
		28 × 10^6
		1·98 × 10^6
		280/341
		10·5 × 10^{-6}
			37
			5820
			
			23 × 10^6
			1·62 × 10^6
			240/297
			13·3 × 10^{-6}

APPENDIX 15.2

PISTON RING FREE SHAPE FORMULAE

This appendix, which relates to ring free shape to give uniform pressure distribution, is from the paper by Professor H. W. Swift listed in the Bibliography.

The increase in radial ordinate allowing for change in angular position of any point on ring periphery when the ring is compressed is given by:

$$R_c + U + \delta u$$

when

R_c = bore radius

$$U = \frac{PR^4}{EI}(1 - \cos\alpha + \tfrac{1}{2}\alpha \sin\alpha)$$

$$u = \frac{R}{2}\left(\frac{PR^3}{EI}\right)^2 (\alpha - \tfrac{1}{2}\alpha\cos\alpha - \tfrac{1}{2}\sin\alpha)(3\sin\alpha + \alpha\cos\alpha)$$

where U is the increase in radial ordinate in excess of the cylinder bore, P the mean wall pressure × ring width (pressure at the neutral axis), R the mean radius of the ring (at the neutral axis), I the moment of inertia about the neutral axis, E is Young's modulus, and α the angle measured from the back of the ring (radius).

This is shown in Fig. 15.34.

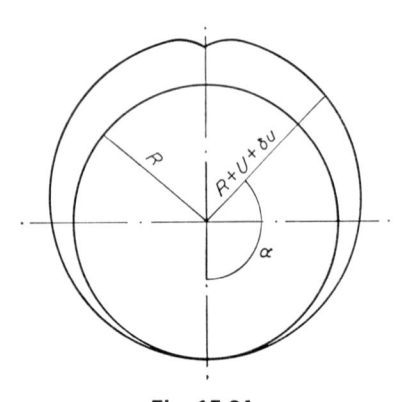

Fig. 15.34

Figs 15.35 to 15.37 show typical modern ring packs.

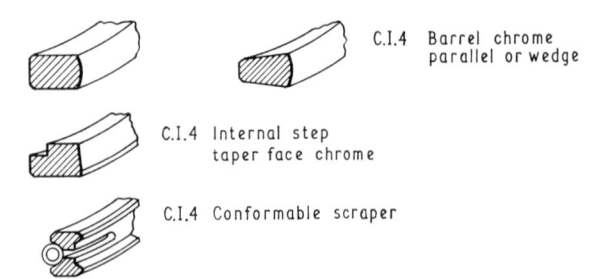

Fig. 15.35. Four-stroke NA or TC automotive engine, all rings above pin

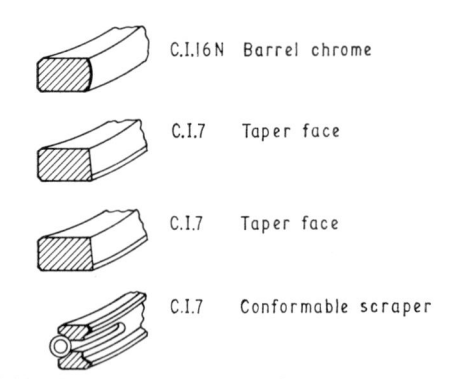

Fig. 15.36. Four-stroke NA or TC industrial and marine engines, all rings above pin

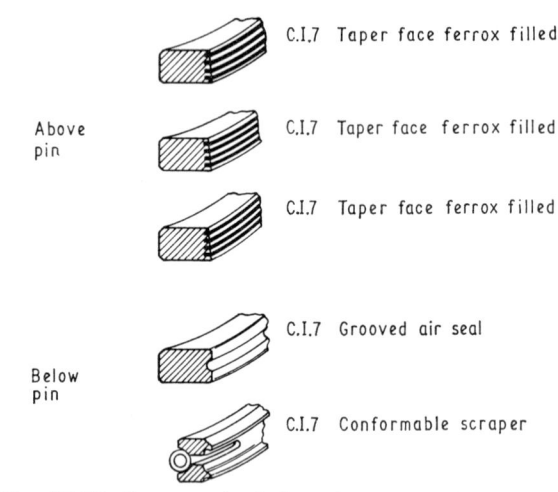

Fig. 15.37. Two-stroke industrial and marine engines

APPENDIX 15.3

BIBLIOGRAPHY

AUE, G. K. and STEIGER, A. 'The influence of the thermal loading criterion on the design of turbo-charged two-stroke diesel engines', *Symp. on Thermal Loading of Diesel Engines, Proc. Instn mech. Engrs* 1964–65 **179** (Pt 3C), 68.

CAMPBELL, J. C. 'Cylinder bore surface roughness in internal-combustion engines; its appreciation and control', *Bull. Mot. Ind. Res. Ass.* 1969 (September–October).

HESLING, D. M. 'A study of typical bore finishes and their effect upon engine performance', *Lubric. Engng* 1963 (October) **19**, 414.

HYDE, G. F. et al. 'Piston ring coatings for high-performance diesel engines', *S.A.E. Paper 670935*, 1967 (November).

LAW, D. A. 'Further developments in cylinder bore finishes', *S.A.E. Paper 690751, National Power Plant Meeting*, Cleveland, Ohio, 1969 (27th–29th October).

PRASSE, H. F. et al. 'Heavy-duty piston rings', *S.A.E. Paper 680238*, 1968 (February).

PRASSE, H. F. 'New developments in piston rings for high b.m.e.p. engines', *S.A.E. Paper 690753*, 1969 (October).

SWIFT, H. W. 'Elastic deformation of piston rings', *Engineering* 1947 (7th March), 161.

WESTBROOK, M. H. and MUNRO, R. 'Electronic instrumentation techniques in the development of pistons and rings', *Symp. on Accuracy of Electronic Measurements in Internal Combustion Engine Development, Proc. Instn mech. Engrs* 1965–66 **180** (Pt 3G), 54.

Paper 16

CRITICAL FACTORS IN THE APPLICATION OF DIESEL ENGINES TO FIGHTING VEHICLES

D. H. Millar*

When diesel engines are installed in fighting vehicles they are subjected to various conditions which may critically affect their performance, reliability, or durability. These conditions can be classified as installational or environmental, and as they are interdependent they cannot be considered in isolation. This is best illustrated by engine rating, since not only does it dictate the performance level but also has a major influence on reliability and life and is itself determined by a number of installational and environmental factors. It is essential to view the problem of engine application as a whole and to apply the solutions not merely to the engine but to all its subsystems constituting the power pack, which must be regarded as an integrated unit.

INTRODUCTION

IN THE ARMY THERE IS A SAYING, 'There are no bad units, only bad officers.' After considerable experience in the field of fighting vehicle design and development there may be a temptation to adapt this to the form, 'There are no bad engines, only bad installations.' While experience proves this to be a somewhat over-optimistic view of 'the engine', the situation could probably be more accurately expressed by paraphrasing another common expression to 'An engine is only as good as its installation', since no matter how well designed, developed, or proved the engine may be, it can be rendered quite unacceptable by faulty installation. Although applicable to many types of installation, this is probably exemplified more clearly in the fighting vehicle field than in many others. In this context 'installation' is applied in its widest sense, covering engine size relative to available space, mounting, rating, cooling, air cleaning, and exhaust. However, in considering these it is essential to relate them to various environmental factors, the principal being high and low extremes of ambient temperature, high dust concentration, duty cycle, and altitude variation.

While it is widely recognized that all design is a compromise, the acceptable compromise is more difficult to achieve in the fighting vehicle field because of the many conflicting requirements. It is even difficult to allot priorities to these requirements in some instances. However, the principal requirements are usually considered to be high performance, small installed volume, low weight, reliability under adverse conditions of environment, maintenance, and fuels and lubricants.

The only factor that can be offset against these is 'life', the requirements for which, by comparison with many commercial applications, is extremely short in terms of mileage. For heavy vehicles this can be as low as 2 per cent, but in terms of time as much as 200 per cent.

The principal factors which critically affect the performance, reliability, and durability of engines in the military vehicle role, and to only a lesser degree in other military roles, are:

(1) *Rating*.
(2) *Installational features*. Cooling system, fuel and oil system, air cleaner, exhaust system.
(3) *Environment*. Ambient temperatures, dust, altitude, conditions during maintenance and repair.
(4) *Fuels and lubricants*. High sulphur diesel fuel; low additive level lubricant.

Although these factors have been separately classified, in practice they are so interrelated that they cannot be considered in isolation.

RATING

Rating is defined as the maximum power and speed of an engine at which a satisfactory standard of reliability and endurance can reasonably be expected under the stated conditions of duty, installation, and environment, and when operating on specified fuels and lubricants.

The MS. of this paper was received at the Institution on 2nd March 1970 and accepted for publication on 7th May 1970. 42
* Assistant Director, Fighting Vehicles R & D Establishment, Chobham Lane, Chertsey, Surrey.

Apart from design and quality control, which are the responsibility of the engine manufacturer, rating is influenced by four principal factors which, although they may be subject to recommendations made by the engine manufacturer, are seldom within his control. These are (a) duty, (b) installation, (c) environment, and (d) fuels and lubricants.

The desire to rate the engine as highly as possible, within the limit imposed by the acceptable levels of reliability and life, is obvious; that is, to provide the most favourable power/volume and power/weight ratios. In the fighting vehicle role the former is the more significant since the volume of the engine and its auxiliaries must be enclosed in armour of high specific weight, while the weight of the complete power pack represents only about 4 per cent of that of a heavy tracked vehicle. There is, therefore, limited scope in effecting an appreciable vehicle weight saving by any reduction in engine weight. Although the duty of the engine, in relation to its rating, has probably a greater influence on reliability and durability than any other factor, the load cycle is most difficult to determine. Unlike some other military applications—e.g. in the general purpose field where, for instance, in a pumping set the load, speed, and time relationship is predictable with a considerable degree of accuracy—the corresponding relationship for a fighting vehicle is largely indeterminate. With a heavy tracked vehicle of relatively low power/weight ratio, where the driving technique is that of 'foot down', the engine operates on the governor for a high percentage of the time and the load is dictated by conditions of terrain and controlled by the engine governor.

On the other hand, the engine of a wheeled vehicle of high power/weight ratio, such as that used for reconnaissance, is subject to constant variation of speed since the vehicle suspension system and crew comfort become the limiting factors when operating under cross-country conditions. The engine of such a vehicle is exposed to frequent variations of load and speed with infrequent maximum load demands, and these of brief duration (Fig. 16.1).

Currently there is a wide variation of opinion on which type of test is most appropriate for the engine. These vary from running the engine on the bed under constant maximum power at its automotive rating to the other extreme of running on successive cycles of 10 min, during which the maximum power demand is of three peaks only, each of 5 s duration. During the remainder of the cycle the engine is subjected to frequent changes of load and speed. Not only does this more nearly simulate vehicle conditions in general but it also subjects the thermally critical components of pistons and valves to thermal cycling so necessary in testing a highly rated automotive engine.

To resolve these doubts and to achieve a form of test which could be applied with the greatest degree of confidence would require the establishment of a test based on an agreed standard duty cycle synthesized from the analysis of a sufficiently wide range of conditions of operation, terrain, and environment. Once established such a test would be operated by automatic test bed control but would require to fully simulate such features as vehicle inertia and transient conditions, both mechanical and thermal.

Investigations to establish such a test are currently being undertaken. The most significant effect of different duty cycles is heat. Where the load remains substantially constant over lengthy periods, thermal stability will be achieved throughout the engine and its cooling and lubrication systems, and hence the level of engine rating must be such that the temperatures of all critical components remain within acceptable limits. Where the load is subject to frequent variation and the peak is of momentary duration only, the thermal inertia of individual

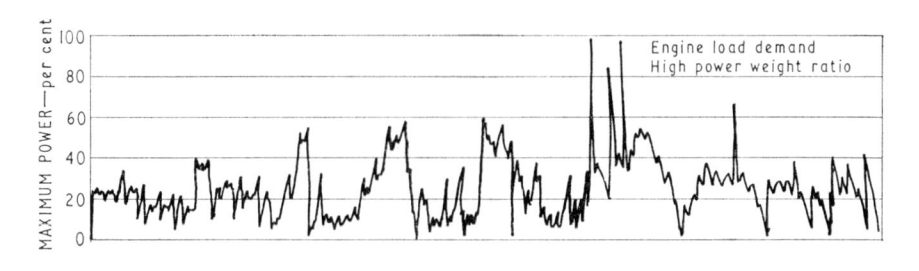

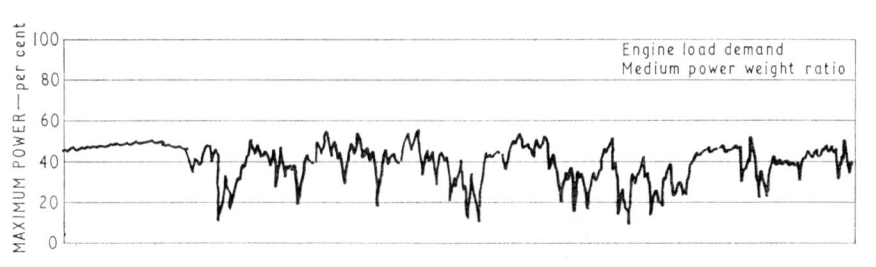

Fig. 16.1. Duty graph

components and of the relevant systems will be sufficient to prevent the occurrence of excessive temperatures. In such cases the rating of the engine may be raised. An example of the effect of this is seen when considering the military tropical requirement, where equipment must be able to operate satisfactorily in an ambient temperature of 125°F. The engine of a generator or pumping set must be so rated that it can sustain its rated output in the full 125°F temperature, whereas that of a high power/weight ratio vehicle to satisfy the same field condition may only require to be tested under sustained conditions at its automotive rated power in an appreciably lower ambient temperature.

As the effects are thermal, they particularly influence piston crowns, piston rings, and injector nozzles, the first two of these being also influenced by the quality of the oil in use.

A useful relationship has been established between ratings of the same basic engine for use in various army applications. The maximum power curve is determined by running the engine on the bed up to the maximum speed specified by the manufacturer and the rack stop adjusted to give 'just visible' exhaust or maximum permitted exhaust temperature. This power is considered to be 10 per cent higher than the 10 per cent overload on the 12-h continuous rating, and hence the latter is 82 per cent, or nine-elevenths, of the 'just visible' power. For continuous operation the appropriate speed is selected, and by applying it to the curve determined by the above procedure the 12-h rating is determined.

As the tropical requirement is to operate in an ambient temperature of 125°F, the tropical rating is determined by correcting the 12-h rating according to the accepted formula.

While this constitutes a useful guide it must be applied with due regard to which aspect of the engine design is critical to ambient temperature, e.g. how ring or injector temperature reacts to variations in ambient temperature. A type test based on the appropriate load–speed cycle is then undertaken on the basis of these values and, if necessary, an adjustment is made in accordance with the engine condition at the conclusion of the test.

The automotive rating is agreed as the result of a separate automotive type test which more nearly reflects the varying speed and load conditions experienced in this class of service, although this may be based initially on the rating at the 'just visible' or maximum exhaust temperature condition.

In addition to their obvious direct effects on engine reliability some installational features also have secondary effects which can influence the acceptable rating for particular installations and duties. An example of this latter condition is the effect which the pressure drop across the engine, largely determined by restriction of air cleaners and exhaust silencers, can have on the trapped air charge and hence on the temperature of such items as piston crowns, piston rings, and injector nozzles for a given fuelling rate.

INSTALLATIONAL FEATURES

Cooling system

In a fighting vehicle the cooling system presents the most difficult design and development problem, while its performance can critically affect the engine reliability (see Fig. 16.2). Despite the apparent advantages of air cooling, this system has not been used for fighting vehicles in the British army since the days of the Vickers Medium Tank in the mid-1920s, for the following reasons:

(1) The lack of established design techniques in British industry applicable to high specific output engines for vehicle service.

(2) More severe development problems resulting in a more lengthy and costly development programme. These arise from the more critical nature of temperature distribution resulting from the use of the less dense cooling medium and affect particularly the cylinder barrel, piston assembly, and the cylinder head where the close proximity of valves and injector presents a particularly difficult problem.

(3) Unreliability in service resulting from fouling of the cooling fins caused by dust adhering to the inevitable oil film on these components and from the much greater difficulty experienced by vehicle crews in cleaning them, whereas the matrices of radiators are designed to be accessible for this purpose.

However, the design of a liquid cooling system (Fig. 16.3) presents its own difficulties, two of the major problems being:

(1) The difficulty of determining the degree to which thermal inertia of the system can be relied upon to cope with transient heat loads and, hence, the steady-state condition that can be selected as the basis of design.

(2) The difficulty of selecting the appropriate value of air restriction through the system. In a fighting vehicle this is high since all the cooling air has to pass

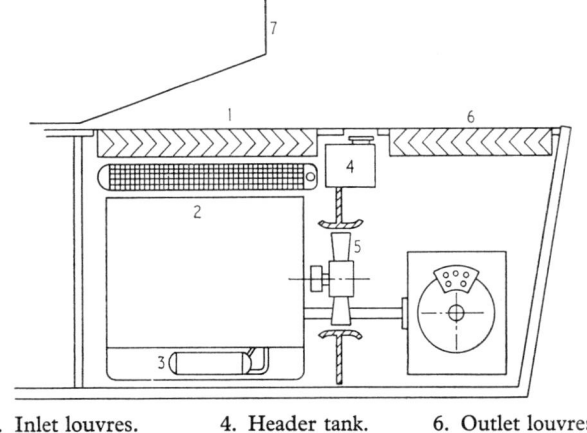

1. Inlet louvres.   4. Header tank.   6. Outlet louvres.
2. Radiator.   5. Fan.   7. Turret bustle.
3. Heat exchanger.

Fig. 16.2. Layout of engine and transmission compartments

1. Inlet louvres.
2. Radiator.
3. Heat exchanger.
4. Header tank.
5. Fan.
6. Bleed line.
7. Make-up line.
8. Overflow tank.
9. Outlet louvres.

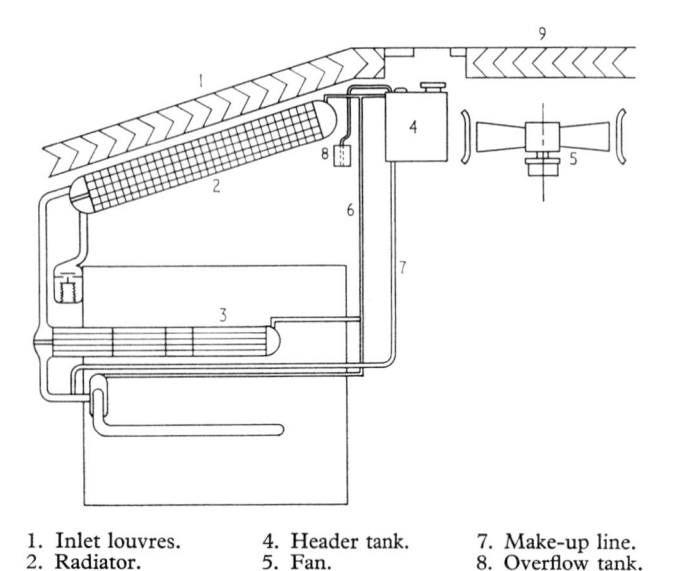

Fig. 16.3. Layout of cooling system

in series through very restrictive air inlet louvres, a dense radiator matrix, and a very confined engine compartment, finally to be discharged through outlet louvres.

These high values of restriction, or pressure drop, are due to the severe space limitation dictated by the requirements to maintain a low vehicle silhouette and to minimize vehicle weight. The difficulty of selecting the appropriate steady-state ambient temperature level for the design peculiar to the vehicle application has already been explained, together with its dependence on the load cycle. This is of even greater significance when considering the design of the cooling system.

Frequent reference is made in this paper to heat exchangers; in this context the term is applied to the coolers in the various oil circuits. Frequently there are four such circuits linked into the engine cooling system which deal with engine, gearbox, steering unit, and hydraulic oils. Because of the different characteristics of these oils and of the temperature sensitivity of the systems which they lubricate, the limiting temperatures differ from system to system. Considering that the boiling point of water when pressurized to 10 lb/in^2 is 239°F there is but limited temperature difference and, hence, the difficulties in designing a complete cooling system with one limiting ambient temperature will be appreciated. Further complicating factors are the different heat rejections and flow rates in the various oil systems.

One important aspect of some vehicle power packs is that if the design is to provide the compromise between being operationally acceptable and economically viable it is unlikely to meet the most adverse combination of requirements. However, the more specialized fighting vehicle, in which performance is more important than cost, must be designed to satisfy the full requirement.

For a given design of engine the rating governs its heat rejection, and the performance of the cooling system will therefore react on the rating which the engine can carry reliably in that installation. The direct effect which the cooling system has on engine reliability is thus a function of the reliability of the cooling system.

Owing to space limitations on the installation, the size of radiator, header tank, and heat exchangers must be minimized. To effect this, the maximum acceptable temperature difference must be employed, which results in the use of pressures considerably higher than in normal commercial practice. However, such increases in pressure necessitate additional design considerations to avoid the tendencies to increased leakage and distortion. As these inevitably result in some increase in weight, and certainly in cost, the optimum pressure must be selected with due prominence being given to these aspects.

One further factor which must be assessed at an early stage is the influence on metal temperature of cylinder heads and barrels. Since any increase in coolant operating temperature results in an increase in the gas side metal temperature, the latter may reach a critical value at the heat flux resulting from operating the engine at its chosen rating. This factor may thus determine the selected boiling point and, hence, degree of pressurization to be adopted.

Another important aspect of the operation of the cooling system is the overall pressure drop on the air side, since this will dictate the power absorption of the fan. Some impression of the problem can be gained by considering the pressure drop distribution across the system. Typically this is approximately: inlet louvres 33 per cent, radiator matrix 30 per cent, engine compartment 1 per cent, and outlet louvres 36 per cent (Fig. 16.4). The overall pressure drop of the system can be as high as 9–10 in w.g., depending on the configuration of the individual vehicle.

Since the heat dissipation, for a given form of matrix, is a function of the pressure drop, the distribution should approximate as closely as possible to 100 per cent attributable to the radiator, which is the case most nearly approached in the orthodox commercial vehicle.

The cause of the high pressure drop is clearly seen as being the armour louvres necessary to protect the engine compartment from shell fragment penetration. Consequently, it is incumbent on the designer to choose the most efficient fan for the installation.

With the increasing tendency for vehicle performance level to rise, and for size to diminish, the problem of high pressure drop becomes progressively more severe. One effect of this is that owing to the characteristic of the axial flow fan—the size of which is limited by the installation—the operating zone tends to be close to the stall point and, hence, the cooling performance of the system may well be critical.

This is particularly undesirable because of the difficulty of allocating a pressure drop value with sufficient accuracy in the design stage, and also because of the variation of pressure drop dependent on the degree of fouling of

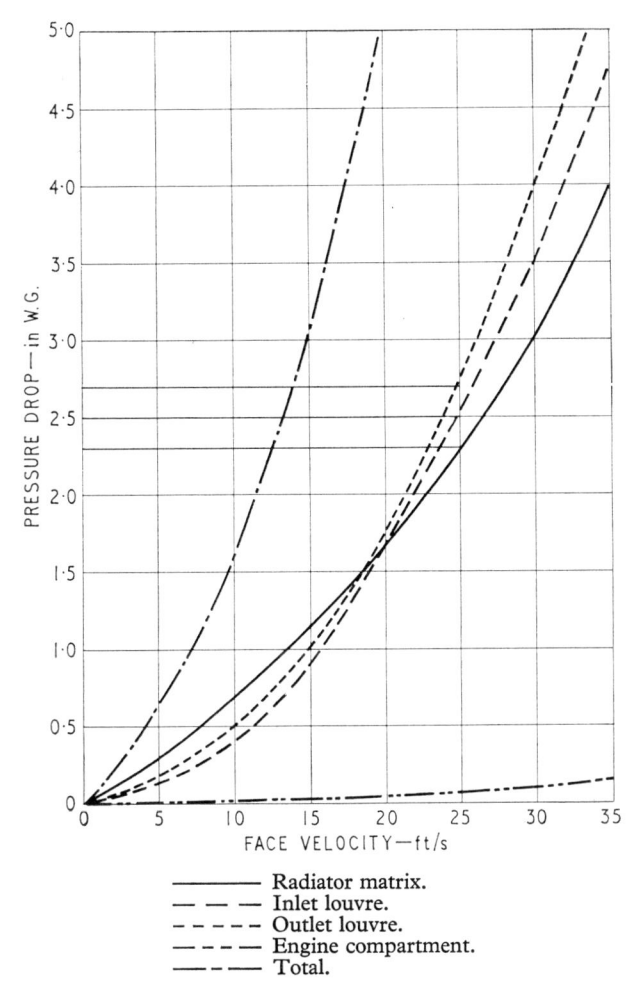

Fig. 16.4. Distribution of pressure drop

— Radiator matrix.
— — — Inlet louvre.
- - - - - Outlet louvre.
— — — — Engine compartment.
— · — · — Total.

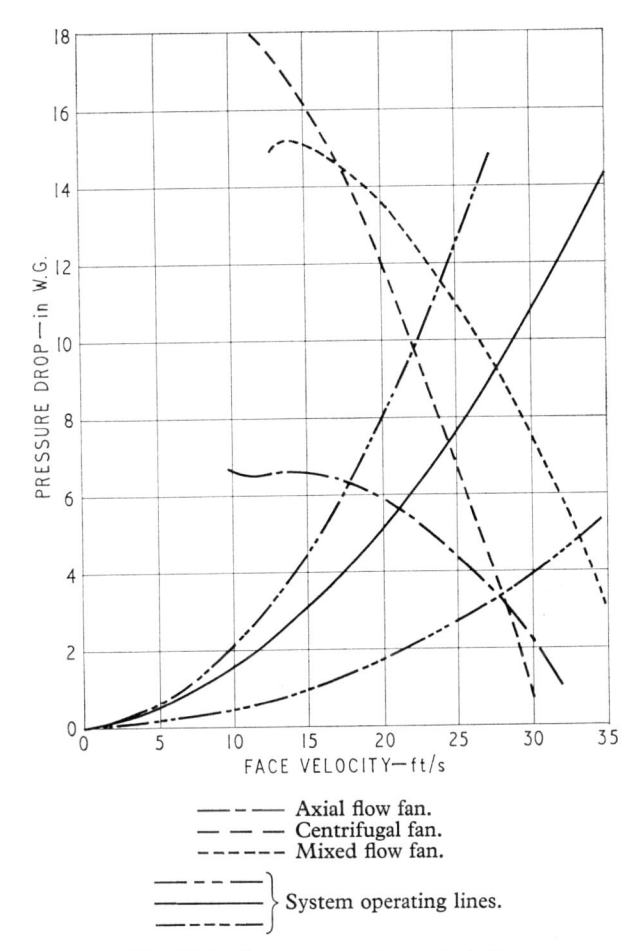

Fig. 16.5. Fan system characteristics

— · · — Axial flow fan.
— — — Centrifugal fan.
- - - - - Mixed flow fan.
⎫
⎬ System operating lines.
⎭

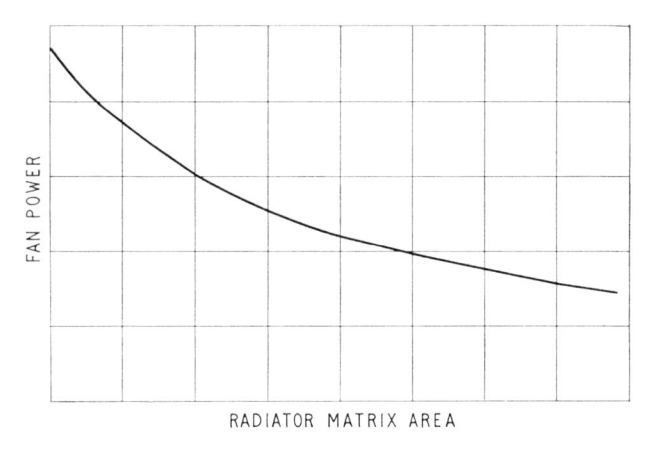

Fig. 16.6. Radiator/fan relationship

louvres and radiator matrix in service by dust, leaves, and other debris.

The requirement is thus for a fan which will operate in a non-critical range under all conditions likely to be encountered in service. For this reason there has been a tendency to adopt the mixed flow fan in preference to the axial as for the same tip speed it develops a considerably higher pressure. This effect is even more apparent with the centrifugal fan, but it introduces installational difficulties so that, for most installations, the mixed flow fan presents the best compromise solution (Fig. 16.5).

A theoretical relationship has been established between radiator size and fan power to assist the designer in his initial choice of parameters and to enable him to optimize the system more readily (Fig. 16.6).

It is clearly desirable, at all times, to limit the fan speed to the value dictated by the cooling requirement. This has led to the adoption of the hydrostatic drive system controlled by an element sensing coolant temperature, and this has the additional advantage that it enables the engine to reach normal running temperature more rapidly, particularly in arctic conditions.

Environment also introduces complications in cooling system design owing to the possible use of aggressive waters and to the high proportion of sand in the cooling air under sandstorm conditions. In certain theatres the most adverse combinations of these two conditions can occur with the further factor of the sand being salt laden.

It is for these reasons that the adoption of light alloy construction has not yet been adopted, although production techniques that are intended to minimize these effects are in course of development and evaluation.

Air cleaner

Reference has already been made to the indirect effect which the air cleaner can have on determining the rating of the engine and, by inference, the dire consequences on engine reliability if the pressure drop through the cleaner rises appreciably above its maximum design value.

The most immediate effect of the cleaner on the engine, however, is due to its efficiency. Under desert conditions of high dust concentration, a drop in efficiency from, say, 99 to 98 per cent will double the dust throughput, a simple fact not always appreciated by the user.

The difficulties in ensuring adequate protection for the engine are due to the severe space limitation and to ensuring that the cleaner remains within its operating limits while in service. Some impression may be gained of the installational difficulties from the following approximate relationships between rate of air throughput and installed volume for various types of air cleaner:

Oil bath air cleaner	250 ft^3 min/ft^3
Cascade cyclone air cleaner	350 ft^3 min/ft^3
Cyclone and paper element air cleaner	200 ft^3 min/ft^3

To ensure that the cleaner always operates within acceptable limits of pressure drop requires either reliance on the vehicle crew for adequately servicing the cleaner or on the provision of automatic dust extraction—the latter being suitable only for use with an inertial and not with a barrier or oil bath type of unit.

Failure of the cleaner in service can thus produce two different types of engine failure:

(1) heavy abrasive wear of piston rings, piston skirt, and bearings;
(2) piston assembly failures due to excessive combustion temperature arising from air starvation

The tendency today is towards adoption of the cleaner with automatic dust extraction.

An assessment must be made of the probable reliability of the crew in servicing the more orthodox cleaner against that of an automatic extraction system. However, one such system based on an ejector principle has operated with complete reliability in a range of vehicles in many theatres over high aggregate mileages.

Exhaust system

As with the air cleaner, the exhaust system has two effects: the first, obvious, one of influencing exhaust noise is more critical in the tactical sense, but the second can be critical on engine reliability due to the back pressure on the engine and its limiting influence on combustion air flow and, hence, on temperatures.

Fig. 16.7. Power pack, drive end

Owing to the restricted size and total enclosure, the engine compartment temperature may rise to as high as 300°F locally, which constitutes a difficult environment for such components as starter motor, control solenoids, and cable insulation.

ENVIRONMENT

Ambient temperature

The ambient temperature extremes at which it is necessary to start and operate impose some critical conditions on the engine. These fall into two categories: those within the engine, and those external to it.

The first of these can affect engine design, since heat distribution and flow within the combustion chamber determine the heat rejection to coolant and, hence, the thermal load on the cooling system. For the same reason, the low-temperature startability is influenced by heat loss from the air charge to the chamber walls and, consequently, can be a factor in the choice of type of combustion chamber, stroke/bore ratio, and compression ratio.

Unfortunately, the latter is in conflict with the requirements for low engine weight and reliability, since peak pressure rises with compression ratio—hence the attraction of the variable compression ratio principle.

The external influences are those of high temperature on the starter motor and on the fuel injection pump. The starter tends to be buried in the bottom of the hull where it is in a pocket of stagnant high-temperature air and is often in close proximity to an exhaust manifold. It is difficult to duct air separately to the vicinity of the starter because of the very confined installation, while if the starter is developed to provide reliable cranking at a sufficiently high speed then rather exotic insulation and commutator soldering techniques are required which result in a costly machine, particularly as the production quantities are limited.

Thus, starter temperature can have a critical effect on engine reliability. Injection pump temperature is influenced (*a*) by the use of positive cambox lubrication tapped off the engine gallery, (*b*) by constant circulation of fuel from the tank through the pump gallery back to the tank in the case of the multi-fuel engine, and (*c*) by the high ambient temperature in the vicinity of the pump. The effect of high pump temperature is particularly significant when operating on the more volatile fuels.

Fig. 16.8. Power pack, free end

Altitude extremes

The effect of variation in altitude in most commercial applications is not critical, but in the case of army vehicles a much more critical situation arises. For the reasons already presented, the engine tends to be rated to give maximum performance, with due regard to the reliability and endurance requirements, under normal conditions (i.e. at altitudes below 5000 ft) and in temperate conditions.

When such an engine is required to operate at high altitude, and when coupled with tropical ambient conditions, the reduction in mass air flow through the engine can produce such excessive temperatures in critical components as to lead to early failure, while producing quite unacceptable exhaust smoke. These conditions can be avoided, at least in part, by fitting an aneroid controlled rack stop on the fuel injection pump to avoid overfuelling resulting from the altitude effect. No corresponding automatic adjustment exists to compensate for the high ambient temperature. The only alternatives are to derate the engine or to adjust such critically rated engines when they arrive in the theatre where these conditions exist.

Conditions during maintenance and repair

One further condition with considerable bearing on the suitability of an engine, and which can also critically affect its rating, is the operating and maintenance environment.

An army vehicle in the field does not come back to base for replenishment, but under active service conditions it will be refuelled, and the cooling and lubrication systems topped up, from containers by crew members. As this operation is frequently carried out under most adverse conditions it will be appreciated why considerable quantities of water, sand, and organic matter are often found in fuel, oil, and cooling systems. Similar effects arise from maintenance and repair procedures carried out under field conditions.

Because the installational factors can have a major effect on the engine, it is essential that an acceptance level common to both engine and the auxiliary systems be adopted, and that this be reflected in the design and development effort required. To enable this to be done effectively and economically the concept of the power pack is introduced rather than that of an engine to which are connected a number of subsystems. The power pack consists of the engine, induction, cooling, lubrication, fuel, and exhaust systems, together with any necessary framing (Figs 16.7 and 16.8). One further benefit is that the whole can be removed from the vehicle with a minimum of disconnection, thus avoiding the necessity to break the oil, hydraulic, and cooling circuits. Apart from the

Fig. 16.9. Power pack removal

reduction of effort in the field, this removes a frequent source of trouble due to the introduction of dirt and to faulty reconnection (Fig. 16.9).

However, to gain the full benefit from the introduction of this policy it is essential that the power pack be designed as one integrated unit with one authority responsible for its design and development. One of the first major decisions to be made during the design stage is the choice of specific loading as shown by the brake mean effective pressure. This critically affects reliability and the installation volumes of the engine and of the power pack. Fig. 16.10 illustrates the relationship between brake mean effective pressure and these two volumes and shows clearly the diminishing return above about 250 lb/in²; it also emphasises the increasing discrepancy between the volumes of engine and power pack at very high loadings.

FUELS AND LUBRICANTS

The fuel and lubricants policy is based on the probability that in war, fuels and lubricants for the services may be restricted to those which are produced by simple refinery processes and hence are likely to be held to relatively low quality levels. It is therefore essential that engines which may be compelled to operate on these are approval tested on representative grades. As these are not readily available in peace time, special reference grades are specified. In the case of fuels these are deliberately held, by specification limits, to the most adverse end of the tolerance band on such critical items as octane number and lead content in the case of gasolines, and cetane number and sulphur in the case of diesel fuel.

Reference grades of engine oils of the OMD series, however, are held within limits not necessarily at the most adverse end of the band. In this respect they are of more consistent, rather than of minimum, quality.

The engine lubricants have more effect on critically affecting the diesel engine than do the fuels. The influence they have on determining the maximum acceptable rating has already been stressed, and this constitutes one of the main factors which necessitate that the engine approval procedure be based on a special services type test, whether the engine is intended for automotive, marine, or general purpose application.

The specifications of fuels and lubricants are constantly under review to ensure that they are related to commercial practice while still satisfying the requirement to reflect quality levels likely to be appropriate in war. Thus, it is to

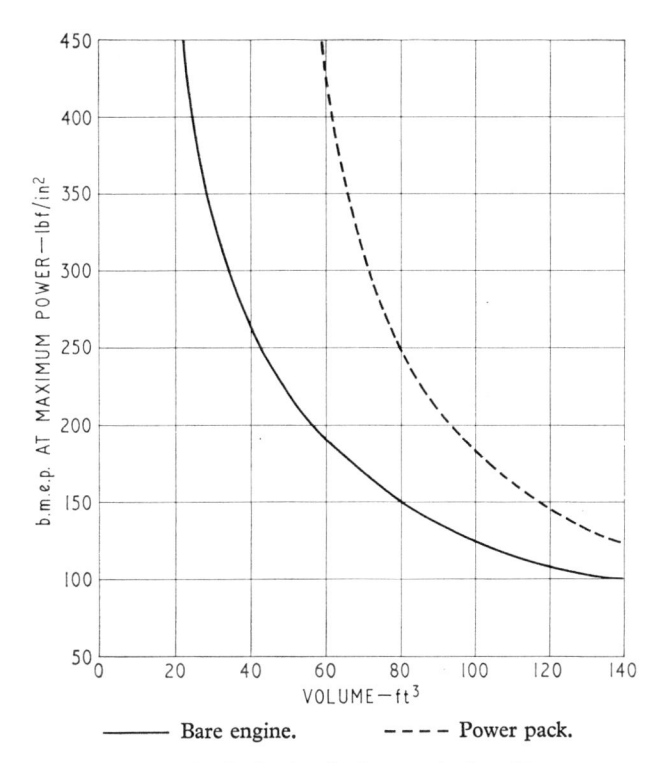

Fig. 16.10. Rating/volume relationship
——— Bare engine. – – – Power pack.

be expected that a rising tendency in quality levels may be anticipated in the future.

CONCLUSIONS

The constant theme throughout this paper has been that because of the complex relationship between the engine and its subsystems, engine functioning can be critically affected by these systems and, consequently, it is essential to regard the power plant as one unified concept.

This stage has already been reached in the fighting vehicle field, but with the general advance in the level of performance demand this trend will progressively extend into the realm of other engine applications. However, if design is to be optimized over all relevant aspects—technical, production, operating, and cost—a unified policy based on the principle of the 'whole' must be applied, not merely to design and development, but also to that of administration and financial control. Only thus can vehicles capable of satisfying future requirements be provided within the limits of acceptable cost.

Paper 17

FURTHER CONSIDERATIONS IN INJECTOR DESIGN FOR HIGH SPECIFIC OUTPUT DIESEL ENGINES

J. P. S. Curran*

Although the complete fuel injection system is a critical feature for the reliable operation of all diesel engines, various applications place a different emphasis on service requirements; particularly with respect to nozzle and injector life. This paper deals with a number of rail traction problems which have arisen in the field where improvements in nozzle/injector life have been needed for both technical and economic requirements. One specific problem—that of injector spring failure—is covered in more detail. A study of spring dynamic behaviour has been made which shows that normal spring 'surge' theory does not adequately explain spring failure. A simple expression is derived indicating that injector spring dynamic stresses are independent of spring design parameters and are a function of nozzle needle velocity only.

INTRODUCTION

THE INJECTOR IN A DIESEL ENGINE is subjected to relatively high mechanical, thermal, and hydraulic loads irrespective of engine application and size. In the case of rail traction diesels, however, the rapid increase in ratings over recent years has probably posed the most serious problems with the injector, calling for special design features to ensure reliable and economic operation.

Diesel engines—particularly in the rail traction field—have been subjected to intensive development to improve their specific output. Engine brake horsepower per litre has approximately doubled over the last two decades, and fuel injection equipment has been required to match this improvement without incurring an increase in size or loss of reliability. Considerable development has been necessary to fulfil these requirements and to overcome various service hazards.

Typical problems encountered on injectors under traction conditions have included nozzle spray deterioration and carbonization, gas blowback and associated needle sticking, and nozzle spring relaxation and failure. Cavitation erosion has also been evident on certain applications and has been the subject of previous papers (1)† (2).

While nozzle spray deterioration and carbonization depend on many factors, they usually result from the compromise in spray hole size to cater for a wide speed and load range. Optimization of the nozzle for idling produces unacceptably high injection pressures at full load. Conversely, the appropriate nozzle match for the high load condition usually produces inferior spray characteristics at idling. Where an engine runs on a fixed load/speed characteristic, improved low-speed performance has been obtained by the use of an upper helix on the pump plunger (Fig. 17.1). This gives an increased rate of injection due to the point of cut-off occurring further up the cam flank.

The problem of nozzle gas blowback/needle sticking is associated with higher fuel throughput at an increased injection rate. The mechanism of gas blowback is well known and is clearly influenced by the response rate of the injector during needle closure. Considerable improvements in reducing nozzle blowback have been made by changing from the conventional high spring injector to the 'low inertia' (low spring) design (Fig. 17.2).

In the low inertia design, the location of the spring close to the nozzle poses considerable design problems due to the confined space available between the cylinder head valve ports. A further complication is the need for increased spring load to match increased cylinder firing pressures at higher engine speeds.

On one particular rail traction application, spring failures—which could not be explained at the time—had

---

The MS. of this paper was received at the Institution on 2nd July 1970 and accepted for publication on 10th July 1970. 23
* Senior Product Engineer, Bryce Berger Limited, Hucclecote, Gloucester.
† References are given in Appendix 17.1.

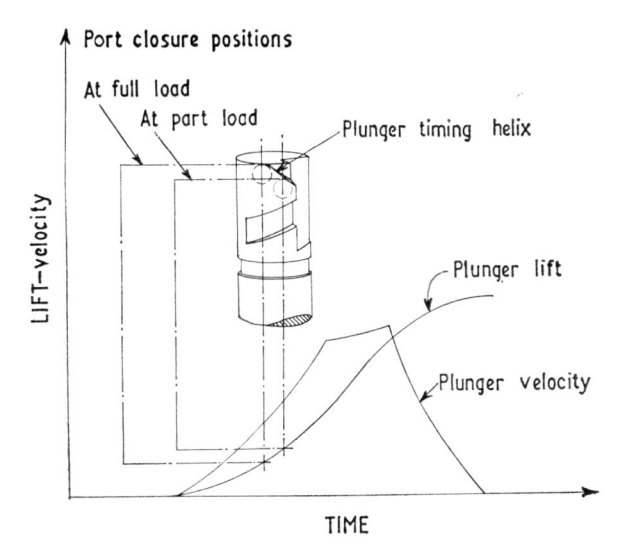

Fig. 17.1. Diagram illustrating effect of pump plunger timing helix

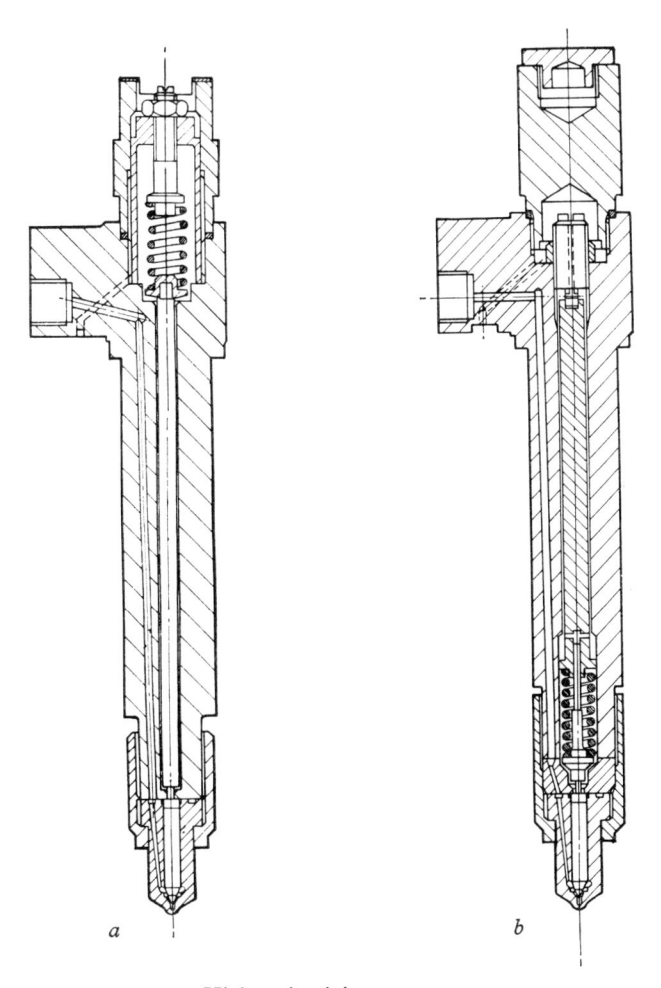

already occurred in service on conventional injectors operating at nominally conservative stress levels. An experimental analysis of spring dynamic operation has suggested that alternative methods of calculation were required to explain the failures. The remainder of this paper deals with a more comprehensive method of considering spring criteria in injector design to overcome premature deterioration or failure.

Notation

A	Cross-sectional area.
c	Spring diameter/wire diameter ratio.
c_s	Spring stress wave velocity.
D	Spring mean diameter.
D_n	Nozzle needle shank diameter.
D_s	Nozzle needle seat diameter.
D_o	Spring outside diameter.
d	Wire diameter.
E	Young's modulus.
f	Acceleration.
G	Modulus of rigidity.
h	Needle lift.
k	Wahl spring constant.
L	Spring length.
M_s	Spring mass.
M_T	Total mass of moving parts.
n	Number of turns.
P_0	Nozzle opening pressure.
R	Spring rate.
S_I	Initial spring stress.
S_R	Stress range.
S_v	Vibration stress.
V	Velocity.
V_F	Final needle velocity.
W_I	Initial spring load.
W_F	Final spring load.
x	Spring spindle overshoot.
ρ	Density.

INJECTOR SPRING DYNAMIC BEHAVIOUR

Conventionally, injector springs have been designed for infinite life when operating between the two fixed stress levels corresponding to the needle on its seat and against its full lift stop. The springs usually have such a high rate that they satisfy the normal dynamic criteria for engine valve springs, which states that the spring natural frequency must be at least 15 times the drive frequency to prevent spring 'surge'. This criterion is to avoid resonance between the harmonics of the cam drive system and the spring natural frequency, and can only occur when the spring is driven by a continuously reciprocating device. High dynamic spring stresses are sometimes encountered in situations not necessarily related to continual cycling, and can be induced by coil action resulting from a single deflection. The difference between the various modes of operation is as follows:

a High spring injector.
b Low inertia (low spring) injector.

Fig. 17.2

Normal spring operation

Fig. 17.3a is a simplified representation of a valve spring driven at low speed where (a) the spring natural frequency does not coincide with any of the lower order harmonics of the drive system and (b) damping is ignored. After the valve lifts the small amount of vibration has a negligible effect on the spring stress.

Continuous reciprocation—spring resonance

Fig. 17.3b is similar to Fig. 17.3a except that the ninth harmonic of the cam drive system now coincides with the spring natural frequency.

During the first cam cycle an oscillation of amplitude x_1, proportional to the velocity, is set up. On the second cycle the valve motion complements the existing vibration and the amplitude increases to x_2, where $x_2 = 2x_1$. This effect continues until the amplitude of the nth vibration equals nx_1, and so on to infinity. In practice, because of damping, the vibration amplitude will reach a finite limit, which for normal valve spring systems could be as high as 500 times the initial amplitude x_1. The situation shown in Fig. 17.3b is the classical valve spring resonance effect.

Single operation—high-velocity coil stress

Fig. 17.3c also represents a similar system to that in Fig. 17.3a but the instantaneous cam velocity is considerably higher and the effect of damping is shown. Here, the initial amplitude, x_1, causes an appreciable increase in stress above the maximum static value. The situation shown in Fig. 17.3c is analogous to injector spring operation where the nozzle needle can lift to its maximum stop in less than 1 ms.

The two effects illustrated in Fig. 17.3b and c cause the spring to vibrate at its natural frequency; that is, all the spring coils are moving in the same direction at the same time. If the coils of a spring are not all moving in the same direction at the same time, the spring is said to be in 'surge'.

Spring surge

Surge occurs when the spring coil velocity is of the same order as the velocity of propagation of a stress wave in the spring. If the end of a spring (A in Fig. 17.4) is suddenly moved at a velocity V, a compressive stress wave will travel around the spring wire at the velocity of sound in the spring, V_s. The time for the stress wave to travel from A to B is $\pi D/V_s$, thus point B will begin to move $\pi D/V_s$ seconds after point A, and the stress wave velocity along the axis of the spring, c_s, is equal to

$$\frac{V_s}{\pi D} \cdot \frac{L}{n}$$

In Fig. 17.3b the amplitude, and hence the coil velocity, increases during successive cycles and the spring motion will change from its simple harmonic form to a concertina-like surge form as the coil velocity approaches the stress wave velocity. In Fig. 17.3c surge will only occur if the initial needle lift velocity is close to the stress wave velocity along the spring axis.

On most injector applications the stress wave velocity is considerably higher than the needle velocity and surge does not occur. However, because practical considerations require injector springs to work at very high static stress levels, the initial amplitude of vibration may be of sufficient magnitude to overstress the spring. It was,

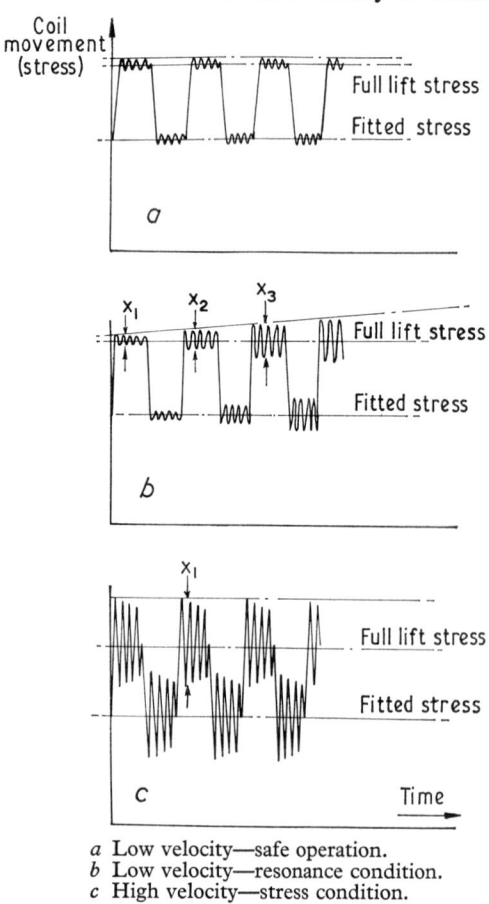

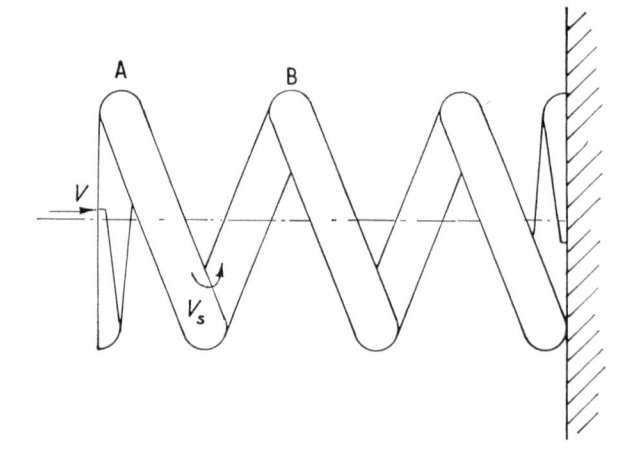

a Low velocity—safe operation.
b Low velocity—resonance condition.
c High velocity—stress condition.

Fig. 17.3

Velocity of sound along spring axis, $c_s = \dfrac{L}{\pi D n} V_s$.

Fig. 17.4

therefore, necessary to obtain theoretical expressions to calculate the dynamic stresses and to evolve experimental techniques to validate the theoretical expressions.

OPTIMIZATION OF SPRING STRESSES

Static stress calculations

The initial static stress in the spring with the needle seated can be calculated from:

$$S_I = \frac{8ck}{\pi d^2} W_I \quad \ldots \quad (17.1)$$

and

$$W_I = \frac{\pi}{4} P_0 (D_n^2 - D_s^2)$$

thus

$$S_I = \frac{2ck}{d^2} P_0 (D_n^2 - D_s^2) \quad . . \quad (17.1a)$$

The maximum static stress at maximum needle lift is equal to $S_I + S_R$ where

$$S_R = \frac{8ck}{\pi d^2} Rh = \frac{Gkh}{\pi n c^2 d} \quad . . \quad (17.2)$$

For a fixed injector application, equations (17.1a) and (17.2) reduce to

$$S_I = B_1 \frac{kc}{d^2} = B_1 \frac{(D_o - d)k}{d^3}$$

$$S_R = B_2 \frac{k}{nc^2 d}$$

where B_1 and B_2 are constants, for a particular injector application.

The maximum spring outside diameter, D_o, is limited by the overall injector size; thus for minimum fitted stress, S_I, d must be as large as possible. The minimum value of c (D/d) recommended by spring manufacturers is 3. Therefore

$$\frac{D}{d} = 3 = \frac{D_o - d}{d} \quad . . . \quad (17.3)$$

and $d = D_o/4$; thus for optimum spring design $d \simeq \frac{1}{4} \times$ (spring chamber diameter). Having fixed d and c (and hence k), the value of the stress range, S_R, can only be reduced by:

(a) increasing the number of active coils, n, to the maximum possible from stability consideration (usually in the range of 10–15);

(b) reducing the needle lift to the minimum necessary for adequate seat flow area.

So for minimum static stress levels, maximum wire diameter and maximum number of active coils are required.

Dynamic stress calculations

Stress wave effects

An expression for the amplitude of the compressive stress wave described previously can be determined as follows. Assume that the spring is a rod of low stiffness and density with one end fixed and that the other end is moved at velocity V; the load in the spring, W_v, will be given by

$$W_v = \frac{EAV}{c_s} \quad . . . \quad (17.4)$$

Then, using the expressions,

$$c_s = L \sqrt{\frac{R}{M_s}}, \qquad E = \frac{RL}{A}$$

$$R = \frac{Gd}{8nc^3}, \qquad M_s = \frac{\pi^2 d^3 cn\rho}{4}$$

in equation (17.4) for the maximum needle velocity condition

$$W_v = V_F \frac{\pi d^2}{8c} \sqrt{(2G\rho)}$$

Then substituting for W in equation (17.1), the stress, S_v, due to velocity, V, is given by

$$S_v = Vk \sqrt{(2G\rho)} \quad . . . \quad (17.5)$$

Equation (17.5) shows, perhaps surprisingly, that for a fixed D/d ratio spring, the amplitude of the stress wave is a function of needle velocity only, and otherwise is not a function of the spring dimensions.

Equation (17.5) was derived assuming that the spring was a rod of low stiffness. When a stress wave travelling along a rod meets a large change in stiffness/sectional area the wave is completely reflected, causing a stress doubling effect. Thus, the instantaneous dynamic stress at the end of the injector spring may be up to twice the value given by equation (17.5).

Spindle overshoot

When the nozzle needle hits the lift stop, the spring spindle can overshoot, causing further compression of the spring. The amount of overshoot, x, is given approximately by

$$x = \frac{M_T V_F^2}{2W_F} \quad . . . \quad (17.6)$$

The increase in stress due to overshoot is then obtained by combining equations (17.2) and (17.4) with $h = x$.

Spindle overshoot [equation (17.6)] is proportional to

mass of moving parts × (velocity)^2

and is inversely proportional to spring load. Needle lift velocity is not normally controlled but again will be minimized by ensuring minimum needle lift. The minimum mass of moving parts will be achieved with a low spring injector design (this type of design is also preferred for minimum needle closing time and prevention of gas blowback).

THE EXPERIMENTAL INVESTIGATION

Three methods of recording spring dynamic behaviour were used:

(a) a high-speed 16-mm camera (15 000 frames/s);

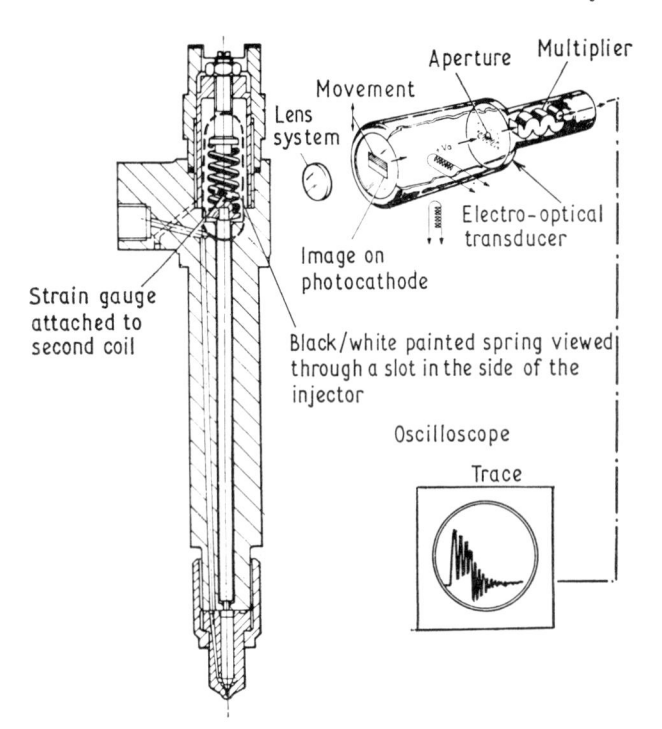

Fig. 17.5

(b) a strain gauge attached to a spring coil;
(c) an electro-optical device for observing the motion of individual spring coils during the injection period.

Fig. 17.5 illustrates the injector used for the tests; the spring was painted half black and half white along its axis and observed through a slot in the side of the injector for methods (a) and (c). The strain gauge was attached to the second coil of the spring for method (b).

High-speed camera method

When the high-speed films were projected at one-thousandth normal speed, severe vibration was observed qualitatively. Unfortunately, owing to the extremely small physical movements involved, it was not possible to obtain quantitative data from the films.

Strain gauge method

In order to check the preceding theoretical analysis, it was necessary to have an accurate record of the motion of each spring coil. To do this, it would have been necessary to attach a strain gauge to each coil. While this was theoretically possible, it proved difficult in practice to reliably mount a multiplicity of strain gauges correctly on the spring and to have them operating simultaneously.

Electro-optical device method

A more practical and satisfactory solution was obtained with an electro-optical device, as yet little used in the U.K. (Figs 17.5 and 17.6). The cross wires of the optical device were focused on the black/white paint junction on one of the spring coils. When the spring coil moved, the change

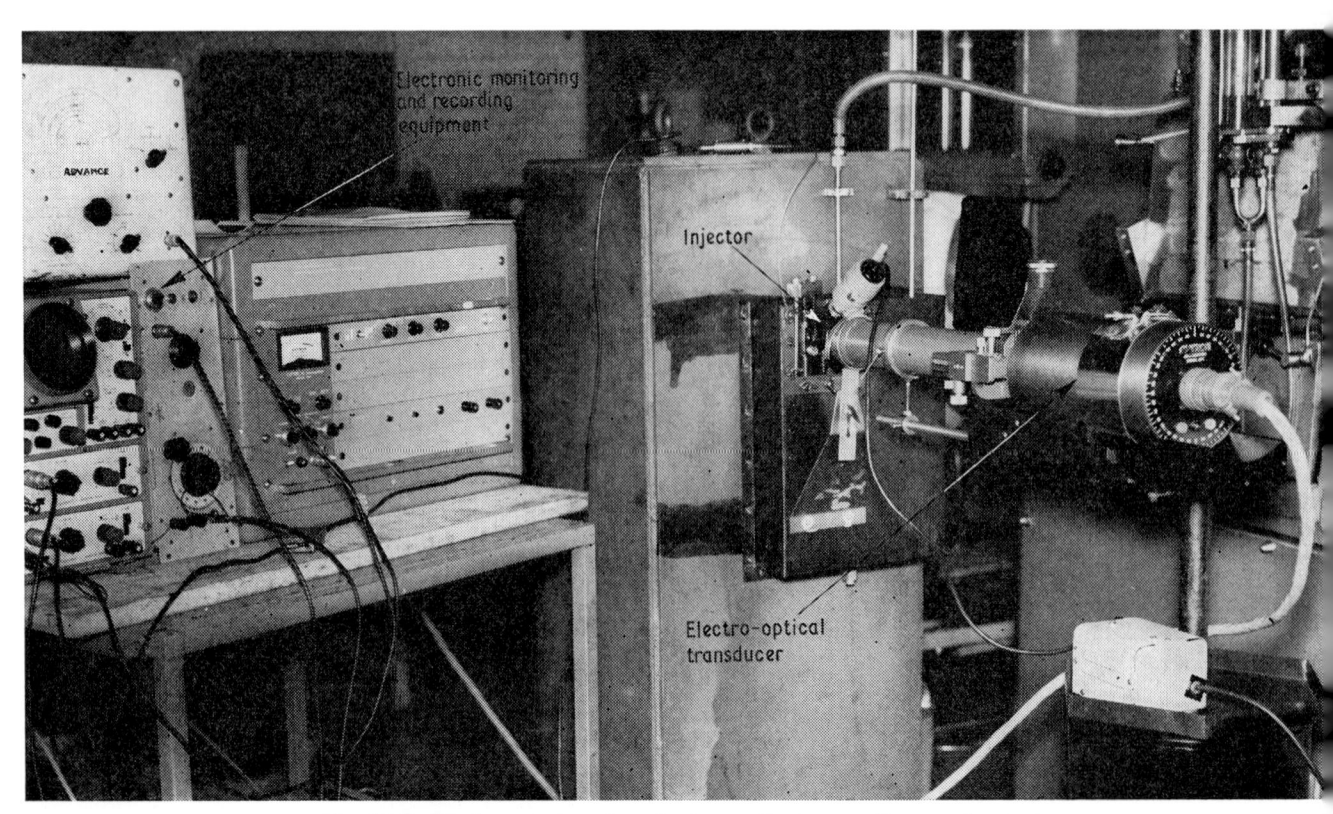

Fig. 17.6. Rig layout showing electro-optical recording equipment

in contrast was sensed on the photocathode and a proportional displacement was registered on an oscilloscope screen. By focusing on each coil in turn, it was possible to obtain a series of oscilloscope traces giving a complete picture of spring coil movement.

The displacement diagram of the last moving coil at a point exactly one turn from the fixed end was used to obtain a value of spring stress. As spring stress is directly proportional to spring movement, the mean distance moved by the end coil was equivalent to the static stress range. The dynamic stress was determined from the ratio of the amplitude of vibration to the mean distance moved.

In addition, one calibrated strain gauge was attached to the second coil of the spring to provide a reference stress value. Fig. 17.7 shows a typical set of spring coil movement diagrams recorded by the electro-optical transducer and the strain measurement recorded on the second coil. From these records the following observations can be made:

(1) The maximum stresses obtained from the electro-optical transducer and strain gauge agree to within 2 per cent; in all the tests carried out the two values agree to within 10 per cent.

(2) The minimum stress value obtained from the strain gauge is considerably less than the value given by the electro-optical diagram. This is most probably due to the high needle seating velocity causing short duration stress pulses in the spring spindle and hence in the spring.

(3) Before injection the spring coils are completely at rest, previous vibration having been completely damped out.

(4) The oscillation occurs at the spring's principal natural frequency, all coils moving in the same direction at the same time.

Fig. 17.8 shows the first, third, and sixth coil movement for similar conditions to those in Fig. 17.7, but with different nozzle opening pressures.

Although the nozzle opening pressures varied by 3:1, the maximum needle lift velocities are within 10 per cent of each other. The spindle overshoot is much greater with the lower nozzle opening pressure due to the reduced spring load [equation (17.6)] but the vibration stresses during lift are virtually the same.

The needle closing velocity with the higher spring load

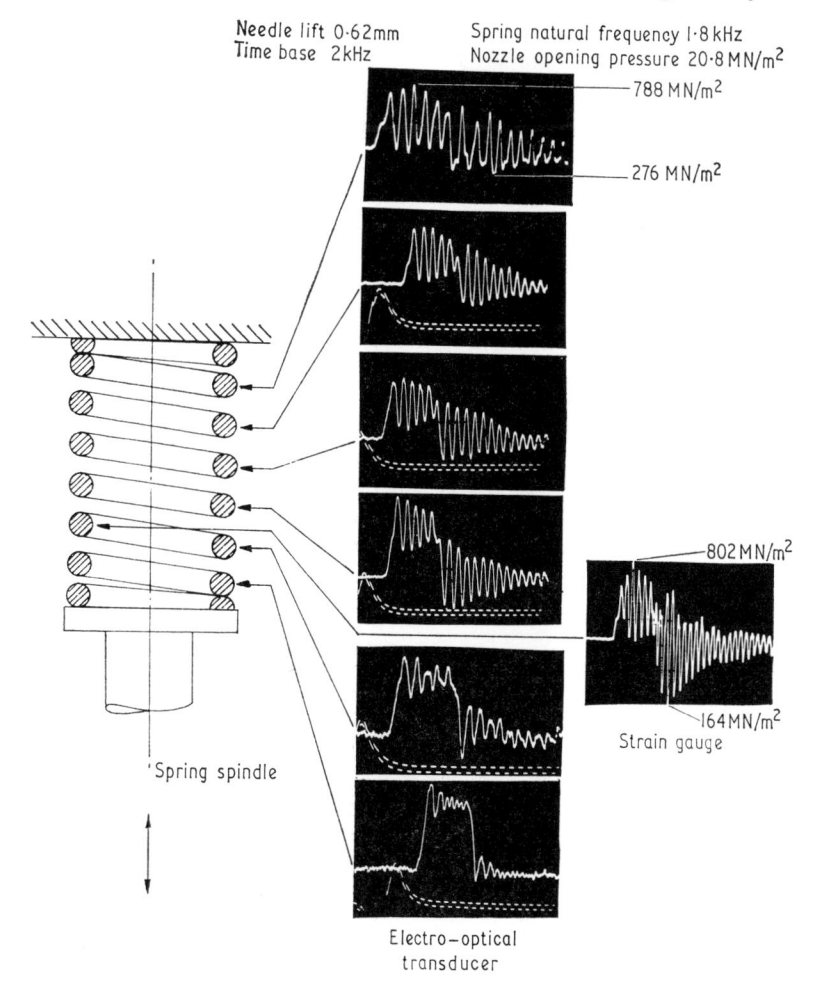

Fig. 17.7. Electro-optical movement diagrams of each spring coil and strain gauge record for the second coil

Nozzle opening pressure = 10·4 MN/m²   Nozzle opening pressure = 31·2 MN/m²
  a Sixth coil.   b Third coil.   c First coil.

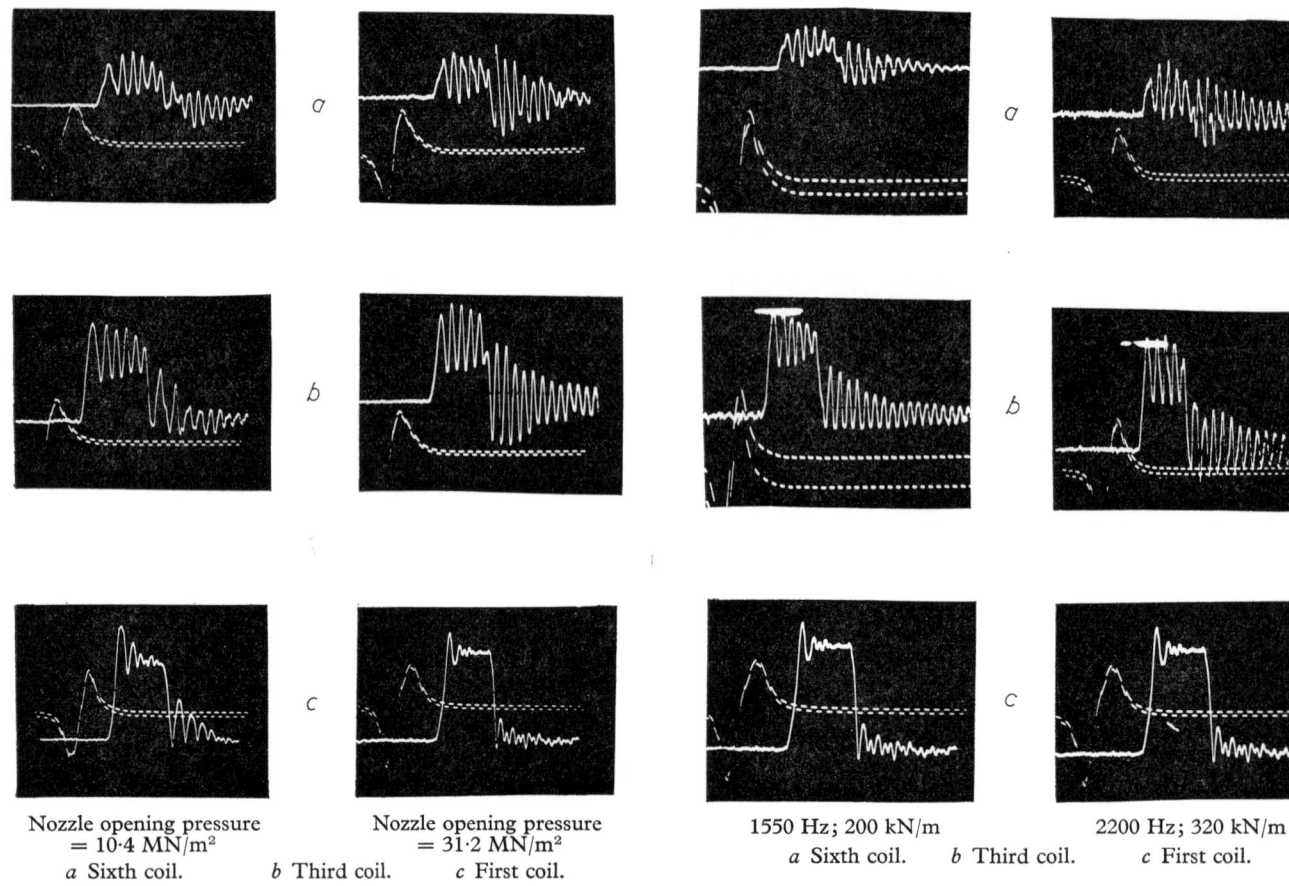

Fig. 17.8. Electro-optical diagrams—conditions as Fig. 17.7 with the exception of nozzle opening pressure

1550 Hz; 200 kN/m   2200 Hz; 320 kN/m
  a Sixth coil.   b Third coil.   c First coil.

Fig. 17.9. Electro-optical diagrams with different frequency/rate springs

is almost twice the value of that with the lower spring load. The vibration is much more severe with the higher closing velocity despite the larger secondary injections at the lower opening pressure. Thus the vibration stress is a function of needle velocity and not fitted spring load.

A comparison between springs of different rate (200 and 320 kN/m) and natural frequency (1550 and 2200 Hz) is made in Fig. 17.9. Again, needle lift velocities are within 10 per cent and the measured overstress is exactly the same in each case. This indicates that the dynamic stress is unaffected by either spring rate or natural frequency.

CONCLUSIONS

While the practical work to date has been restricted to considering a limited range of variables, it closely supports the theoretical analysis.

Other workers in this field have investigated similar phenomena (3). Their conclusions advocate that spring natural frequency should be maximized to reduce spring vibration stresses. This argument is untenable, since the optimum static spring stress is achieved with maximum wire diameter and a minimum D/d ratio of $c = 3$.

$$\text{Spring natural frequency} = \frac{\text{constant}}{c^2 dn}$$

Therefore, the natural frequency can only be increased by reducing d and/or n. Any reduction in either will increase the static stress levels and adversely affect the margin of safety—which is clearly undesirable.

Our analysis shows that for minimum static stresses the wire diameter and number of coils must be maximized [equations (17.1) and (17.2)] and, more important, equation (17.5) indicates that dynamic stresses are essentially independent of spring design. They can only be reduced by minimizing the nozzle needle opening/closing velocities and spindle mass.

ACKNOWLEDGEMENTS

The author wishes to thank the Directors of Bryce Berger Limited for permission to publish this paper, and extends his gratitude to colleagues who assisted in its preparation.

APPENDIX 17.1

REFERENCES

(1) GROTH, L. 'Cavitation damage to fuel injection systems and its prevention', *Symp. Mechanical Design of Diesel Engines, Proc. Instn mech. Engrs* 1966–67 **181** (Pt 3H), 130.

(2) CLIFTON, W. L. and GRATZMULLER, C. A. 'Injection diesel dans les domaines industriel, marin et ferroviaire', *Jl S.I.A.* 1970 **43** (March).

(3) BOSCH, W. 'Untersuchungen über die Schwingungsbeanspruchung von Einspritzventilfedern', *Bosch Tech. Ber.* 1965 **1** (No. 4, September).

SUMMING-UP

W. P. Mansfield*

THE PAPERS IN THIS SYMPOSIUM present a great deal of information on a wide variety of subjects. In order to make a brief summing-up possible, it is necessary to limit consideration to that information which relates directly to the symposium theme. Accordingly, an attempt has been made to extract from the papers the answers to the question, 'What are the critical factors in the application of diesel engines?'

The results are shown in the table. The group headings, 1 to 8, which form a logical sequence, have been chosen to cover all the possible critical factors, the factors covered in each group being as follows.

Group 1

This covers the critical factors in the design and development of the engine and installation. Since the design is largely determined by service requirements and since the choices made will greatly influence service performance, design factors can be critical factors in the application of engines.

Group 2

Before, during, and after the design and development of the engine and installation, communication is necessary between engine maker, component makers, makers of any equipment or vehicle of which the engine forms part, and the ultimate users. The effectiveness of such communications can be critical.

Group 3

When the engine enters service, the fuel and lubricant actually used, which may differ from those recommended, may prove to be critical factors, as may also the composition of the coolant.

Group 4

(a) The environment is clearly a major factor affecting the behaviour of the engine in service, and the term is broad enough to include not only atmospheric conditions but also factors affecting the application of external forces to the engine, e.g. mounting conditions and the degree of alignment with the driven machine.

(b) The engine's environment may also be a critical factor by affecting reaction to the engine, particularly in respect of noise and emissions.

Group 5

Clearly the actual engine duty, which may differ from that envisaged during the design and development stages, is very important. The expression 'combinations of load, speed, and time' includes such conditions as prolonged idling, rapid speed and load changes, frequent speed and load changes, and even running-in schedules when running-in is carried out by the user.

Group 6

The extent and quality of maintenance received during service is another factor of great importance, and this in turn is affected by such factors as the quality of design in respect of ease of maintenance, and by the effectiveness of the instruction imparted to the personnel responsible.

Group 7

Any of the factors mentioned so far can be critical in determining the engine's behaviour in service, as defined by performance, reliability, life, noise, vibration, and exhaust emissions. Any of these factors may, in turn, prove a critical factor in the application.

Group 8

The success or failure of all the efforts expended is judged by the comparison of actual service behaviour with that required. Since the requirements are very extensive, Group 8 in the table is confined to any outstanding requirements, especially those defined by standards or legislation.

If the actual behaviour falls seriously short of that required, further design and development work is needed. The requirements grow continuously more exacting, so the whole process is continuous.

The columns at the left of the table, headed 'Paper classification' and 'Author(s)', show, above the double horizontal line, a classification of the papers according to the specific application with which the authors are concerned. The remaining papers all relate to the more general

---

The MS. of this Address was received at the Institution on 3rd August 1970.
* BICERI Ltd, Buckingham Avenue, Slough, Bucks.

Paper classification	Author(s)	Group 1 Design and development of engine and installation	Group 2 Communications: makers of engine, components, equipment, and final user	Group 3 Fuels, lubricants, and coolants
RAIL TRACTION	PETROOK STEWART	Accommodation of adequate cooling systems and exhaust silencers		
FIGHTING VEHICLES	MILLAR	High specific power, problem of housing adequate air, exhaust and cooling systems	Need for feedback from testing and field experience	Multi-fuel operation. Low-quality fuels. Low additive lubricants
FIGHTING SHIPS	SAMPSON	Shock and vibration mountings. Acoustic cladding or enclosures. Oil-cushioned piston. Crankshaft		Need for single oil. Test kit for oil checks. Composition of antifreeze for coolant
ROAD	HOLMÉR HÄGGH	T/C, exhaust brake, inlet valves, CR, head gaskets, starting, radiator size. Cost	Market research re gasket failure	Series 3 oil to eliminate scuffing
	RUSSELL	Modifications to reduce noise: stiffening, isolating, damping of walls		
TRANSPORT	WATERS LALOR PRIEDE	Paramount effect of speed and bore on noise intensity ($I \propto N^3 B^5$)		
ELECTRICITY	RATCLIFFE	Design for reliability using cheap fuel		Residual fuels, treatment effect on components. Low cost lubricants, treatment, monitoring
GENERATION	MACKENZIE	Need for normally aspirated engine with high inertia. Provision of heat recovery for desalination	Difficulties of remote site	
MINES	LUNNON	Combustion system. Flame-traps. Exhaust conditioner. Injector timing		Low sulphur fuel for low nitrogen oxides. Fuel additives
INDUSTRIAL (SMALL ENGINES)	SENIOR CLARKE	Design of base engine and finishing parts for diverse applications with low cost. Testing	Requirement survey. Data from equipment maker. Continuous liaison with customer	
FUEL INJECTION EQUIPMENT	HOWES	Injection rate and timing for optimum noise, smoke, and economy. Fuel setting procedures	Co-operation of fuel injection equipment maker and engine makers in diverse applications	Fuel viscosity: 1–7 cS, impurities
PISTON RINGS	LAW	Ring materials, wall pressures, finish, coatings, cooling, rig testing, quality control	Full co-operation between engine designer and ring designer	Lubricant limits permissible ring temperature
CRANKSHAFT	CASTLE	Design of support structure. Alignment of seatings. Accuracy of jigs	Co-operation of engine-builder and shipbuilder at design stage	Effect on alignment of temperature of lubricant and cooling water
FUEL OIL	HOWELLS WALKER	Need for improved fuel system design for low-temperature operation		Fuel factors affecting smoke, carbon monoxide, hydrocarbons, oxides, nitrogen, and sulphur
FUEL: NATURAL GAS	EKE WALKER WILLIAMS	Dual-fuel or spark. Gas/air control. Installation for liquid natural gas		Low-knock (92% methane). High heat value. 'Heat sink' value. Less detergent oil
AMBIENT CONDITIONS	LOWE	Choice of turbo-chargers and charge air coolers for site conditions		

Group 4	Group 5	Group 6	Group 7	Group 8
Environment (a) affecting engine (b) affecting reaction to engine	Service duty: combinations of load, speed, and time	Maintenance: ease, schedules, instruction fulfilment	Engine behaviour: performance, life, reliability, noise, vibration, emissions	Special behaviour requirements: standards, law
(a) Air temperature. Air contamination	'On–off' operation (thermal cycling). Determination of appropriate rating	Need for ease of maintenance	Component failures. Water leaks. Turbo-charger fouling. Engine and exhaust noise	High reliability
(a) Temperature extremes. Dust. Altitude change. Air flow restriction	Various load cycles (effect of thermal inertia)	Adverse conditions of maintenance. Use of power pack	Cooling system unreliable. Air cleaner failure affects pistons etc. Noise and emissions	High reliability but only short life necessary
	Appropriate rating	Need for accessibility. Schedules. Evaluation at design stage	Problems with liners, crankshafts, bearings, viscous dampers, fuel injection equipment, noise, and vibration	
				Reliability, low cost, low noise and smoke (limited by Swedish law)
			Noise	B.S. noise measurement method. M.O.T. noise regulations
			Noise	U.K. laws regarding vehicle size, weight, speed, smoke, and noise
(a) Use of external air intake and filter to avoid air contamination	Rapid starting and loading	Cost reduction. Cylinder balance. Monitoring pistons, bearings. Replacement system	Good economy and reliability (calc.). Various outage causes including early vibration	Reliable starting and running. Low fuel and maintenance cost
(a) Lack of natural resources. Dust problem, particularly during assembly	Frequent and rapid load changes	Use of single engine type and size to simplify maintenance	Good reliability. Initial carbon deposits. Importance of control components	Reliable automatic starting. Close regulation. Reliable heat recovery system
(b) Confined spaces make emissions of special importance	15% derating for gas flow restrictions	Flame-traps cleaned daily. Exhaust gas checks	Exhaust emissions. Fire risk	Statutory requirements for CO, NO_2 in mines. Minimum smoke
				Early 'bed-in' of rings. Flame-proofing, low-temperature starting, cyclic regularity, inversion
(a) Ambient: 40 to 100°C	Various torque curves. Various governor run-outs, etc.			Various requirements for starting. Legal limits on noise, smoke, speed
			Gas pressure and rise rates. Ring, groove, and liner temperatures	
(a) Effect on alignment of ship loading and state of sea			Thermal deformation, main bearing failure, crankshaft breakage	
(a) Low-temperature operation	Effect of running conditions on emissions		Exhaust emissions	Air pollution legislation
		Less with natural gas. Fuel injection equipment less than spark. Carbon deposit minimum. Less cylinder wear	No smoke or smell. Low noise level. High part-load s.f.c.	Ability to switch to full diesel
(a) Atmospheric temperature and pressure. Cooling water temperature	Calculation of appropriate rating		Output may be limited by air/fuel ratio or exhaust or metal temperature, etc.	

application of engines and are mainly concerned with components, fuels, and ambient conditions.

Reading across the table shows the critical factors discussed or mentioned in each of the papers. Reading columns 1 to 8 vertically shows to what extent the critical factors are common to several applications. The main points that emerge are as follows.

Group 1. Design

In rail traction and fighting vehicle applications, where a high power per unit volume of available space is needed, the accommodation of the cooling systems, and, to a somewhat smaller extent, the air and exhaust systems, is a major problem that calls for much more research and development effort. The problem of housing adequate cooling systems is also of growing importance in road transport applications, because of the need for larger vehicles with engines of higher specific power.

Group 2. Communications

The engine and component makers lay heavy emphasis on the need for co-operation with those to whom they supply their products, and there are indications that such co-operation is sometimes lacking. It appears that more should be done to improve communications.

Group 3. Fuels, lubricants, and coolants

A variety of considerations is indicated according to the application or component considered. No common critical factor of outstanding importance is revealed.

Group 4. Environment

As would be expected for air-breathing engines, the most important environmental factors shown are the pressure, temperature, and cleanliness of the air entering the engine.

Group 5. Service duty

Years ago the performance of an engine was considered to be largely defined by curves of specific fuel consumption against load for several different engine speeds. Simple load–speed cycling tests were later introduced and today increasing efforts are being made to reproduce service conditions closely during development work, to ensure that the engine's response to the various kinds of load changes will be satisfactory, both as regards meeting the power requirements and as regards the resultant thermal and mechanical conditions in the engine and the effect on emissions. The entries in this group suggest that this attention is very necessary.

Comments regarding the choice of engine speed and load as a rating are included in this group. Rating is an obvious critical factor. The more thorough testing under various simulated service conditions now being undertaken by engine makers should enable them to determine safe ratings for the various applications.

Group 6. Maintenance

The most strongly emphasized critical factor in this connection is ease of maintenance, which is largely dependent on accessibility. Other important points are the need for evaluation of maintenance procedures at the design stage and the system of replacing components after a known safe life. Where reliability is of great importance, this system seems essential. The more thorough development testing should enable engine makers to make more effective recommendations regarding maintenance generally. None of the papers mentions instruction, or maintenance schemes run by engine makers.

Groups 7 and 8. Engine behaviour and special requirements

These two columns are best considered together, as they are comparable. Engine makers will be pleased to note that engine behaviour is not always the exact opposite of the special requirement. Moreover, a closer study of the entries taking account of circumstances, reveals that when reliability was not achieved, the failures were due in the majority of cases to unsatisfactory features of the installation rather than to defects in the engine itself. The importance of installation factors was stressed in several of the papers.

No mention was made of crankcase explosions, which are perhaps the most unpleasant type of engine behaviour. There have been several damaging crankcase explosions recently, showing that not all engine makers have yet taken the simple steps necessary to ensure safety.

In Group 8, several entries refer to the increasing legislation regarding noise and exhaust emissions. With the recent sudden awakening to the need for conservation, such legislation has already become a very important critical factor as far as vehicle engines are concerned.

The information presented in the papers and summed up in the table reminds us of the many critical factors of various types on which the success of an engine in service depends. Some of these, such as the accommodation and reliability of cooling systems, call for further research and development. Others can readily be dealt with by existing methods provided that the need for action is recognized. Overlooking just one critical factor can be disastrous. Perhaps the most important conclusion to be drawn is that we should take a more comprehensive view of the situation and adopt more thorough and systematic means of dealing with it. As a step in this direction, engine makers might find it helpful to draw up tables on the lines of that used here, covering all the various applications of their particular engines and including all the additional critical factors indicated by their own knowledge and experience. Such tables distributed to all the staff concerned, and perhaps also to users, should help to avoid costly oversights.

It may be that some engine and component makers already deal with this matter comprehensively and systematically. If so, the symposium will have been of interest and value to them as a check on their methods.

We must all have gained a somewhat better appreciation of the extent and importance of critical factors in the application of diesel engines.

Discussion

Mr K. J. Bresser (Shrewsbury)—Paper 2 sets out clearly the work which has to be done by independent engine manufacturers to win, and to keep, that most desirable person, a satisfied customer. It would seem that out of this work the situation will arise where the authors' company will be in a strong position to recommend rationalized packs for all sorts of duties, and this influences customers' choices of ancillary equipment, on the grounds of economics, reliability, and availability.

A somewhat narrower range of options seems also to be likely, which will reflect in the radically different installation paying a relatively higher price.

Paper 7 has defined the authors' views of a probable future road transport engine, and has pointed out the critical factors encountered in trying to meet the requirements of this particular market.

It is interesting that the engine preferred is a six-cylinder unit or an alternative 12-cylinder unit. It appears that engine length is less important to vehicle builders these days, if the recent introduction of a straight eight 240 b.h.p. engine is any indication. One might have expected height to be the drawback of this type of engine, but Swedish vehicle manufacturers have already shown the virtues of high cabs, and made them more popular, so perhaps we may look forward to an even longer life for the straight six vehicle engine.

The lowered compression ratio, aimed at increased reliability, carries the obvious penalties of poor startability, and light load smoke. It is the writer's experience that, whilst starting problems can be mitigated fairly readily, the light load smoke, either white and/or blue, is more difficult to overcome. Palliatives in the form of a modified nozzle and air movement have helped, but there is a paramount demand for increased cylinder temperatures.

As remarked by the authors, variable compression ratio is a solution which commends itself.

The authors' expressed wish for an efficient turbocharger with a pressure ratio capability of 3/1 and turbine inlet temperatures of 800°C is shared, together with the view that as wide a compressor map as possible is needed. Our experience with charge air cooled engines, where the engine has low rotational inertia, is that the turbocharger is forced into surge when the engine is throttled back. Ducting changes and the use of minimum inertia turbochargers have eased this problem, but obviously the higher the operating pressure ratio, the more difficult it becomes 'to obtain' sufficient operating margin to avoid surge.

Two small points raised in the paper are of interest. The use of paper air cleaners is said to result in no reduction of turbocharger compressor output by dirt. I am not sure where the air cleaner is situated on a Volvo vehicle, but in the past there has been experience of high inlet restriction with paper element filters, due to rain-water being drawn into the intake. Has this been the authors' experience?

Regarding the more flexible valve employed to reduce seat wear, would the authors be kind enough to give a little more detail about this. Is the head as a whole more flexible, or are they referring to the junction of the head and stem?

Mr W. R. Dingle, C.Eng., F.I.Mech.E.—I would like to comment on the remarks contained in Paper 4 concerning the cooling system. I agree that engine builders are reluctant to increase coolant temperatures as this tends to increase metal temperatures and may entail more sophisticated cooling methods for the piston and head to maintain a given temperature level, adding to the complication and cost of the engine.

It is now common practice to put the engine oil coolers in series with the engine cooling water, which means that the oil temperature will be just above the water outlet temperature, giving temperatures of approximately 240°F at the authors' proposed cooling water temperatures.

The authors state the coolant system faults already form a major source of trouble in locomotives. I would suggest these troubles could increase if temperatures are raised. I doubt whether the proposed 'sealed for life' system would really contribute, as this would not overcome the major fault, which is stated as being leakage and hose failure.

The authors refer to water washing turbochargers every 2000 h. A short time ago we carried out some tests on our development test beds and found that if maximum turbocharger efficiency was to be maintained it was necessary to water wash at approximately 200 h intervals. If a thick layer is allowed to build up on the diffuser, normal water washing will not remove it, but more frequent washing prevents this build up.

Referring to the points made in Paper 5 concerning the reliability of fuel injection equipment, we find that vibration is the most frequent cause of fuel injector pipe breakage and this in turn will show up any faults in the quality of the connections, etc. It is essential that fuel injection pipes are well clipped and this should be considered in the initial design stage so that fixed positions can be provided for the clips. This principle was applied to our Ventura range where clamps with nylon ferrules hold each individual pipe at fixed positions on the air manifold; see Fig. 5.4.

The incidence of pipe failure when correctly clipped is low. Unfortunately, fuel injector pipes are disturbed during maintenance; the clips are not always refitted correctly and this is where the trouble starts. I think the importance of fuel injector pipe clips is not always appreciated by the operating personnel.

With regard to the reliability of fuel injection equipment, I agree with the author that a fuel pump should last to the major overhaul period of the engine and the injectors should only require periodic maintenance to the nozzle and release pressure. I suggest, however, that the author underestimates the collaboration which exists between the engine builder and fuel injection manufacturer. For instance, a pump and injector we currently have under joint development with a fuel injection equipment supplier has recently completed a 10 000 h rig test at our works under overload conditions, in addition to 8000 h on development engines. Meanwhile, the fuel injection equipment supplier is also running performance and proving tests.

Mr T. A. C. Dulley, M.A., and **Mr B. Wood**, M.A., C.Eng., F.I.Mech.E.—The dynamic response of a supercharged diesel engine must, of course, be worse than that of the same engine unsupercharged. Firstly, because the flywheel effect per megawatt of rating is less, unless the flywheel is altered. Secondly, increase of output level calls for speed-up of the charging gas turbine compressor, which takes time.

Looking back on the whole issue as reported in Paper 11 it appears that the end result was generally acceptable and the difficulties were more apparent than real. Drawers-up of specifications who have no practical experience, in seeking to cover imagined difficulties, may often ask the impossible.

Available remedies from experience of variable load plant of various types are:

(1) For speed control. Initial excursion and time lag in getting back to correct speed inherent in normal governing arises from the speed having to drop in order to inform a speed governor that more fuel is required. This can be avoided by putting the fuel supply directly under control of a kilowatt meter. Transient response can be obtained by directly actuating the fuel rack in terms of kilowatts. The normal flybolt governor arrangement can be retained with its normal droop. Alternatively, more complicated electric governors (called frequency load regulators) with an adjustable speed kilowatts characteristic can be purchased. To obtain parallel running stability, the normal governor droop should be retained and should be the same for all units. In order to correct for the static droop, the speed setting of all governors must be modulated as load changes by means of the speeder gear. Formerly, this was always done by hand even in public supply stations running alone, and is still done by hand in the great majority of large networks. However, it can be done automatically by a frequency regulator to say 0·1 per cent. No frequency regulator is accurate enough to maintain correct frequency time by clocks. Such a frequency regulator will ordinarily pass equal impulses to speeder motors of all governors but can quite readily be arranged to distribute load in any desired proportions.

(2) Voltage control. The author's estimate of voltage drop from the generator transient reactance is valid only at about zero power factor which is hardly a likely condition. At unity power factor the reactance drop from no-load to full load with 15 per cent reactance will be

$$\sqrt{100^2+15^2}-100$$
$$=\sqrt{10\ 225}-100 \approx 101 \cdot 1-100 \approx 1 \cdot 1 \text{ per cent}$$

There will be a further drop of the order of 1 per cent due to resistance. Hence it would appear that voltage regulation difficulty has been grossly exaggerated. However good the voltage regulator (and very good ones have been available for 50 years or more) the overall response is largely dictated by the exciter. Stable high speed response can be obtained by a separately excited exciter using a constant voltage source which means a pilot exciter or battery excitation. It is unlikely that distinctly better results are to be obtained with static excitation arrangements since these may be faced with the difficulty of producing increasing amperes at the moment when their supply voltage is falling. A possibility is to use the same technique as mentioned above on speed control, namely to inform the voltage regulator of the increased kVA demand a stage earlier, rather than having it wait until the voltage has fallen before it can respond, even though admittedly in this case nothing can be done about the instantaneous reactive drop. Some crude attempts in the past to do this by compounding have failed because transformer coupling between exciter windings simply led to flux remaining constant when the series winding passed a higher load current.

Mr J. G. Fowler, C.Eng., M.I.Mech.E.—In Paper 15 Mr Law states that a typical engine test to establish fatigue levels on piston rings can be made on an accelerated basis by increasing the top ring to groove side clearance and advancing injection. We agree that this type of test is costly and can better be performed on a rig. However, could the author give any recommendations for an accelerated engine test that would permit blow by, oil consumption, groove/ring wear and compatibility etc. to be assessed? Such an acceptance test should be representative of an engine life cycle and of shorter duration than the 1000 h endurance test generally used at present.

From the paper it would appear that the wedge type top ring is just a palliative for ring stick and should by now be somewhat dated. With the more sophisticated materials and techniques available now, what design does the author recommend to replace the wedge ring, assuming no improved temperature condition?

With reference to Fig. 15.15 on piston ring wear tests, perhaps the author could give a correlation between cast iron bearer wear and that of aluminium alloy, say SAE 332.

Would the author care to give maximum piston operating conditions for a tractor diesel engine that would preclude fitment of a top ring groove austenitic iron insert and still give competitive life?

Mr C. C. J. French, M.Sc.(Eng.), C.Eng., F.I.Mech.E.—The response of turbocharged diesel engines under transient conditions has been mentioned in several papers including those by Holmér and Häggh (Paper 7) and by Mackenzie (Paper 11). While black smoke under increasing load conditions can be suppressed by employing a boost responsive maximum fuel stop, there are some operating conditions, e.g. tug boat manoeuvring, or excavators, where the engine will stall, since the torque on the drive exceeds that which can be supplied by the engine which is at this time effectively unboosted. Further study should be given to ways of quickly running up the turbocharger or to the provision of external supplies of air for combustion or supplying energy to the system, since this problem could be a bar to further increases in turbocharging of diesel engines in certain applications.

Paper 1 by Eke, Walker and Williams advocates a wider use of natural gas as an engine fuel. The Americans are beginning to be concerned about natural gas supplies by 1980 and beyond and are talking of going over to coal gas in the long term! Can the authors confirm that natural gas supplies in Britain over the next 20–30 years are going to be such that we can consider gas as an alternative to liquid fuels for engines.

I would certainly agree with Mr Petrook and Cdr Stewart (Paper 4), and with Mr Millar (Paper 16) on the importance of load cycling for endurance testing of thermally stressed components. Continuous full load must also be used, however, where it will exist in service, since with component wear and lubricating oil life for example, continuous full load would be expected to be a more severe duty.

Operating at higher water temperature with pressurized systems would give no increase in thermal stresses and could in fact lead to a reduction of thermal stress in the centre of cylinder heads since, with nucleate boiling at least partially established, the temperature of the metal in this region would rise less than that of the cooler outer areas of restraining metal. The increase in liner and piston ring groove temperature must be held down by improvements in cooling. It might, in fact, be possible to use a dual circuit, allowing the cylinder head coolant temperature to rise but holding down that of the cylinder jacket coolant, although care would have to be taken that the relative expansion of head and block did not accentuate gasket troubles.

Since in locomotives the provision of adequate radiator area is a major problem it would be worth considering the use of insulating liners in the exhaust manifold. Our tests have indicated that something like half the heat from the cylinder head to the coolant is from the exhaust bend and a worthwhile reduction in radiator requirement should therefore be possible by fitting such liners.

It is interesting that there has been so much emphasis on leaks in cooling systems. We might expect troubles with internal combustion engines as we have had only about 100 years of development but plumbers and their problems have been with us for 3000–4000 years! Surely we can do a lot better than we do, even if additional expense is involved.

Mr Sampson has described with great conviction in Paper 5 the advantages of the oil cushioned piston, not the least of which should be a considerable extension in lubricating oil life due to reduction in gas blow-by. I should be interested in any information which he may be able to give as to the actual improvement in such life and also to know why the Ministry of Defence does not specify this panacea for all evils for all their engines!

The tests on the effect on fatigue strength of thick chrome platings are suspect due to deposition faults. It would obviously be of considerable value if these tests could be repeated. The cheap torsional vibration tester sounds as though it should have wider use. Would it be possible for us to have more details?

Other speakers have talked about the difficulties of ensuring that fuel pump pipe clamps are replaced in service. From Mr Sampson's description of pipe failures close to the nipples it sounds as though these are of the solid forged variety. Would it not be better to use brazed-on nipples?

Paper 10 by Howells and Walker gives some interesting data on what is becoming, and will become, an even more vital topic—namely gaseous emissions from diesel engines. California will have legislation by 1973 and the rest of America probably in 1974 or 1975 and, since we live in an era of environmental control, it must surely only be a matter of time before we have similar legislation in Europe.

I am afraid that I do not really agree with the authors that the diesel engine is only slightly better than the gasoline engine for CO and HC emissions. Even with mass emissions over a load–speed cycle, we can obtain with the diesel engine values which the gasoline engine needs a further 10/1 reduction to attain, even though the present day gasoline engines are already down by a considerable factor from what they were five to ten years ago.

The key emission for the diesel engine is of course oxides of nitrogen and we are faced with the prospect of legislative limits which may be very difficult to attain. The indirect injection engine gives values which are about one half or one third of the direct injection engine but we are likely to need very substantial further reductions to meet the values which have been suggested by some authorities as possible limits for America in 1975–1980.

Two possible methods of approach are by the use of cooled exhaust gas recirculation or by water injection into the charge air. Fig. D1 shows the results of such tests on a swirl chamber diesel engine. Some initial reduction in NO_x emissions is obtained by retarding the injection

162   DISCUSSION

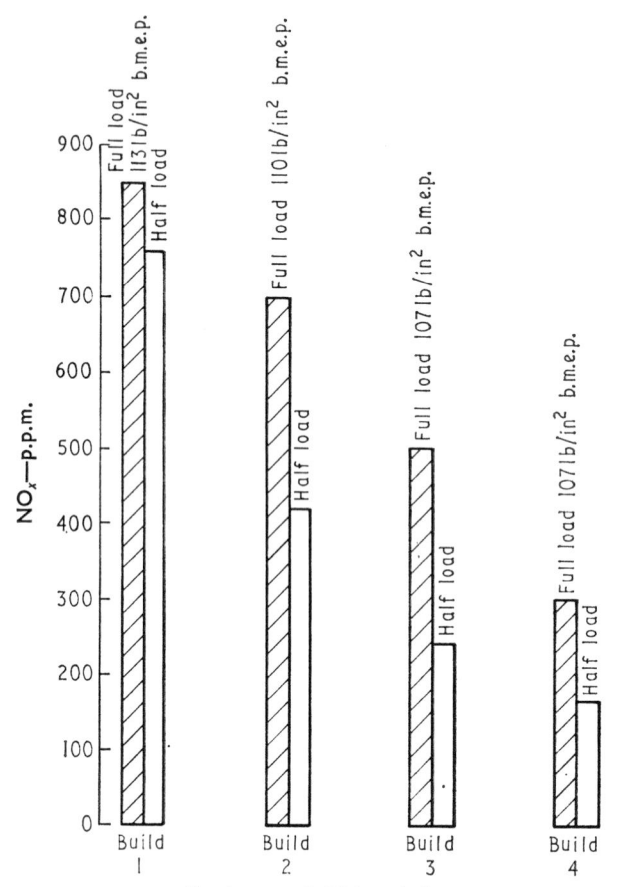

Build 1. Optimum injection.
Build 2. 4° retard injection.
Build 3. 15 per cent by volume cold exhaust recirculation, 4° retard injection
Build 4. 2/1 water/fuel injection into inlet, 4° retard injection

Fig. D1. Emissions of oxides of nitrogen from a 4·75 × 5·5 in Comet Mk V swirl chamber engine, illustrating effect of injection timing, exhaust recirculation and water injection into inlet manifold

timing as shown, with further improvements from recirculation or water injection.

Even with all this, however, the NO_x levels are such that we believe that they may exceed by a factor of two or three possible American limits for 1980. Clearly much more work is necessary.

Mr J. V. Garside (Dursley, Glos.)—Referring to Paper 6, it would be of interest to know the extent of stiffening to the engine seatings and if the discontinuity of the girders was corrected on the ships under discussion; this point is not made clear in the paper.

The fact that the port seating has a different deflection from the starboard, and the difference between inboard and outboard longitudinals from the port engine of the *Manchester Progress* are facts that I am sure the shipbuilding industry will find somewhat perturbing, particularly as twin input–single output installations are gaining favour in the smaller, specialized unclassed type of vessel where there is a tendency to reduce the scantlings to a minimum, and yet the powers specified are ever on the increase.

All the rules of procedure that have been proposed by the author when installing medium speed propulsion engines should be rigorously carried out and, I stress again, particularly on the smaller class of vessels.

In conclusion, the results of this investigation which the author has placed before us, I am sure will prove of value in the future for those concerned with engine seatings for medium-speed propulsion engines, so that one may be wise before the event.

Mr A. J. Glasspoole, B.Sc.(Eng.), M.A., C.Eng., F.I.Mech.E.—I found myself in general agreement with the majority of the views expressed in Papers 4, 5 and 16 and feel that these will be of particular value to engine builders.

Despite the fact that two are in the defence field and only one is commercial, I made an attempt to correlate these three papers. The only single common problem area related to cooling systems. There were a number of problems common to two groups such as noise, rating, overhauls and maintenance, reliability and lubricating oil, but again only one of these was common between the two defence groups.

Both British Rail (Paper 4) and the Navy (Paper 5) indicated a need to relate maintenance periods to the type of service and I take it that the authors from British Rail are suggesting using either a measure of kilowatt hours or total fuel consumed as a part of the criterion. To my mind these are synonymous but, as written, the paper suggests they have in mind the use of both.

British Rail asks whether failures of 'tried and tested' diesel engine components at conditions 'slightly different' from those giving satisfactory reliability are 'due to changes in manufacturing techniques, insufficient inspection or just a case of engine manufacturers resting on their laurels'. In connection with the serious aspects of the above I would include other factors in the list to cover the difficulties facing an engine builder in having to develop a product to meet an almost infinite variety of uses and conditions. To test under all of these is too expensive, even if they were all known precisely beforehand. While the severest may be chosen in good faith there can be insufficient knowledge to ensure that it is in fact the severest and also there may be different severe conditions for different components. Another idea to consider is that, to allow progress to be made in reducing areas of dissatisfaction with the product, the user should conduct a discussion with the manufacturer with a view to specifying in precise probability terms the degree of reliability he requires. The manufacturer can then give an indication of the effect on cost. Once a figure for reliability has been agreed the manufacturer can allow for this in his product, by, for instance, allowing for variations that can occur in material properties etc. For this process to produce a useful outcome much more data will be required than is currently available.

Mr Sampson for the Navy (Paper 5) suggests a joint programme of research between engine builders and fuel injection manufacturers aimed at improving the reliability of fuel injection equipment. This raises the problems of this symposium by substituting 'fuel injection equipment' for 'diesel engines' in the theme. To my way of thinking it is detailed specific design defects that can cause troubles and not an absence of fundamental knowledge. To avoid this sort of failure, we have tended to develop and endurance test the fuel injection system on rigs away from the engine and under severe conditions in order that delays in the engine development programme do not delay the development of the fuel injection system. In this way many more hours can be obtained on the fuel injection equipment and it can complement the results of engine testing. I think it might be worth the author elaborating further on any specific research items which he has in mind.

Mr Millar of the Army (Paper 16) mentions four principal factors seldom within a manufacturer's control which can affect rating. I would like to point out that the manufacturer can quite often cater for these factors by carrying out his engine development work under the most adverse conditions likely to be encountered in service. For instance, one can test at high charge air temperatures out of the intercooler, using a low grade lubricant and fuel, a higher overload than is anticipated and load cycling tests by rapid stopping and starting. As many auxiliaries as possible should be fitted during these tests.

I trust the authors will forgive me for not commenting in more detail on many of the interesting points raised in their papers but I hope that other contributors will make up for my omissions.

The authors of Paper 10 state that 'the relatively slow rate of combustion of residual fuel presents formidable problems for the engine designer who must ensure that combustion goes to completion in order to minimize the deposition of carbon'.

The overall objective of the designer of an engine which has to burn heavy fuel is to keep all economic and social factors the same as for a distillate fuel in order that the user can obtain the full benefit of the lower cost of the residual fuel. That is to say that the same overhaul periods, component life, fuel consumption, exhaust clarity and ability to run at part loads etc. are required. These certainly present an engine manufacturer and user with problems but I am not clear that these are due to a great extent to the relatively slow rate of combustion of the fuel. Exhaust valve life, wear rates, and maintenance of the fuel injection system in top condition can become critical, but the problems are due to other critical properties of the fuel.

In my experience those parameters such as thermal efficiency and exhaust clarity, that might have been expected to show problems due to the slow speed of combustion of residual fuels preventing combustion going to completion, are solved by the normal methods used for distillate fuel, although the avoidance of secondary injections does become more vital. When a serious development programme of work on heavy fuel was started over 10 years ago at Lincoln there was some concern that the slow combustion properties of the fuel would raise serious problems. Initially, using a $12\frac{1}{2}$ in bore engine at 160 lb/in^2 b.m.e.p. and 500 rev/min, these did not materialize and testing has now proceeded as far as an 8 in bore engine at 1000 rev/min at 250 lb/in^2 b.m.e.p. without obvious trouble from the speed of combustion point of view. 1500 seconds fuel has chiefly been used in this work with occasional tests with 3500 s, 950 s or 200 s fuel.

There is some uncertainty as to whether there will be a lower limit to bore size or upper limit to engine speed with regard to the possible combustion of heavy fuel, and any views the authors have on this would be welcome.

It would also be appreciated if the authors could indicate the sort of problems the designer is faced with in their experience due to the slow rate of combustion of heavy fuel, and if in their views these problems are ones that are being overcome using existing knowledge or whether more data are required. Some sort of specification of the heavy fuel causing these problems may be necessary.

The authors raise the subject of exhaust valve life and I suspect they have deliberately over-simplified the mechanisms of failure which can limit this. At Lincoln at least three separate mechanisms of failure are identified. More important, however, no indication is given of the critical constituents in the fuel which can affect exhaust valve life and it would be appreciated if the authors could summarize these.

It should be recorded that despite co-operation from oil companies, engine development in the manufacturers' works can be limited to an extent when heavy fuel is considered because of difficulty of obtaining fuels of a selected composition in Britain. This is sometimes desired by the manufacturers to allow them to carry out development testing using a worse fuel than is likely to be encountered in worldwide service.

Referring now to Paper 15, I have always found Mr Law's company most helpful in dealing with problems and would be grateful for his comments on the following points of his paper.

Although a section of the paper is headed 'Piston ring groove temperatures' the text appears to deal with piston rings and it is stated that these 'must operate at a temperature of between 220° and 240°C'. I would like confirmation that the author intended these temperatures to apply to the ring groove. While I think a figure of 230°C for the maximum operating temperature at the centre of the back of the top ring groove is accepted for medium speed engines, although manufacturers try to operate well below this figure, I am aware of higher temperatures being permissible on small high speed engines for the same quality of oil and would be pleased to learn of any explanation that can be given for this.

I feel that the author's statement 'the goal of any piston ring manufacturer' should be extended to include the requirement that the ring must not wear the liner or piston so as to cause the life of these components to be less than the life of the engine.

Confining my remarks to prototype engines under development, I would agree with the author's statement that scuff can in certain circumstances be independent of lubricating oil consumption and would also suggest that this can apply to wear as well. The author says that values of oil consumption 'lower than 0·25 per cent can give rise to excessive ring–liner wear' but I would also feel that, equally, excessive ring–liner wear can occur at higher oil consumption values and consequently it seems that oil consumption in itself is not a basic factor affecting wear.

I have been aware of certain installations running at less than 0·25 per cent oil consumption with low wear rates also, which seems to provide a challenge to engine design and development engineers to attain these figures consistently. Naturally the assistance of firms like the author's would be required and also there would have to be a market requirement to justify the work. I would be interested in the author's views of the possibility and desirability of providing ring arrangements which can consistently run at oil consumption values of between 0 and 0·1 per cent.

I was disappointed that Mr Law did not feel able to make specific recommendations in his paper with regard to the limits to be set on the nine specific factors affecting lubricating oil consumption. I would welcome a statement from him of at least the ideal arrangement that he would like to see adopted by medium speed engine manufacturers using currently available knowledge for new engines now being designed. Even a list of basic 'do's' and 'dont's' would be helpful.

One such critical principle that seems very important to me but is rarely put on paper is that the major and final oil control should be carried out by the top piston ring. This can then ensure that adequate oil is present up to the top ring to result in low wear rates and scuff-free running provided other conditions are acceptable.

Mr M. T. Hall (Associate)—I would like to add to, and comment on, Paper 13 by Mr Lunnon and my comments to some extent bear on Paper 10 by Mr Howells and Mr Walker. These papers are most interesting and a valuable contribution to the literature of this particular aspect of diesel engine operation. Whilst by no means wishing to belittle toxic risks, in my experience when one is asked to investigate a pollution problem arising from the use of a diesel engine, the answer often concerns objectionable materials or conditions which are not necessarily of a toxic nature, but are annoying or disagreeable in some way. They may, however, constitute a safety risk, although not always from an aspect of toxicity.

One would not at the present time completely ban the use of diesel engines because of their exhaust emissions. However, it is essential that adequate measures to control their use are provided, and as a guide to industrial hygiene 'threshold limit values' (I) may be used. These refer to airborne concentrations of substances and represent conditions under which it is believed that nearly all workers may be repeatedly exposed day after day without adverse effect. Threshold limit values refer to time-weighted concentrations for a 7 or 8 h working day and 40 h working week. They are only guide limits and should not be used as fine lines between safe and dangerous conditions, except in the case of ceiling values (designated with a 'C') which should not be exceeded at any time.

In connection with the use of diesel engines the following table of threshold limit values is of interest:

Substance	Threshold Limit Value (TLV)	
	p.p.m.	mg/m³
Carbon monoxide	50	55
Carbon dioxide	5000	9000
Sulphur dioxide	5	13
Nitric oxide	25	30
'C' Nitrogen dioxide	5	9
'C' Formaldehyde	5	6
Oil mist (particulate)	—	5
Oil vapours	depends upon content of aromatics and additives.	
Carbon black	—	3·5
'Inert' or nuisance particulates	—	15 (or 50 m.p.p.c.f.)

(1) p.p.m.—parts of vapour or gas per million parts of contaminated air by volume at 25°C and 760 mm pressure.
(2) mg/m³—approximate milligrammes of particulate per cubic metre of air.
(3) m.p.p.c.f.—millions of particles per cubic foot of air.

On railways one may be faced with toxic risks from diesel engines running in the confines of a locomotive maintenance depot, but as far as I am aware the only case akin to mining operations investigated in any detail on British Rail is that concerning the operation of the Severn Tunnel. The tunnel has a length of 4¼ miles and is force ventilated by a single fan which delivers air into the tunnel near to the half-way point.

Tests carried out to establish the pollution rise in the absence of fan ventilation but with normal train service have shown that there is little general toxic risk in the short term. The nitrogen dioxide levels did not exceed 30 per cent of the TLV, averaging 1 per cent of this. Likewise the total nitrogen oxides did not exceed 38 per cent of the TLV, averaging 16 per cent; carbon monoxide was similar. There was a significant degree of ventilation due to the piston effect of passing trains. Dust levels were about 2–3 mg/m³. This result was comparable with similar values obtained by Katz et al. in the St Claire tunnel in America—also a submerged rail tunnel, unventilated (2).

On the papers themselves, what does Mr Lunnon mean by low or medium levels of oxides of nitrogen bearing in mind the TLV for nitrogen dioxide is 9 mg/m³—a 'C' value? The author considers that any error in the analytical techniques due to oxidizing all the nitrogen oxides to NO_2 is not significant because of behaviour of the gases in the lungs. If this is so, why then is an oxides of nitrogen strength of 27 mg/m³ (15 p.p.m.) regarded as being just acceptable when it is three times the TLV for NO_2, a value which from hygiene standards, should not be exceeded at any time? One of the insidious hazards of NO_2 is that one could get a fatal dose and not feel any ill effects until hours later, by which time it could be too late, particularly if the symptoms were not recognized (3).

The analytical difficulties recorded are noted and to some extent confirmed by our experience in the field. One of the American defence laboratories has produced a nitrogen dioxide dosimeter (4) which is very similar to the radiation dosimeter (or film badge) used for radio-active substances. Have the authors any experience of these? I am reminded of a comment by a guide at an atomic energy establishment about the radiation film badges, 'They won't protect you if anything goes wrong, but at least we will know what killed you!'

There are references to aldehydes in both papers. Have the authors any further information on aldehyde nature, as well as the concentrations which have been referred to in diesel exhausts in the field from a ventilation aspect, bearing in mind that the TLV figure for formaldehyde is 5 mg/m³—a 'C' value? The limits for other aldehydes, however, may be higher.

According to Mr Lunnon it would appear that local concentrations of exhausts are considered unlikely. However, a serious situation arose during tests with a locomotive, although powered by a gas turbine, in a tunnel due to exhaust staying with the locomotive and the first vehicle.

When a diesel engine is in a good state of repair the only toxic risks should be those from its exhaust. However, risks can arise due to the poor condition of various components. This may result in significant amounts of oil mist and oil vapour being formed, and it should not be overlooked that the TLV for oil mist is 5 mg/m³, whilst that for the vapour depends upon the nature and content of aromatic constituents and additives.

Finally it should be borne in mind that when two or more hazardous substances are present, their combined effect, rather than that of either individually, should be given primary consideration. In the absence of information to the contrary, the effects of different hazards should be considered as additive. That is, if the sum of the series $C1/T1 + C2/T2 + \cdots + Cn/Tn > 1$ then the threshold limit of the mixture should be considered as being exceeded. $C1$ indicates the measured atmospheric concentration and $T1$ the respective threshold limit value.

REFERENCES

(1) 'Threshold limit values for 1969', Technical Data Note 2/69 (H.M. Factory Inspectorate).

(2) KATZ et al. 'Air pollution hazards from diesel locomotive traffic in a railway tunnel', *A.M.A. Arch. Ind. Health* 1959 **20**, 493.

(3) 'Methods for the detection of toxic substances in air, No. 5 Nitrous fumes', 1939, D.S.I.R. (H.M. Factory Inspectorate).

(4) 'Development of a colorimetric personal dosimeter for nitrogen dioxide', USAF Aerospace Medical Research Laboratory Report AMRL-TR-68-104.

Mr S. H. Henshall, B.Sc.(Eng.), C.Eng., M.I.Mech.E. —Marine engineers have always been aware that the machinery seatings provided in ships distort and flex for various reasons and the case which Mr Castle has closely investigated in Paper 6 throws an interesting light on some of the causes of this distortion and the effects that they can have. In spite of the thoroughness of the investigation, the different shapes taken up by the seatings for three engines, the starboard and port engines of one vessel and the port engine of a sister vessel, are largely unexplained.

In comparison with large bore crosshead engines, the medium speed engine is short and on this account should suffer less from flexing of the hull, but on the other hand the crankshaft of the medium speed engine is very much stiffer, the high output designs frequently having considerable overlap of pins and journals. The stiff crank is more vulnerable to the distortion effects when they arise and, as the paper has shown, advantage must be taken of the shortness of the engine and other components to avoid straining the stiff crankshaft unduly. It would seem highly desirable to provide flexibility in shafting systems; between engine and gearbox, for instance, so that the components of the installation can be arranged to give flexibility of the whole with comparative rigidity of the parts, although it must be admitted that this would not have prevented the major part of the distortions in the case quoted.

In view of the likely development of engines to even higher outputs with crankshafts of greater stiffness in proportion to their length, it would seem desirable that future generations of medium speed engines should be designed with more rigid frames capable of supporting their crankshafts in accurate alignment, whilst in turn being supported themselves at the minimum number of points, preferably four points only or perhaps even three, so as to avoid all strains arising from hull deformation.

Dr W. P. Mansfield, B.Sc.(Eng.), C.Eng., F.I.Mech.E. —Paper 3 by Mr Lowe is well suited to the symposium theme. The pressure and temperature of the air the engine breathes and the temperature of the available cooling water are obviously important critical factors in the application of diesel engines, and the paper indicates the nature and extent of their effects.

The paper is largely concerned with the provision of standard methods of derating engines in accordance with these ambient conditions, and it is on this aspect that I would like to comment.

I have been involved from time to time in discussions on this problem of derating turbocharged engines ever since the BICERI pressure charging panel was asked by the B.S.I. to make recommendations for the revision of B.S. 649. That was many years ago: the problem is of long standing. The demand is for a simple yet generally applicable formula which will enable the engine user to check that the derating proposed by the engine maker is adequate.

I have always held that this requirement cannot be met. Mr Lowe has made clear some of the main reasons why it cannot. His suggested solution of the problem is to use a simple empirical formula of the form of equation (3.12) in the paper to derate to the extent necessary to maintain a constant trapped-air/fuel ratio, but to apply this formula only up to limiting values of air pressure and temperature and coolant temperature specified by the engine maker, who also is to specify the steeper derating to be applied thereafter.

This seems to me very reasonable, and probably the best that can be done. However, the method does not appear to give the user very much assurance that the proposed derating is adequate, since the user is entirely dependent on the maker's assertions as to the appropriate steep derating and the point at which it should begin to apply, and also on the maker's word that the turbocharger rematching carried out does give a result up to this point in accordance with the standard formula.

If the maker has all the information necessary to apply the proposed method, i.e. he has curves such as those in Figs 3.2 and 3.3, he could show the doubtful user these curves with the proposed rating marked on them. This would seem to be more realistic and convincing than the use of the proposed standard formula.

A further indication that equation (3.12) is of somewhat limited value is provided by Figs 3.2a, b and c and 3.3a and b, which are discussed in relation to a turbine inlet temperature of 980°F. All these data are presumably fairly representative. Taking the five sets of curves in order, the greatest derating applied according to equation (3.12) would be, with variable turbocharger match, 0·8 per cent for air temperature, 0·5 per cent for pressure and 0·8 per cent for coolant temperature, and, with fixed turbocharger match, 1·1 per cent for temperature and 0·7 per cent for pressure. Earlier in the paper, in connection with the effect of mechanical efficiency, it is implied that a difference of 1 per cent in rating is negligible. Hence, when a substantial derating of the engine in this example is necessary, the proposed standard formula provides a check of only a small part of it. Of course, for other engines, equation (3.12) may require greater values of the indices so that the standard values agreed might be somewhat greater.

The method may be the best possible, even though in many cases its use might be almost equivalent to simply accepting the maker's recommendation. Since there appears still to be a widespread belief that some sort of standard formula is essential, the proposed method may serve the useful purpose of satisfying those who hold this belief, thus saving still further expenditure of effort in this direction.

Mr W. H. Sampson, C.Eng., M.I.Mech.E.—The information gathered during the investigation described by the author of Paper 6 is of great interest to the marine engineer. The Royal Navy has had problems with the alignment of items of machinery and shafting and experience has shown that with the continuous type of support illustrated in Fig. 6.2a poor workmanship during the initial lining-up can result in distortion of a piece of equipment and this can be worsened by a change in ship loading and temperature variations in the hull. For these reasons the Royal Navy prefers, when mounting a large item directly on to a seating, to use the two or three point method of support. The PC2 engine is sufficiently stiff to be mounted on the type of mounting shown in Fig. 5.1 and I consider it a mistake to install it on a seating 'sufficiently stiff to accept the hogging of the engine due to thermal expansion' (rule 5 in Paper 6). Is the author certain that the differences shown in Fig. 6.4 between the engine hot and engine cold conditions are due to the engine? The Royal Navy has conducted a number of tests on resiliently mounted engines including one with a generating set running at 1200 rev/min, on load with 50 per cent of the mountings removed from beneath the centre half of the set, and in no case found evidence of main bearing clearances being absorbed due to an engine bending because of temperature or its weight.

The failure in the starboard engine of the *Manchester Port* was due to poor installation. It is our practice to take crankweb deflections on all new installations and after every 1000 hours' running. Was this done in the *Manchester Port*?

Finally, the first rule of procedure infers that when an engine is cold its main bearing housings are deliberately set out of line. This, if true, must be difficult to assess in the design stage as the movement of an engine after installation will be dependent on the type of ship.

Regarding the author's remarks under 'Application classification' in Paper 9, it is Royal Navy practice with the smaller standard engines, e.g. Perkins 6·354, to use the same combined fuel pump and governor for all applications of the engine, i.e. propulsion, generator and general auxiliary duties. Does the author consider this to be good practice?

On the question of rating engines, a limiting rating is set for each standard engine and this is set by adjusting the fuel stop when the engine is carrying out its acceptance tests. However, it is possible that a fuel pump has to be changed in service and the fuel stop of the replacement pump would be set to the setting of the stop on the defective pump. It would appear from the author's comments under 'Production results' that this could mean a difference in output between the two engines.

Referring now to Paper 17 the Royal Navy has in use a large number of high spring injectors as shown in Fig. 17.2a and has had a number of spring failures. It is our practice to carry out maintenance routines in accordance with a schedule and this means that the injector in question is replaced every 500 hours by a reconditioned one. From his work, is it possible for the author to estimate the life of a spring under the conditions described, as the Royal Navy would prefer to replace a spring during a specified injector change rather than let it fail and then replace it?

Mr T. E. Scott, C.Eng., M.I.Mech.E.—The authors of Paper 4 state that the total diesel engine weight for a six-axle locomotive should not exceed 35 000lb, and later in the paper say that the cooling equipment for the charge air of an intercooled engine is as big, and presumably as heavy, as that for the engine cylinder jacket, turbo-blower jacket and lubricating oil cooling system itself.

In view of this I would like to ask if any design studies have been carried out to see if charge air cooled engines are in fact the most suitable type of engine for main line locomotive use?

Would it not be better to lay down a total weight and possibly volume for the complete engine and cooling system including all the pipe work and the water?

It would seem that a slightly larger engine (in terms of swept volume), with a relatively low pressure turbo-blower without charge air cooling, might be more suitable and also give greater reliability. This type of installation would possibly be more attractive if the cooling water was slightly pressurized and could possibly run at similar piston and lubricating oil temperatures to those of a

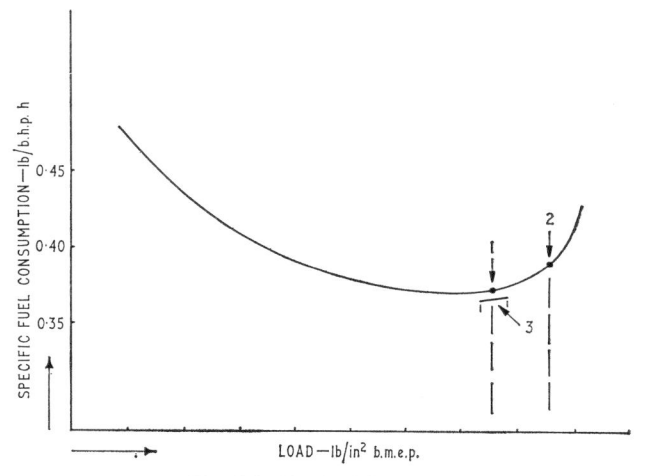

Fig. D2. Typical fuel loop

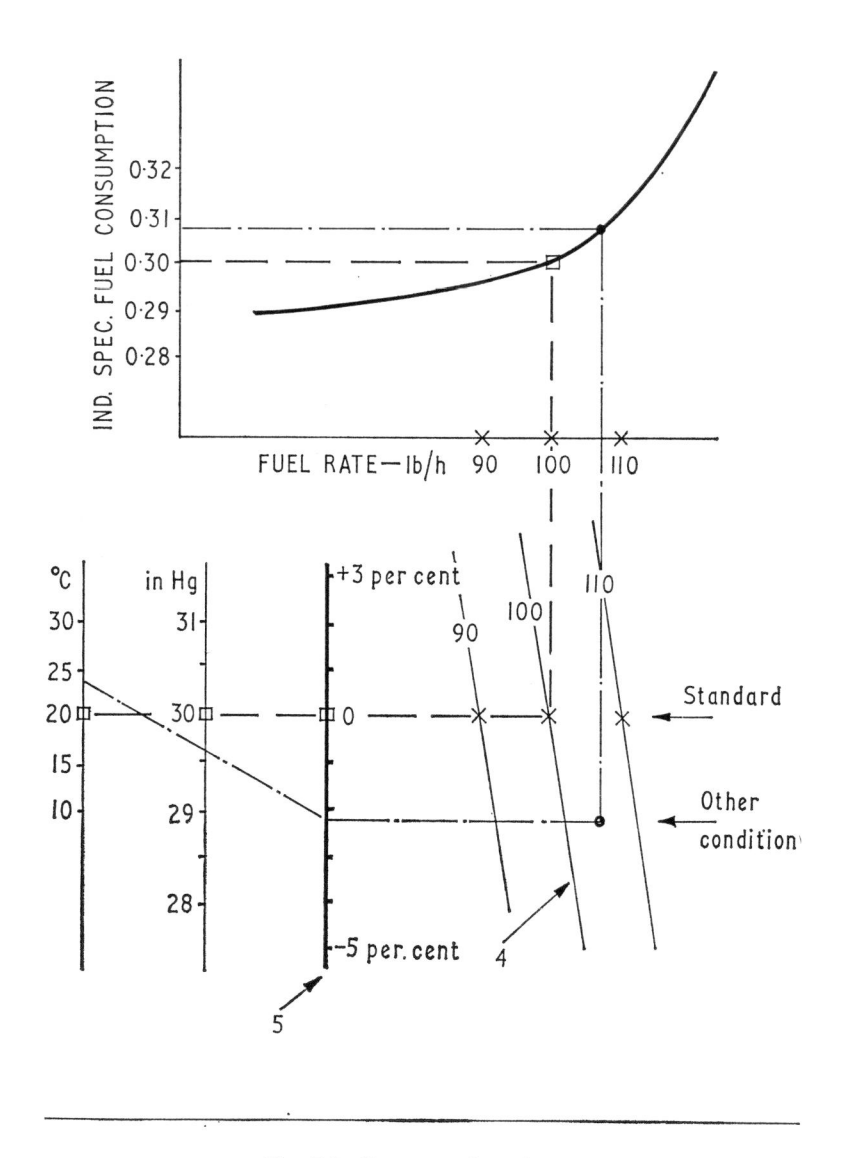

Fig. D3. Construction chart

charge air cooled engine with a non-pressurized cooling water system.

The authors have confined their remarks mainly to high power locomotives. I would be grateful to have their comments on diesel engines for shunting duties. In this application, with even more engine idling, is the turbo-blower friend or foe?

With reference to the rating and use of traction diesel engines, whereas in the early days of conversion to diesel on British Rail the driver's controller had notches and the engine speed control was driven by an electric motor, which took a few seconds to run up or down, nowadays we use a notchless air control and a driver can, if he so wishes, open up or shut down an engine very quickly, as shown in Fig. 4.4 of the paper. Is this detrimental to engine life, bearing in mind the fact that the cooling water pump is engine driven, and is there a case for having a separate electrically driven water pump as well as the normal one?

Mr D. W. Tryhorn, B.Sc., C.Eng., F.I.Mech.E.—Mr Millar has referred in Paper 16 to the problems of choosing a power rating for an engine which spends practically its whole working life below maximum power but frequently uses peak torque at the part speeds. This leads to an engine with a fuel setting set for high torque and which is therefore capable of a high top power at top speed. By high top power I mean running closer to the smoke limit than normal commercial engines.

This high top power must be checked on engines coming off production and the problem has been that, because they are running close to the limit, the actual brake power is very sensitive to the test cell ambient conditions and the particular full rack fuel pump setting.

Fig. D2 shows a normal diesel engine fuel loop on a load or b.m.e.p. base. A normal industrial rating would be at 1 whereas a high output engine is rated at 2. The important difference is that the gradient of the curve is relatively constant at 1 but changing rapidly at 2. The

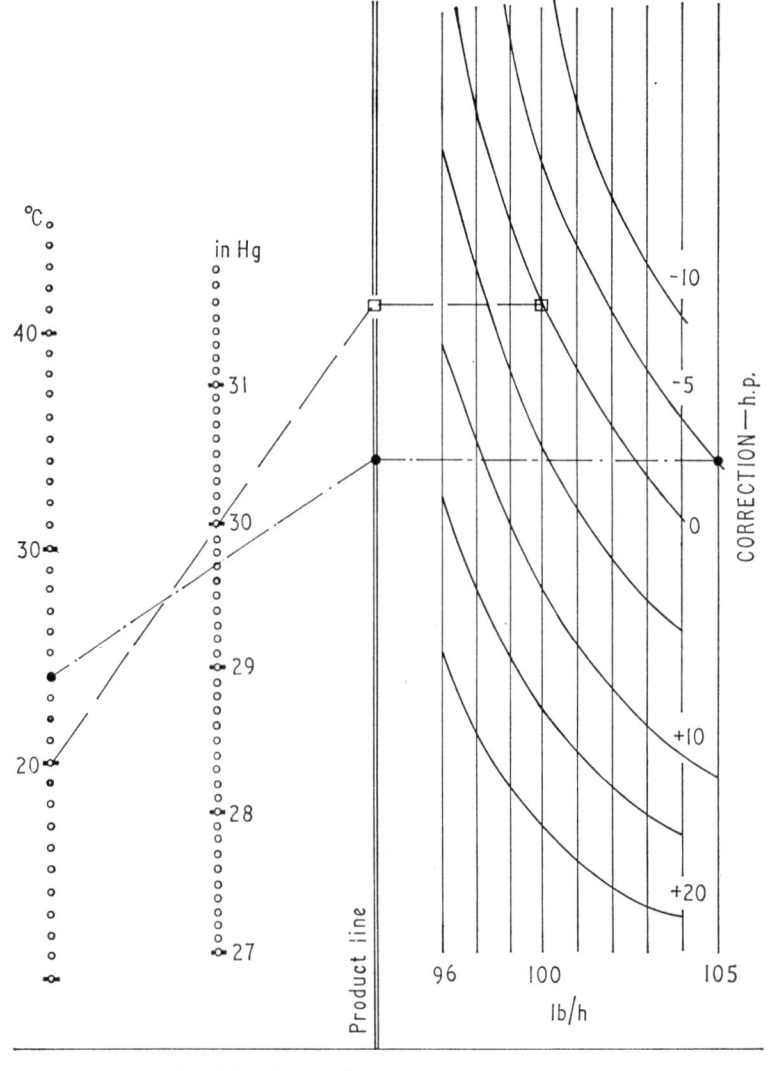

Fig. D4. Correction curves, not to scale

effect of changing ambient conditions is to shrink or expand the curve in length relative to points 1 and 2, low air density moving the turn up of the curve closer to 1, as shown in practice by the increase in fuel consumption, exhaust temperature and smoke. If at 1 the straight line 3 represents the range of ambient conditions, then because the inclination can be parallel to the fuel loop the normal correction factors of x per cent/degree used and y per cent/inHg, can be used. At 2, however, the curve is too far from linear for such factors to be accurate, and the fuel loop does not have to shrink far before there is a big drop in power accompanied by a voluminous black smoke.

The loop describes the engine but it is difficult to use it for rating purposes. So what we do is to replot it in the form of fuel consumption in terms of lb/i.h.p. h against the fuelling rate in lb/h, as shown in Fig. D3. The i.s.f.c. depends upon the fuel to air ratio, so it changes the same no matter whether it is the fuel increased or the air decreased. Inclined lines, 4, can therefore be drawn which define either a percentage change in fuelling rate or in air trapped in the cylinder. This percentage change scale is shown as 5 and used for the air density changes. To the left of it, scales of temperature and pressure project across to the line to give the air density change from the given standard, shown as 20°C and 30 inHg for an engine of 250 b.h.p. and 83 f.h.p. This chart describes the whole of the engine performance at one speed, and the example condition shown is for 24°C, 29·7 inHg projecting across to a fuelling rate of 105 lb/h. Then up to the i.s.f.c. curve and across to read off the i.s.f.c. as 0·307, so giving the output as 342 i.h.p. or 255 b.h.p. This means that the standard engine would have been 5 hp up in output, and 5 hp must be subtracted from the power of the test engine to obtain its output at standard conditions.

For simplification the data from this chart is then used to produce correction curves, as shown in Fig. D4 which shows the same example and the standard condition as a spot and square respectively, and gives the power correction directly. At this stage the scales are drawn to cover the expected range of test cell conditions, but there is no reason why they could not be extended to permit prediction of altitude performance, and curves added to give fuel consumption and smoke figures.

Only when these correction curves turn out to be parallel straight lines can simple correction factors, as say the B.S. factors, be used with accuracy.

Mr R. A. Vaughan, C.Eng., F.I.Mech.E.—The principal outlet today for base load diesel engines is overseas and the information given in Paper 14 will give the potential user confidence in this form of prime mover.

The author has referred to the Working Cost Reports published by the Diesel Engineers and Users Association and from this source I note that these engines have averaged 3800 hours per annum over the nine years to 1968. Although the annual load factor has only been between 20 and 30 per cent, on many occasions the maximum demand has equalled the installed capacity and the running plant load factor has varied from 87 to 95 per cent. To run at 95 per cent load factor for 5000 hours as was done in 1962 reflects great credit on the C.E.G.B. staff at Ashford and the engine maker.

Turning to Table 14.1 listing the ten most economical stations in the United Kingdom I have re-calculated these total running costs on the basis of a fuel price at all stations equal to that paid at Ashford. I have done this because, although the operator cannot get away from the hard facts of Table 14.1, yet the cost of the fuel is largely fortuitous according to the geographical location of the power station. Bringing these figures back to a uniform basis (omitting Kirkwall and Haverfordwest because they use a Class A fuel) the order of merit changes somewhat and Ashford goes from fifth place to eighth, as follows.

Name of undertaking	Fuel viscosity, s R.1.	Revised total running cost, d./unit	Order of merit
Guernsey	950	0·705	1
Gremista	950	0·810	2
Jersey	200	0·812	3
Douglas	250	0·823	4
Isle of Man: Peel	220	0·861	5
Macclesfield	220	0·925	6
Station C	3000	0·934	7
Ashford 'B'	950	0·977	8

In the paper the author studies the elements of the total running cost. I do not disagree with what he says but I arrive at the result in rather a different way. For this purpose 1968 was an unfortunate example because the maintenance cost of 3 per cent of the total was far away from the 30 per cent recorded in the previous four years. Thus it can be said that normally 40 per cent would represent fuel, 3 per cent lubricating oil stores and water, 27 per cent maintenance, leaving some 30 per cent as the cost to be met whether the station runs or not. Thus I am not quite sure where this part of the author's argument is intended to lead us. However, he finishes by claiming overall efficiencies of 33–35 per cent with which I would not quarrel.

He goes on to say that the 'operational costs' are low enough to offset the substantial maintenance expenses but, from what he has said, I am not clear to what these operational costs are being related when maintenance cost is excluded.

Is the author crying 'Wolf' in his reference to disturbing stories of blended fuels? Does he not mean the quality of the residuals rather than the proportion? I can understand the quality varying but the proportion is that which is required to produce the required viscosity.

I agree as to the importance of air/fuel ratio, but the author deals with only one aspect based on an air intake design which today is seldom adopted. I am sure the author will agree that there are two important elements, overall air/fuel ratio and trapped air/fuel ratio. The former has a bearing on scavenging and exhaust valve temperatures; the latter influences combustion and thus thermal efficiency. In the main, if the air temperature and pressure in the intake manifold are correct then all can be assumed to be well, but how many users monitor this vital feature adequately? If either is wrong then the cause has to be found, such as reduced blower speed, fouled and thus inefficient compressor, intercooler fouled on either the air side or the water side. It is thus clear that monitoring of these two simple parameters covers a number of potential defects, any of which can lead to bad combustion and in turn lead to fouled pistons and rings, contaminated lubricating oil, local overheating with incipient scuffing and seizure. I consider this to be one of the very critical factors in the operation of diesels.

On lubricating oil, the reference to centrifuging as a means of dealing with fuel dilution is new to me. Centrifuging plus complete replacement with new oil every 17 days is calculated to keep the lubricating oil in good condition. In general terms the consumption is about 1·6 per cent of fuel, which is not unusual for an engine of this type although more recent designs use only half the quantity. On a cost basis it is 6 per cent of the fuel cost which is very creditable, but it is encouraging to know that price is by no means everything and that such good results can be obtained with a low cost oil.

Under the heading of overall reliability, availability is included, but the definition does not make provision for operation at reduced load. This is probably an unnecessary refinement in the author's case but nevertheless it is quite frequently met, particularly under severe climatic conditions or with old engines. I assume that in the definitions of reliability the word 'not' has been omitted in each case because one hopes that availability will be 100 per cent whereas the author uses the inverse. Reliability and availability are surely related to load factor. The author's figures represent the achievements under his particular regime, but the higher load factor of the normal 365 day power station makes maintenance more difficult and thus a different set of figures is likely to emerge.

It is of interest that a recent American Army study came up with an availability factor almost identical with that of the author's, i.e. 96 per cent, but the same study gave figures of only 500–700 hours as the mean time between failure. Perhaps the author would say whether his highly creditable figure of 2·3 per cent for unplanned outage represented a large number of short stops, which would support the experience of the American Army.

Since we are discussing critical factors, I would like to mention four. First, I would welcome the author's opinion on foundations, their design and construction. A reference to bedplate defects and crankshaft alignment suggests that he has had his troubles along with an increasing number of users today, a situation aggravated by increasing speeds and ratings. Foundation trouble is always expensive trouble and it is worth spending money to avoid it. Does the author feel satisfied with the conventional method of engine mounting or would he support the Aldershot school of thought where steel bearers are used?

Second, since turbochargers accounted for 27 per cent of the outage time in Table 14.2, I would welcome the author's views on the causes of this outage time and his suggestions for reducing it.

Third, instrumentation. I think insufficient attention is paid to the quality of the instruments supplied with the engine, no doubt because the manufacturer is under economic pressure. This is an area which would repay attention by the buyer. Then subsequently there is an absence of calibration. In spite of years of use under conditions of heat and vibration the user seems to have a touching faith in the continuing accuracy of the instruments. Good instruments coupled with periodic calibration are vital to successful operation.

Finally, does the author agree that the human factor is perhaps the most critical of all? A man is either master of his diesels or his diesels will master him. Having had a hand in the installation of over 100 MW of diesel plant in the last ten years, I have seen many examples of success and failure, and unfortunately far too often the engine is blamed for results which in fact were due to the shortcomings of the operator, but the qualities that make a successful diesel operator are hard to define. Perhaps the author could do so were he not too modest!

The author of Paper 11 has presented a most interesting paper, interesting because of the problems of load fluctuation coupled with the use of desalination which is currently attracting worldwide attention. On desalination some costs would have been helpful.

On the rating of the plant, the engines are rated at the equivalent of 850 kW but the alternator is rated at 930 kW. The reason is possibly to take advantage of low air temperature to obtain more power, but there is always a danger of operating at overload in hot weather. Again a larger alternator may have been chosen to assist regulation.

The reference to the loss of waste heat due to the use of a naturally aspirated engine is not understood. A given load represents a given amount of fuel and thus a given amount of exhaust heat. The lower efficiency of the naturally aspirated engine, of which complaint is made, should result in a small increase in waste heat.

The change to the air filtration system and the subsequent improved running suggests insufficient air in the first place.

It is not understood why 12–15 starts per day were necessary. Examination of the load curve suggests that the unit size was almost ideal and that only six per 24 hours would be required. Presumably, consideration was given to the use of electrode boilers to augment the water supply and to improve the load factor.

I agree that the components of an auxiliary system are always the Achilles' heel. The experience of the C.E.G.B. in this respect with gas turbines is the outstanding example.

Mr M. Vulliamy, C.Eng., M.I.Mech.E.—We must all be intrigued by the method by which the authors of Paper 7 have arrived at 'Pay-off potential', given in

Table 7.1. I am sure many of us would welcome an explanation of their method, involving the factors they have listed.

It is always a relief to find that other engine makers have had gasket trouble. We would agree that extremes of operating temperature are a major influence, and that thermal cycling tests can reveal weaknesses in much less time than the more usual endurance test schedules. It is not clear in the paper just what is meant by 'A defined part of the bolt load was transferred to the cylinder block face, etc.' Please may this be explained, and also can the authors say if gasket problems are reduced, or increased, by the use of individual separate cylinder heads, as on T.D. 100, compared with monobloc heads?

With reference to constant power engines, if we assume such a characteristic is really possible down to a very low engine speed, thus giving enough torque multiplication to avoid the need for a several speed gearbox, an intriguing point is what governs the best choice of final drive ratio. Although an engine may in itself be quite capable of a full load speed over say 2500 rev/min, as noise and fuel consumption should decrease with speed it would be tempting to govern the engine down to 2000 rev/min, or even less, and to use a higher axle ratio. There are doubtless other factors affecting the final choice, and I would ask for the authors' thoughts on this point even if it may be a little academic at this stage.

Little information seems to be available on gaseous emissions from highly turbocharged diesel vehicle engines, so any data the authors can release would be welcomed.

Referring to Papers 8 and 12 I rank the matter of engine noise as the most critical factor of all those reviewed in this symposium. It is critical in the sense that it affects the lives of people. Noise reduction is not technically necessary for the correct functioning of the diesel engine at the present day standard, but we are faced with an obligation to the public and to meet legislation which will be progressively made more difficult to satisfy.

On vehicle noise testing it may be worth stating that with the ISO conditions of test it is often possible for the governor to cut in before the 20 metres are covered. The B.S. 3425:1966, however, insists that a higher gear is used if this happens. This can give different results.

On fan noise we can confirm that the value of A in the formula is about 60, and on certain gearboxes we have had lower results than 84 (moving test), e.g. down to 80, but still quite enough to worry about.

The most important point of the paper is the means of prediction. For a long time we doubted the dependency only on speed and bore. We have tried to find all kinds of other parameters but over quite a number of cases at Perkins, including variations in the number of cylinders of the same bore and stroke, we admit that Professor Priede's formula fits best.

However, it does seem essential to embrace all the terms in the formula. Just to say '$I \propto N^* B^{**}$' may be misleading when comparing say turbocharged with normally aspirated engines.

I want to say more about Paper 12 by Mr Russell as it is particularly timely and it should help to make noise reduction part of routine design and development techniques. We must commend the *aim* expressed at the end of the introduction. These are strong words. The changes advocated must usually involve new or considerably modified patterns and dies. We all know the sort of lead times that are involved and these make us think first of all of the desire of the Ministry of Transport to tighten vehicle noise limits in 1973.

There is also the need to minimize the cost impact, but perhaps this is included in Mr Russell's words 'minimum upset'. There will thus be a temptation to restrict changes, to what may be done using cast iron, though here we have to watch weight, particularly in vehicle engines.

I regard the advice given on improvements to the main crankcase or block structures as particularly valuable, as I think there has been a tendency on the part of designers to go by the old rule 'what looks right is right' (there has been little else to go on), but in noise reduction work I do feel this can lead one astray, or put millstones round our necks. Certainly some of the measures now advocated are not always the obvious ones.

While on the matter of panels, is it right to presume that the vertical rotation of the point of attachment of the crankcase wall and bearing webs (item C) contributes to crankshaft axial movement, and so if this is much reduced there would be no need for treatment of the pulley?

While mentioning the pulley, it looks as if the recommended pulley design will not be easily applied to many diesels requiring multiple belts, and hubs also tend to be large relative to the rims.

I would also like to know more about the relationship between the natural frequency of a panel surrounded by ribs in relation to its principal dimensions. I suppose as its size is reduced the acoustic coupling improves, though the exciting forces will be lower?

On covers, one engine I know well has had an isolated rocker cover (stiffly corrugated and using grommets) in production for about 15 years. I think it has been successful, and a practical approach, but is this approach likely to be worse or no better than a damped flexible cover? I like the slit type cover; for flat surfaces it is a neat solution but its cost and life have still to be investigated.

A word about timing gear rattle. This does not bother us at the speeds involved in vehicle noise testing but it subjectively worries customers. It is not so much the result of the vibration of the crank as interaction of cyclic torque variations in camshaft as well as crankshaft. At low speeds it is not always easy to reduce gear tooth clearances without causing whine problems at higher speeds than idling. Good gear concentricity is difficult to realise in production at an economic cost.

Going back to Fig. 12.1, a small point which may be troublesome is that the qualification about the $\frac{1}{3}$ octave band level being influenced by the filter pass bands does not appear in any way in the picture, which might be used out of context, and thus mislead the unwary.

One wonders also why Figs 12.2 to 12.5 use acceleration as the ordinate, when Fig. 12.1 is based on velocity. As the text says the radiated *sound pressure* is proportional to the *velocity*.

I cannot help having a dig at Mr Russell on fuel injection noise. He does refer to it certainly, but in the briefest way. If all the engine treatments are successful then surely the fuel injection equipment will be noticeable. We want to be sure that when we as engine makers have done our bit, we do not have to wait for something else.

Finally, there seems to be a slight conflict between the predictions of the two papers: the first seems to indicate that short of complete enclosure there is little one can do to basic engines, while the second paper's case is that there is quite a lot—maybe 8 dBA. Perhaps all the authors could get together (peacefully I mean) and come out with a single view, if not a compromise. Of course what can be done to an engine will depend on its noisiness in its initial state.

Mr C. J. Walder, C.Eng., F.I.Mech.E.—I should like to endorse the points made by the authors of Paper 7 in regard to the need for a higher horsepower per ton of vehicle weight, particularly in respect of the larger vehicle. The sooner legislation is introduced in Britain demanding at least 8 hp/ton the better, because of the stimulus it will provide to designers and manufacturers to produce engines of high specific output.

Fig. D5 summarizes a recent survey made of European trucks and it will be clear from this that many manufacturers are still offering vehicles well below the German regulation of 8 hp/ton. In Japan, where they have a habit of doing yesterday what we are thinking about doing tomorrow, they already have a regulation calling for 10 hp/ton+15. By this standard the proposed European 44 ton vehicle will require a minimum engine output of 455 hp. Accepting the fact that vehicle weights and the limiting hp/ton will continue to rise in the future, I do not believe that the authors' 800 hp engine study is at all unreasonable. Indeed, we may well see 800 hp required for vehicles appreciably lighter in weight than 100 ton.

The current big unknown in form of the high output engine proposed by the authors is the level of exhaust

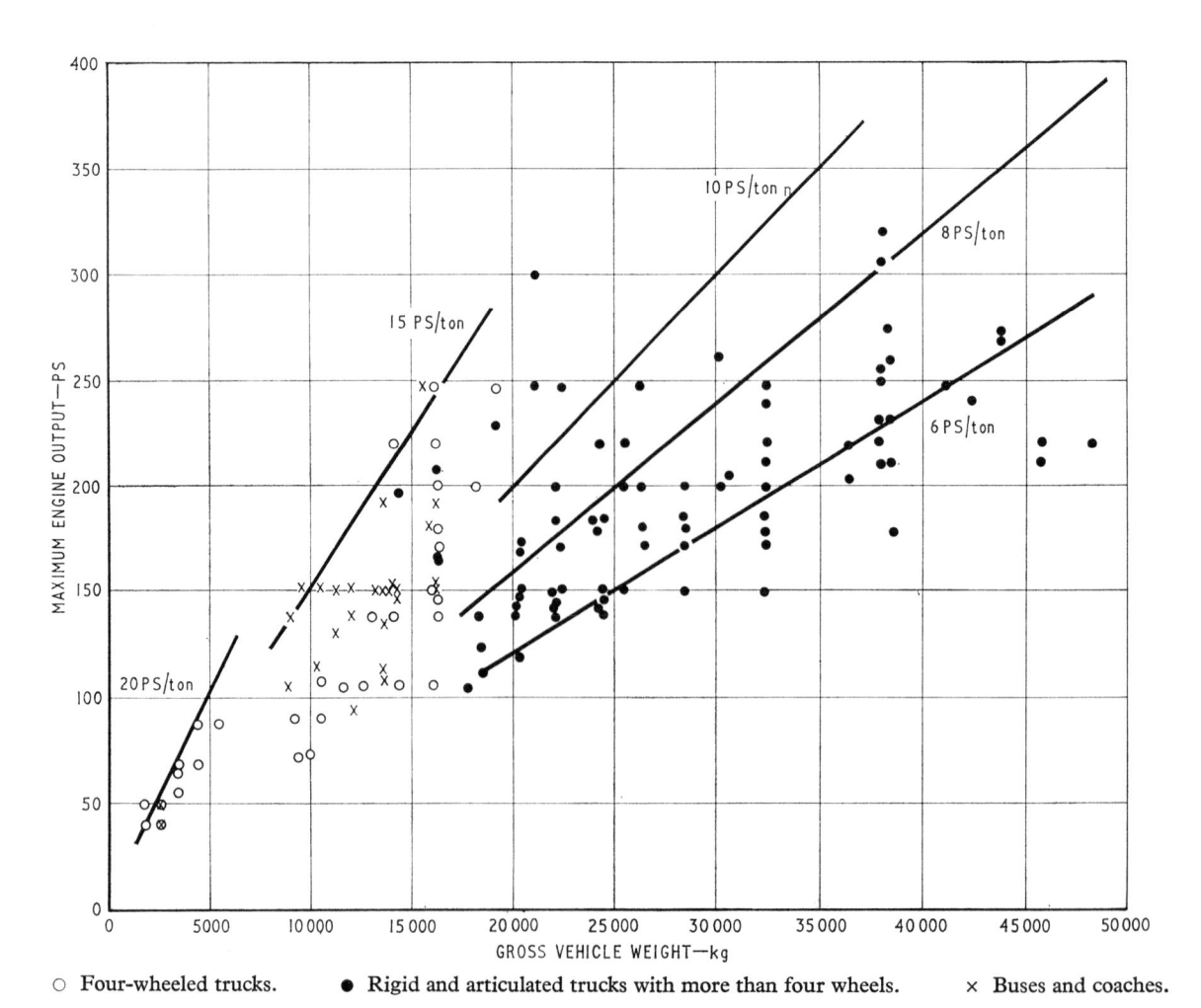

○ Four-wheeled trucks.   ● Rigid and articulated trucks with more than four wheels.   × Buses and coaches.

Fig. D5. Power requirements of diesel engined commercial vehicles produced in Europe (1969)

emission, particularly in respect of oxides of nitrogen. If, however, these very high powers are to be obtained from an acceptable package size, it is clear that high piston speeds and high b.m.e.ps will have to be accepted and measures will then have to be taken to control exhaust emissions. Just how this can be achieved is not clear at the present time.

The piston speeds and b.m.e.ps quoted by the authors in their Table 7.2 are within our current experience, and my personal choice for an engine to meet the 88 hp objective would be for their engine No. 3. Although more expensive to manufacture than engine No. 1, it should be more compact, lighter in weight, as well as being significantly quieter, if rated by the formula quoted in Paper 8 of this symposium.

I am interested in the authors' comments on inlet valve seat wear because in one respect they appear to have gone against accepted practice by introducing flexibility into the valve head. It is usually assumed that the wear that is often experienced on the inlet valves of turbocharged engines is due to the flexing of the valve head on the dry seat. The cure proposed by Professor Zinner and others has been to stiffen up the valve head as well as to adopt the other measures employed by the authors, i.e. 30° valve seats, reinforced valve train, etc. It would therefore be interesting to have details of the design of the inlet valve adopted by the authors and to know if they have introduced lubrication for the seat by the addition of fuel or oil to the inlet manifold.

The problems experienced with cylinder head gaskets reported by the authors are interesting in that they appear to be confident that they were due to thermal deformation associated with low temperature operation. Their solution of doubling the bolt load is classic when faced with gasket trouble but one wonders what initial cover was employed if they could double it without difficulty. We would normally design for a bolt load of some 4–4·5 × the gas loading.

The fact that these gasket troubles were experienced mostly under start-up conditions in periods of intense cold suggests that the possible excessive use of an ether starting aid could be a contributory factor. The rate of pressure rise and peak pressure can be very high indeed if this form of starting aid is used without discretion.

Papers 8 and 12 are very valuable and they will undoubtedly play their part in the design of future diesel engines. The influence of the formula for producing the noise level of current generation diesel engines quoted in Paper 8 will be especially important in relation to the choice of stroke/bore ratio.

One can visualize the scene in some design offices as the full impact of the paper dawns upon the Chief Designer! Out will come the erasers and the cylinder bore of his precious new engine will be reduced and the stroke lengthened. Justification for the move will be found in the shorter overall length of the engine, the increase in height will be ignored and the development engineer will be pleased to find that the new long stroke engine has a higher mechanical efficiency than the fast running, over-square engine so recently the darling of the Chief Designer's eye.

However, it is not every day that new engine designs are produced and many of us will have to continue to live with what we have for some time to come. What can be done to reduce the noise level of these units and particularly the noise under the drive past ISO tests? The authors of both papers have given an indication of what can be achieved by attention to the detail design of the covers, inlet and exhaust systems, fan drives, etc., and it should be possible to introduce such modifications without major alteration to the main components. Furthermore, if the current engine is normally aspirated, the adoption of turbocharging will help.

An alternative or possible complementary approach is to employ an engine enclosure. We at Ricardo's have had considerable experience with this technique, and we have demonstrated in one case at least that the difference in test bed noise and vehicle noise can be as much as 20 dBA by this means. Indeed, it does appear that some form of enclosure or close shielding will be essential even if engines designed in the light of the authors' recommendations and developing more than, say, 300 hp are to meet future noise regulations. It would be interesting to hear the authors' views on this point.

It would be interesting to know the level of noise to be expected from a coasting 44 ton g.v.w. vehicle. If noise increases with g.v.w. as one suspects it must, the noise of such a vehicle is going to be pretty high even without the engine in operation.

The authors of Paper 8 are careful to state and repeat throughout their paper that their predictions have been based on 'current generation commercial vehicles and engines'. It would be nice to peep inside their crystal ball to know what noise level they consider could be achieved with a 300 hp engine if they were given a completely free hand.

Another aspect of noise not covered in Paper 8 is the

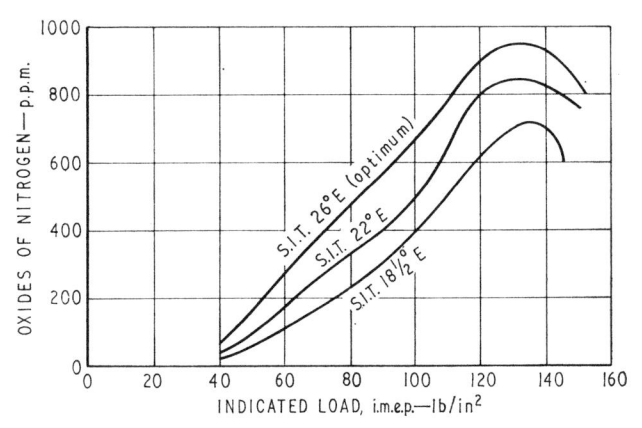

Fig. D6. Emissions of oxides of nitrogen from a 120 × 140 mm Comet Mk V engine at optimum and retarded injection timings, 2000 rev/min

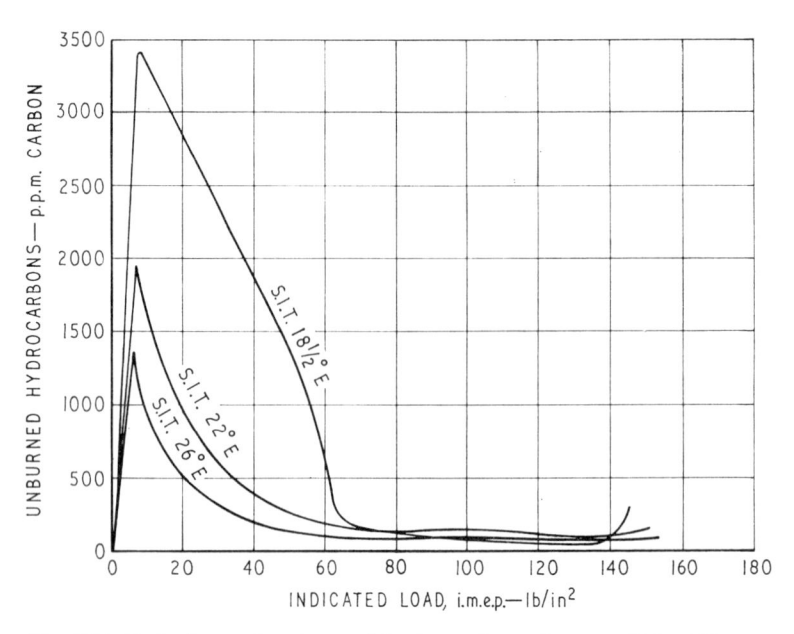

Fig. D7. Emissions of unburned hydrocarbons from a 120 × 140 mm Comet Mk V engine at optimum and retarded injection timings, 2000 rev/min

idling noise of the small high speed diesel engine as used in taxis, delivery vans, etc. In terms of dBs the noise level of these engines under idling conditions is low, but subjectively it can be very annoying. Have the authors attempted any prediction of noise levels to be expected from this form of engine under these conditions?

Finally, I would like to ask the authors if they have any idea of the level of supercharge required to change the index D from 30 to 40. This information would be of value to manufacturers who would like to apply some supercharge in the interests of reducing noise but who may have an engine that is sensitive to an increase in maximum cylinder pressure.

The author of Paper 9 presents an excellent summary of the critical factors affecting the specification of the injection equipment for both indirect and direct injection engines. From personal experience I have much sympathy with his problems, the greatest of which has been the absence of some form of positive control on the fuel delivery over the speed range of the D.P.A. pump. However, there seems to be little prospect of things being any easier in the future because, in addition to the legislation concerning smoke and noise, there is the possible introduction of regulations affecting exhaust emissions in the not too distant future. The work carried out in this field at Ricardo's to date indicates that injection timing plays an important part in deciding the levels of unburnt hydrocarbons and oxides of nitrogen in the exhaust of both direct and indirect combustion chambers. It is possible that we will require precise control of injection timing at all speeds and loads. Too much advance of timing leads to excessive level of NO_x whilst too much retard at light load can lead to a marked increase in unburnt hydrocarbons. An example of the effect of timing is shown in Figs D6 and D7.

One wonders if these exacting timing requirements can be met with the current design of D.P.A. pump or whether, in the long term, some alternative design of pump or even electronic triggering of the injection timing will be required. The timing characteristics that I have in mind are the opposite of those of the current D.P.A. pump; that is to say, it might be advisable to advance the timing with reduction of load.

The author's remarks on the part to be played by the engine manufacturers to obtain a more consistent engine performance must also be endorsed. If timing is to be held to a tight tolerance, all sources of backlash in the drive train to the pump must be minimized, or eliminated. Factors affecting combustion efficiency such as swirl generation by individual inlet ports and the piston head clearance of individual cylinders will have to be rigorously controlled.

Authors' Replies

Mr N. R. Senior and **Mr D. J. Clarke**—We would agree with Mr Bresser's statement that we are achieving rationalized packs, and that this does and will continue to influence customers' choice of ancillary equipment, on the grounds of both economics and availability.

With relation to the installation that has to be treated as special for a variety of reasons, this service can still be given. The logic of the system outlined will ensure that difficult and/or high volume projects can be correctly tailored so that from a cost aspect the end product is economic.

The design, material, method of manufacture and finally method and area of procurement must be studied to ensure this overall cost is held to a minimum level.

Also by reviewing these 'special options' for inclusion into rationalized pack form, both extremities of the market can benefit.

Mr W. Lowe—Dr Mansfield has commented mainly on the difficulty of establishing a simple and yet generally applicable formula for the de-rating of turbocharged engines, and goes so far as to say that the requirement cannot be met. I agree with him wholeheartedly that the problem is complex and that a solution which is acceptable to both manufacturer and customer may be difficult to achieve. I do not, however, accept that the task is impossible, and hope that my paper may have contributed something towards the understanding of the problem and hence help towards a solution, even though the proposal I have put forward goes only part of the way towards an acceptable formula. The theme of the symposium has been '*critical factors* in the application of diesel engines', and it is important that both customer and manufacturer recognize and identify the parameters which limit the rated output of a particular engine design under adverse ambient conditions. The concept of a limitation declared by the manufacturer is not new, since the CIMAC 'A' and 'T' de-rating methods for naturally aspirated engines have been in use for many years, although the methods are actually intended for the correction of test-bed results. It is the manufacturer's responsibility to declare whether the engine output is limited for thermal reasons (Formula T) or by excess air limitation (Formula A) and I can imagine a future in which the manufacturer of a turbocharged engine is willing (and able) to state the limiting factor on which his de-rating proposal is based.

Although we have not had a paper on governing, several earlier speakers have referred to the poor capability of the turbocharged engine to accept sudden load changes compared with the naturally aspirated engine. I have no wish to deny the importance of the ability of an engine to accept a sudden load application, and a highly rated turbocharged engine can seldom accept a rapid load increase of more than about 70 per cent of its rated full load without a large speed reduction. I would point out, however, that percentages are relative and a turbocharged engine rated at, say, twice the power of a naturally aspirated engine of the same physical size is actually accepting a load application 50 per cent greater than the naturally aspirated engine could.

Mr W. Petrook and **Mr W. A. Stewart**—It is significant that Mr Glasspoole found the common factor in the three papers on which he opened the discussion to be the cooling systems, and much of our effort on British Rail to improve reliability is concentrated on the cooling systems.

With regard to the relating of maintenance periods to the type of service, while I agree that kilowatt-hours and total fuel consumed are similar criteria, they are not necessarily synonymous where long periods of engine idling are involved, because of the differences in specific fuel consumption over the engine operating range.

With regard to the customer specifying the degree of reliability which is required, we feel that this should really be specified by the manufacturer. Otherwise, one gets the impression that a manufacturer will supply an engine, but if the customer wants the engine to be reliable then he has to pay more for it. In addition, in many cases it seems a surprise to a manufacturer that when a component is supplied to British Rail, it is put on to a locomotive. As customers, we would expect every component in an engine to last the full life of the engine unless the manufacturer specified otherwise, rather than for the customer to have to specify the life that he expects from every component.

With regard to water washing of turbochargers, Mr Dingle is quite right to query the statement made that on British Rail we water wash turbochargers every 2000 hours. This period is in fact in error; the time period is between 600 and 800 hours (the text of the paper has been altered accordingly). The effect of water washing gives greater improvement on some engines than on others. However, we have not yet found it necessary to water wash every 200 hours.

Mr Scott's contribution and question are very interesting, as it would appear to be an attractive proposition to have a lower-rated engine of the same overall power within the same total weight and size parameters. However, with say a typical 3000 hp charge-cooled engine installation, the total extra cooling system weight is of the order of 2000 lb. Contrary to the inference in Mr Scott's question, the top gas ring temperature is usually higher with a non-charge cooled engine than with a charge cooled engine. In fact it is unusual for a non-charge cooled engine to run at b.m.e.p. of greater than about 130 lbf/in^2, whereas with a charge cooled engine the b.m.e.p. can be 220 lbf/in^2 for the same top gas ring temperature. Thus power for power, the non-charge cooled engine would have to be about 50 per cent larger and hence 50 per cent heavier than a charge cooled engine.

With regard to the question on the operation of diesel engines for shunting duties, in our opinion, the turbocharger would be a disadvantage unless special arrangements were made to keep the seals clear of carbon build-up by means of an external air supply. In short, if it is possible to have a normally aspirated engine for shunting duties, then this would be desirable. We do not feel that the quick opening up or shutting down of an engine is necessarily detrimental to engine life. We are not clear on the question in the last sentence of Mr Scott's contribution, but we think it can be answered with the information that, if an engine with an engine-driven water pump has been running at full power for a considerable period of time and is suddenly shut down, then the metal temperature immediately falls, although the coolant temperature could increase. Therefore, if some pressurization was introduced, purely to stop the water from boiling, no ill effects would be felt by the engine.

Mr D. Castle—Mr Garside raises the question of seating stiffening and continuity.

The design of seating referred to in the paper was subsequently modified by the addition of a longitudinal beam immediately under the top plate and of a longitudinal casing plate along the outside of the web plates, thus creating a box section seating. A section is shown in Fig. D8. The discontinuity was not corrected on the ships under discussion, because it was felt that more harm than good might result from alteration of the magnitude required, with the machinery in position. This did mean, of course, that alignment had to be subject to more frequent checking than is desirable.

With regard to smaller unclassified ships now being fitted with twin-input drives, it is of course important to observe the principles I have outlined, but the danger recedes somewhat as the volume of the engines becomes greater in relation to the size of ship.

Mr Henshall and Mr Sampson both suggest that flexible mounting is the real solution to the problem of reconciliation between engine and ship structural movements and alignment inaccuracies. Certainly there is a strong case for flexible mounting from this point of view. I consider that the number of mountings would have to be

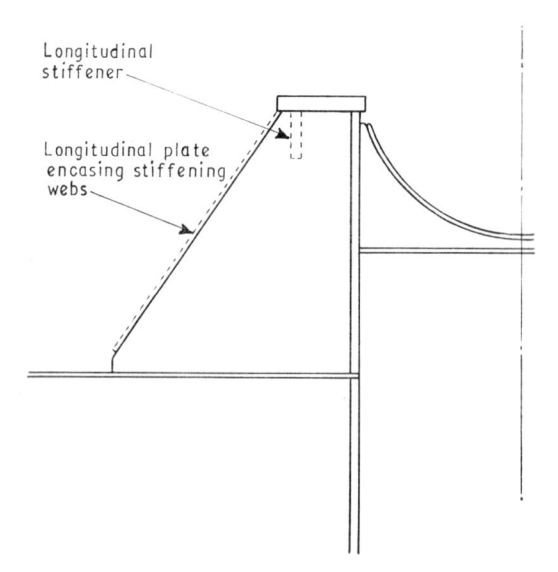

Fig. D8. Modified box section seating

four rather than three, since three would involve a substantial sub-frame to link the mounting points and a central mounting would inevitably lead to a higher engine position in the ship, which is not tolerable in most cases. More than four would begin to defeat the object, although four pairs or four groups would probably be possible.

So far as the absorption of main bearing clearances is concerned, it is pointed out on page 52 of the paper in paragraph (1) that 'For all PC2 engines examined, the loss of alignment of main bearing housing falls within the main bearing clearance.' The cause of failure was the permanent set taken by the seatings during early running of the ship. It was when further thermal deflection was superimposed on this set that trouble developed.

It is my company's policy to recommend to all owners that crankweb deflections are taken every three months for the first year of operation, and subsequently every 2500 h. Deflections had been taken on the *Manchester Port* on the voyage prior to that on which the failure took place, but unfortunately on the starboard engine these were meaningless, as they had been taken by a junior engineer between the balance weights. They were to be taken correctly at the termination of the fatal voyage.

The setting of bearings out-of-line is a standard procedure which represents a mean allowance designed to meet all conditions. It is too little for some ships and too much for others, but it has resulted in an improvement in the hot deflections recorded in service. That it was difficult to fix I will not deny, and I do not necessarily advise others to adopt this method, but I am prepared to be judged by results.

Mr E. Holmér and **Mr B. Häggh**—It is interesting to hear from Mr Walder that Japan already has a regulation calling for 10 hp/ton+15 hp. Information from the U.S.A.

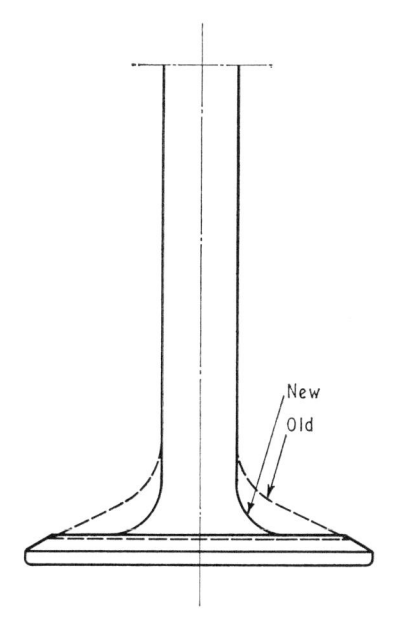

Fig. D9. Inlet valve design

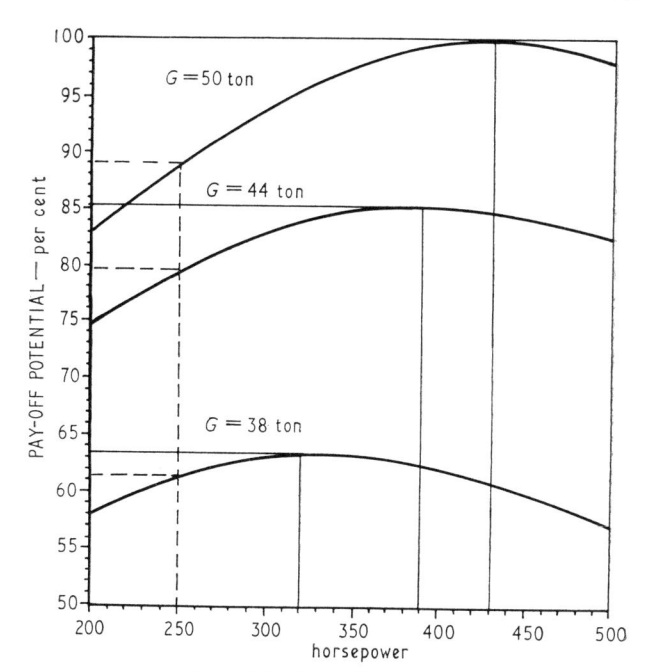

Fig. D10. Pay-off potential

indicates a request for about 11 hp/ton. For special purposes there are vehicles in the U.S.A. with a gross weight of up to 58 tons, which consequently need 640 hp.

Regarding the design of the inlet valve we refer to Fig. D9. Neither oil nor fuel are added to the inlet manifold.

Using the classic method of increasing the bolt load to solve the trouble with cylinder head gaskets was made possible by transferring about one third of the bolt load to the block, instead of having all the load on the liner, and thus avoiding higher stresses.

Extensive investigations regarding the influence of other starting aids on cylinder head gaskets have not been made, since the measured peak pressure is not higher than that at full load, due to the low compression ratio and higher b.m.e.p. on the turbocharged engine. The reason for trouble in cold starting is due to the extra bolt load, resulting from the slower increase in temperature in the bolts than that in the cylinder head.

Mr Bresser is right about white and blue smoke at light loads. We solve that problem by loading the engine through an exhaust pressure governor (see Fig. 7.3 in the paper).

We agree that the surge line can not be defined at steady speed. It is necessary to run the engine until it attains a stabilized condition at constant speed, and then lug it, i.e. reduce the speed of the engine at full load, quickly.

The air intake has to be placed so that water is not forced into it when driving the vehicle on flooded roads. The paper used in the insert must be impregnated to avoid absorption of water.

We cannot answer Mr Vulliamy to the extent we should like, since this would involve writing further papers, but we can say a few words on the matter.

Table 7.1 in the paper is not easy to understand and we therefore showed a diagram, Fig. D10, at the symposium. As engine output goes up, both engine weight and average vehicle speed do so as well. Optimum output is reached when the gain in average speed no longer balances the loss of payload owing to the heavier engine. Lower specific weight (lb/b.h.p.) results in a higher optimal output and is therefore essential to meet the demand for higher brake output per vehicle weight. Whether individual separate cylinder heads reduce or increase gasket problems depends on the gasket material. With the solid steel gaskets we use, it gives a more uniform sealing pressure.

A constant power engine has a limited speed range, due to limited cylinder peak pressure (Fig. 7.6 in the paper). Even with a variable compression ratio we think there is a need to use the speed of which the engine is capable.

To change gear is a problem at vehicle speeds lower than 30 km/h. At higher speeds there is enough time due to momentum to do it under all conditions. We think that constant power down to 70 per cent of full speed and constant torque between 70 and 50 per cent, in combination with a torque converter for moving away from rest to 30 km/h and a 5- or 6-speed gearbox, is the optimal practical solution.

Our investigations regarding exhaust emissions are so far limited. We will be glad to answer this question as soon as we know more.

Mr P. E. Waters, Mr N. Lalor and **Dr T. Priede**—
We agree with Mr Vulliamy that the gear chosen for the B.S. 3425:1966 test can make a difference to the results obtained. When a repeat test in a higher gear is carried out as a result of governor cut-in during the original test the new noise level is not necessarily higher than the original level. We have tested several vehicles which gave the

highest level, due to the corresponding higher engine speed, where the governor cut in. B.S. 3425:1966 implies that the highest level is taken regardless of the gear and not just the level for the repeat high gear test. This is the interpretation used by the Ministry of Transport.

Mr Vulliamy's confirmation of the characteristics of fan noise and the importance of gearbox noise is welcome.

We should like to stress that we feel that the main function of the noise prediction formula is to show the trend in the basic design parameters which must be followed to produce quieter engines rather than to predict absolute levels. Mr Walder's comments in this connection are particularly apt; the height of the engine is the least critical of its three dimensions in most installations. Any noise reductions achieved by attention to details will apply to any engine regardless of the chosen stroke to bore ratio. Thus the noise produced by a long stroke engine will usually be less than that produced by its over-square high speed counterpart within the same design framework.

It should be noted that the noise prediction formula is based on engines where combustion noise predominates. This is the usual case with automotive diesel engines and explains the different form taken by prediction formulae obtained for other types of engine where other noise sources predominate.

We agree with Mr Vulliamy that it is essential to embrace all the terms in the formula and we suggest that this has been taken care of in the general formula on page 71 of the paper by the use of the parameter D. Perhaps this would have been clearer if we had said, on page 69, $I \propto N^{D/10}$. Subsequent work (5) has led to a refinement of this relationship and new values of D have been found as follows:

Conventional D.I. engines (four hole injectors) . . . . $D = 28$ to 30
D.I. engines, pressure charged . $D = 40$
Two-stroke diesel engines . . $D = 40$
M.A.N. combustion system . . $D = 45$
Petrol engines . . . . $D = 45$ to 50

and the prediction formula for two-stroke diesels has been found to be

$$\text{dBA} = 40 \log_{10} N + 50 \log_{10} B - 54 \cdot 5$$

Table D1 is an extended version of Table 8.1 and shows some two-stroke engines in addition to four-strokes.

There is no real conflict between our findings and those of Mr Russell. As Mr Walder has noted our findings are based on 'Current generation engines'. We agree with Mr Russell that improvements can be made to conventional engines by attention to the design of such things as covers and pulleys. We have occasionally met cases in which the covers and crankshaft pulley are the predominant sources of noise and reductions of some 6 dBA can be obtained by attention to them alone. However, in general, our experience is that an 8 dBA reduction could only be achieved by a fairly radical change in the design of the basic engine structure as well. This is borne out by Mr

Table D1. Predicted and measured noise levels of some automotive diesel engines

N.A. = Normally aspirated. T.C. = Turbocharged.

Engine type	Cycle and induction	Bore, in	Rated speed, rev/min	Overall noise level, dBA Calculated	Overall noise level, dBA Measured
V form, 8 cyl.	Four-stroke, N.A.	4·625	3300	107·2	108
		4·625	3300	107·2	109
		4·50	3000	105·5	106·5
		4·25	2800	103·3	103·5
		5·31	2600	107·2	109
		5·5	2100	105·5	106
In line, 6 cyl.		4·65	2600	104·4	102
		4·56	2800	104·9	105
		4·125	2800	102·7	102·5
		3·875	2800	101·3	103
		3·81	2800	101·0	102
In line, 4 cyl.		3·875	2800	101·3	103
		3·688	4000	104·9	107
		3·125	4000	101·3	101·5
		3·50	3500	102·0	101·5
V form, 8 cyl.	Four-stroke, T.C.	5·50	2600	107·3	107
In line, 6 cyl.		5·50	2100	103·3	102
		4·75	2200	101·0	100
		3·62	2200	107·1	102
Opposed piston single crankshaft with rocker arms, 3 cyl.	Two-stroke	3·25	2400	106·3	108·5
Opposed piston double crankshaft in-line 6 cyl.		3·4375	2400	107·5	109·5
		4·625	2100	111·6	113·5

Walder's experience at Ricardo's where 20 dBA difference between test bed and vehicle has been obtained by engine enclosure. If we assume that 16 dBA is due to the difference in measuring distance (he does not say that his measuring distances are different from ours), only 4 dBA has been achieved by this technique. We agree with Mr Walder that for engines around 350 hp some form of shielding will probably be needed in addition to structural modifications if substantial noise reductions are required. Our crystal ball tells us that by employing both these techniques we should be able to achieve 80 dBA maximum vehicle noise level at least up to 350 hp, but this assumes that the engine is then still the predominant source and is, of course, still subject to further practical investigation. For an up-to-date state of the art of vehicle noise reduction we recommend the recent literature survey of the Ministry of Transport working group on road traffic noise (6).

In this connection Mr Walder asks about the level of coasting noise from a vehicle of 44 tons g.v.w. This is about the heaviest type of vehicle for which we have any data, and the indications are that tyre noise would not be an impossible problem in achieving an ISO test level of 80 dBA, but would prevent us from developing a vehicle which did not exceed 80 dBA under all operating conditions. We have insufficient data to give a definite answer to this question, but the coasting noise of such a vehicle is of the order of

$$\mathrm{dBA} = 30 \log_{10} V + 32$$

where V is the road speed in mile/h. We also confirm Mr Walder's suggestion that the rolling noise increases with design g.v.w. Rolling noise is quite an involved subject and it would take most of a full length paper to describe adequately!

Idling noise is not only a problem with small high speed engines, but can also be a problem with most automotive engines. The International Union of Public Transport carried out a survey by questionnaire of 115 member bus undertakings and of these 41 per cent claimed to have had complaints of noise from stationary buses against 47 per cent who had had complaints of noise from moving buses (7). The noise of the diesel engine is particularly noticeable at tick-over compared to the petrol engine because of their different rates of change of noise with speed. This difference gives the greatest difference in levels at tick-over. The actual value of noise at tick-over will be lower than that predicted by the formula by an amount equal to that due to the reduction of speed plus that due to the removal of load.

The reason why turbocharging increases the value of D from 30 to 40 is that the ignition delay is reduced by the increase in compression temperature with a consequent smoothing of the cylinder pressure spectrum. 40 represents the limit to which D can be changed by this means and a 50 per cent increase in b.m.e.p. is certainly sufficient to achieve this limit. However, we have little experience of supercharging to much less than this and it may be that in some cases this effect could be obtained by less.

REFERENCES

(5) ANDERTON, D., GROVER, E. C., LALOR, N. and PRIEDE, T. 'Origins of reciprocating engine noise—its characteristics, prediction and control', A.S.M.E. Paper 70 WA/DGP 3.
(6) 'A review of road traffic noise', Report LR 357, 1970, Road Research Laboratory, Crowthorne, Berkshire.
(7) TAPPERT, H. and LIPPACHER, K. 'Measurement of bus noise', Paper 5b of 37th International Congress of the International Union of Public Transport, Barcelona, 1967.

Mr P. Howes—Mr Walder's understanding of the critical factors regarding the application of F.I.E. is appreciated, and he has raised the new problems of the emission of oxides of nitrogen and unburnt hydrocarbons.

It is of course too early to say what effect the control of these emissions will have on the development of future F.I.E. The two figures contributed by Mr Walder, D6 and D7, showed the response of exhaust emissions given by an indirect injection engine. This response may not be typical for other makes of direct injection engines.

What is clear, however, is that the fuel pump of the future, indeed the whole F.I.E. system, will have to be more sophisticated than at present, giving a more precise control and variability of timing and fuel than at present.

In reply to Mr Sampson, the use of a common pump and governor for *marine* propulsion, generator and general duties can be satisfactory, as the same percentage governor run-out can be adopted for all cases. The speed delivery characteristic of the pump is largely immaterial as the engine is subjected to the propeller law in the case of propulsion, and is usually working on the governor in the other two cases. So unless there is any objection to close governing when under propulsion, the same pump and governor will suffice for all the applications mentioned by Mr Sampson.

On the question of ratings and fuel delivery, the paper deals specifically with pre-setting for production. The fuel setting, once set, is then used also in service.

Setting the fuel delivery to give a fixed power output is to be deplored, as a production technique, as this can mask engine deficiencies, and similar engines can then have widely different fuel levels.

If, however, the pumps are pre-set to whatever the engine rating (and there can be two or more levels to cater for varying power requirements), then the variation in power output between production and service pumps, when tested on the same engine with the same set of nozzles, will be within ± 2 per cent, this being the over-checking tolerance on pump setting.

The power scatter shown in the paper is for new 'green' engines, which includes the scatter due to engine, pump, nozzles, high pressure pipes and test beds.

Mr S. T. Walker and **Mr H. E. Howells**—Mr Glasspoole asks whether there is a lower limit on bore size and upper limit on engine speed beyond which residual fuels cannot be used if acceptable carbon deposit levels are to be maintained.

The authors are very conscious of the efforts made by engine designers to achieve similar engine life, maintenance costs and engine output with heavy fuel compared with distillate. Our own experience has been that a 5 in bore engine can be run at 1000 rev/min certainly for periods of a few hundred hours on residual fuel without any significant build-up of carbon. There is a slight darkening of the exhaust, and this suggests that combustion is not as complete as when using distillate fuel. This might, of course, be overcome by modifications to the fuel injection characteristics.

However, the authors feel that there must be limits to the extent that residual fuel can be used, certainly so far as engine speed is concerned. In broad terms, the time available for combustion is inversely proportional to engine speed. The time required to burn the fuel cannot be substantially changed. It is made up of two components, (1) that necessary to physically prepare it, i.e. bring it into association with oxygen, and (2) that necessary to carry out the chemical process of combustion. Various techniques may be developed to shorten the first of these, but there is a limit if satisfactory performance is to be achieved over the load and speed range. The time for the chemical processes cannot, essentially, be changed.

Some work indicates that the combustion of a residual fuel droplet takes approximately twice the time of that of a gas oil droplet of the same size. Alternatively, the residual fuel has to be atomized to give droplets half the size, in order to give comparable combustion times. However, the pressures necessary to achieve this may not be practical and other effects, such as increased penetration, may be the reverse of what is desired. In any case, it is the larger droplets which are more difficult to burn and therefore more likely to cause carbon formation. Injection pressure does not appear to have much effect on maximum droplet size.

The sum total of the above factors is that there may be sufficient time for combustion to be completed in the combustion chamber in the approximately 0·02 s available at 1000 rev/min. However, the figures available to the authors suggest that this may not be the case at speeds very much above this. To prevent combustion continuing when the gases are passing the exhaust valve, and thus causing excessive valve temperature, it will be necessary to quench the combustion and this will result in the production of carbon in the combustion chamber.

Mr Glasspoole also asks whether we would indicate the fuel constituents which can affect exhaust valve life. To generalize, these are any of the ash forming materials. To be more specific, the more important ones are vanadium and sodium. Vanadium content varies with crude source; that from the Middle East has about 50 p.p.m. whereas that from the Western hemisphere has about 300 p.p.m. Sodium can also be present, sometimes as a result of the presence of sea water. If the ratio of vanadium to sodium is about 4 to 1, the melting point of the mixture is around 400°C, and in its molten form it is highly corrosive. Vanadium pentoxide, on the other hand, which is formed when no sodium is present, has a melting point of 600°C. Water washing will remove the sodium, but this, of course, must be done before centrifuging the fuel.

The final point made by Mr Glasspoole is a plea for the supply of a consistently poor fuel for engine development work. It may be possible to arrange this, but large stocks would have to be laid down and tankage set aside for this purpose. This would undoubtedly make the fuel expensive. Whilst appreciating the reasoning behind this request, it is perhaps worth pointing out that the use of an artificially poor type of fuel might lead to the incorporation of unnecessarily expensive engine design features.

Mr A. K. Mackenzie—Mr Dudley and Mr Wood have offered some interesting comments. First, the author would reply that in his experience it has not been so much a problem of specifiers asking the impossible to cover imagined difficulties, but rather one of them asking for what is ultimately desirable and for this to be balanced against what can be economically provided. As the contributors went on to discuss, electromechanical governing systems which could have provided the closer speed control are indeed available.

On the subject of voltage regulation, it is accepted that the reference in the paper was over-simplified, and thereby misleading. Whilst not disputing the contributors' equation for purely reactive voltage drop at unity power factor (1·0 pf), in fact the further theoretical drop due to alternator resistance would have been closer to 2 per cent, and it must be remembered that at 0·8 pf (a much more likely load condition in this case) the voltage regulation is several times that at 1·0 pf. Detailed theoretical calculations are of little value, however, as the results often do not closely agree, particularly under transient conditions, with those found in practice. During recent works tests on a similar alternator, figures of 17 per cent transient reactance and 10 per cent transient voltage regulation at 0·9 pf were derived by the manufacturers, and the application of full load at 0·8 pf was estimated to give a transient voltage drop of 24 per cent. It is considered that the approximate 15 per cent voltage regulation estimated for the Ascension alternators, with an anticipated load between 0·8 and 0·9 pf, was in fact quite realistic in practice.

In defence of the choice of static excitation, it should be noted that the exciters used were capable of providing the full forcing field current within 2 cycles ($\frac{1}{25}$ s) of the application of a load increase to the alternator, and that this current level was effectively independent of the alternator transient voltage drop. This resulted from the control being based on a separate voltage reference circuit, which did not rely on a stable external power supply. For accuracy and stability of voltage holding under arduous and cyclic conditions, and in particular for speed of response, it is doubted if alternate forms of excitation could have made any improvement on the performance obtained.

The author would like to thank Mr Vaughan for his comments on the paper. The subject of costing of the desalination process had been omitted as the only figure currently available, which was approaching £1·25 per 1000 gallons for most consumers, would have been misleading because it included several factors peculiar to the Ascension installation. For example, the running and maintenance costs of a pumping, storage and distribution system 1000 ft above sea level were included, as was the cost of running an oil fired boiler to boost the steam supply (to meet a subsequent large increase in the water demand). Interest and depreciation on the plant value, however, were excluded.

The rating of the diesel engines had been incorrectly quoted, and the manufacturers confirm that the official site rating of these engines is 1280 b.h.p. The design rating of the alternators was 900 kW, excluding the energy for excitation. As static exciters were used, this energy was produced by the main alternators, and 930 kW was taken as their gross rating.

It is agreed that the quantity of waste heat in the exhaust of a naturally aspirated diesel slightly exceeds that of a turbo-charged engine of the same rating, but the problem lies in the temperature of the gases leaving the engine. In the case of the naturally aspirated set, particularly when on part load, this temperature is considerably lower than the equivalent from the blown engine. The capacity of these gases for raising steam in the exhaust boiler is thus lower, and in effect less usable waste heat is available.

The more frequent engine starts came about in the early days when the daily load curve was far from the anticipated one, and often included short spells of additional running sets for transmitter testing. The load curve has been further improved now by the subsequent inclusion in the design of a 1 MW electrode boiler. Consideration had been given in the initial design study for the inclusion of such an item, but, surprisingly, calculations showed that on economic grounds if additional heat was required this could best be met by the use of a direct oil-fired boiler, which had been provided anyway for use during the construction period and for any subsequent emergency. However, at a later stage it became apparent that the demand for water would rise to a high level well before the electrical load was available to provide waste heat, or even to test the generators, and for this reason the electrode boiler was provided.

Mr M. F. Russell—Mr Vulliamy is better qualified than most to comment on methods to reduce the noise of conventional diesel engines, since some of the engines with which he has been associated have noise-reducing features built into them already. The stiffly-corrugated rocker cover of which he speaks is in fact isolated from the high frequency components of vibration by the combined flexibility of gasket and rubber grommets under the securing nuts. It must be admitted that it is difficult to improve on this particular cover, but a 5 to 10 dB reduction in vibration has been obtained by damping in addition to this isolation. The vibration levels on the damped cover in the paper are 10 dB less than those on the corrugated and isolated cover above 630 Hz, and the excitation is greater between 1 kHz and 2·5 kHz.

Mr Vulliamy has correctly interpreted the sentence on the aims of the investigation as being an attempt to achieve a worthwhile noise reduction with minimum changes to the structure of existing engines, and perhaps more important, with minimal changes to expensive production plant. The modifications to crankcases were made by adding ribs to development patterns in such a way that the casting will pass down the transfer machine which machines the cylinder blocks. Slit-isolation is amenable to die-casting the slits in position. The quiet pulley is intended to be built up from cheap stampings.

These modifications have not been optimized for weight or cost as it is felt that this is the forte of the engine manufacturers. While on the subject of aims, the conflict in the predictions of the two papers on engine noise mentioned by Mr Vulliamy is less real than one might suppose from reading the conclusions of these papers. The 7 to 10 dBA reductions in noise achieved by the structure research engines made by Professor Priede place these engines far outside the scatter of the results quoted in Table 8.1. The results in that table show that there is surprisingly little difference between the noise-

Table D2

Engine	3-cylinder tractor		6-cylinder truck	
	Inlet side	Exhaust side	Inlet side	Exhaust side
Calculated noise level	98·5 dBA	98·5 dBA	105·5 dBA	105·5 dBA
Actual noise level of untreated engine	97 dBA	97 dBA	109 dBA	107 dBA
Noise level of quietened engine	91 dBA	91·5 dBA	100·5 dBA	101·5 dBA

attenuating properties of various engines of different makes, but this does not mean that improvement on the norm is impossible. The noise levels for the two conventional engines which have been treated, calculated according to the formula in Paper 8, and the actual noise levels are compared in Table D2. (The truck engine was measured on a test bed lined with less acoustically absorptive material than most.)

The above noise levels were achieved with the standard fuel injection equipment fitted to the engines. For many engines, the noise level drops off sharply above 5 kHz. This allows the injection equipment to be distinguishable as a high frequency clicking, particularly at light loads and low speeds. However, the engine noise in the middle and lower frequency range is much louder, and this controls the overall noise levels. We have not met an engine yet where the injection equipment was the major noise source. Despite this encouraging state of affairs, the author is actively engaged in reducing the noise of fuel injection equipment against the day when diesel engines are quiet enough to warrant it.

To deal with some of the more particular queries, multiple belt pulleys should be easier to isolate than single row pulleys, as they have heavier rims. The simple design involving a pair of stampings will not be suitable, but leaf springs attached to the axial extremes of the rim might be designed to carry the required loads. A reduction in the axial vibration of the crankshaft, thus obviating the need for an isolated pulley, may be difficult in view of the small linear displacements involved. Restricting the crankshaft movement with the main bearing diaphragms may carry a penalty in reduced damping of those crankcase panel modes which caused vertical rotation of the point of attachment of the crankcase wall and bearing webs.

The natural frequency of a particular normal mode of a built-in panel is proportional to the panel thickness and inversely proportional to the square of its linear dimensions; hence smaller and/or thicker panels are introduced to increase the natural frequency of a particular normal mode. Sound is radiated efficiently by the sizes of panels encountered on conventional automotive and tractor diesel engines. Reducing the size of the panels, and increasing the natural frequency of their normal modes in the ratio above, tends to leave them as efficient radiators of sound, although, as Mr Vulliamy says, the exciting forces will be lower and therefore they will make less noise.

The one-third octave band filters used for analysis of both noise and cylinder pressure signals have a bandwidth of 26 per cent of their centre frequency, and en-encompass a variable number of Fourier harmonics of the repetitive signals depending on the centre frequency. (A 1·2 per cent narrow band analyser cannot separate engine harmonics above 1 to 2·5 kHz either.) A measurement of the peak value of the collection of harmonics in any one-third octave band has little significance, and accordingly the root mean square of all the Fourier components in each filter band has been measured. This provides a direct comparison between noise and cylinder pressure components, but it is a higher level than that indicated by a narrow-band analyser. The measurement of cylinder pressure spectra presents a particular problem, in that the true dynamic range between peak cylinder pressure and the peak value of an individual Fourier harmonic can exceed 90 dB.

This is greater than the filter skirt attenuation of most analysers, and it has become standard practice in the author's company to use two filters and an intermediate amplifier even with a narrow band analyser to avoid such errors, and those due to clipping and excessive instrument self-noise. The panel vibration spectra of Figs 12.2 to 12.5 were displayed as acceleration to provide a direct comparison with future measurements, for use by

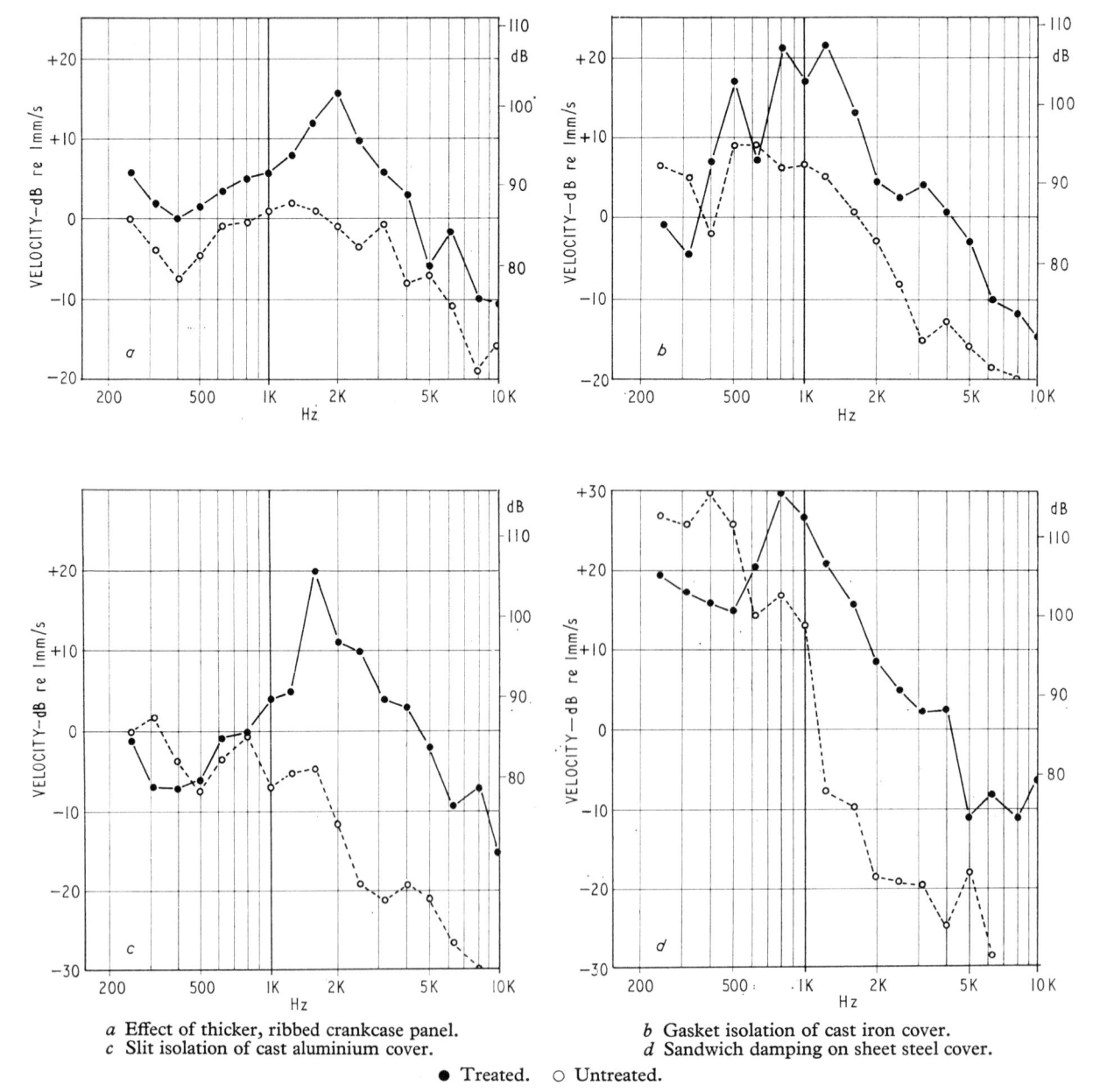

a Effect of thicker, ribbed crankcase panel.
b Gasket isolation of cast iron cover.
c Slit isolation of cast aluminium cover.
d Sandwich damping on sheet steel cover.

● Treated. ○ Untreated.

Fig. D11. Reduction in vibration of panels and covers achieved by various treatments, expressed in velocity units

engineers developing the quieter engines of the future. These measurements are just as they came off the engines. It is almost as easy to quote these same levels as velocities or as sound pressure levels equivalent to velocities as in Fig. D11. This is a useful presentation when comparing the noise from the air intake and pulley with the noise radiating potential of a vibrating panel or cover as in Fig. 12.1, which is a comparison of sound pressure levels at the important points in the generator, on the radiators and at the receiver of the noise from a running engine.

Mr C. Lunnon—Mr Hall asked why a concentration of oxides of nitrogen of 15 parts per million is regarded as just acceptable when this figure is three times the total limit (*C* value) for nitrogen dioxide (i.e. 5 p.p.m.). There are a number of points to bear in mind. First, the figure of 15 p.p.m. is a calculated figure for the total oxides of nitrogen under the worst operating conditions envisaged, and this figure includes other oxides of nitrogen besides nitrogen dioxide. There is no published total limit value (TLV) for mixed nitrogen oxides, and the TLV will vary according to the proportion of the various oxides of nitrogen present.

Both the Safety in Mines Research Establishment (S.M.R.E.) and the National Coal Board have done very extensive work on actual concentrations produced in underground locomotive roadways, including circumstances in which the diesel locomotive is moving at only very low speeds with respect to the ventilation. Recently S.M.R.E. have published a report (8) from which the following quotation relates to points raised by Mr Hall:

'3. EXPERIMENTAL RESULTS

Almost 70 per cent of the samples were taken at the engine-driver's position, a little over 10 per cent at loading points and nearly 20 per cent at other locations. ... Nearly all the samples contained less than 25 p.p.m. of "total nitrogen oxides" and less than 5 p.p.m. nitrogen dioxide (the ceiling limit referred to in Section 1.1). The proportion of nitrogen dioxide in the "total nitrogen oxides" varied from 0 to 100 per cent: however, only one value in 40 exceeded 40 per cent and the average value was 25 per cent.'

It is fair to conclude from this data that overall conditions in N.C.B. underground diesel roadways with respect to oxides of nitrogen are almost always well within the appropriate total limit values (including ceiling limits where appropriate).

We have not tried the nitrogen dioxide dosimeter and are most grateful to Mr Hall for drawing our attention to it. We hope to be able to try this or a similar type in the future.

We are regrettably unable to add to the information on aldehydes, save to remark that the nose is an extremely sensitive detector of aromatic aldehydes and men are likely to complain of the strong smell before the concentration reaches dangerous levels. The amounts of total aldehydes found by Commins, Waller and Lawther in two bus garages when the buses were moving about were never more than 0·4 p.p.m., well below the aldehyde value given by Mr Hall.

Mr Hall's comment about the gas turbine locomotive moving with the air-stream in a tunnel would be more significant if he could indicate how the output of oxides of nitrogen of a gas turbine compares with that of a diesel of equivalent power.

With regard to oil mist and vapour, it is interesting to note that the latest requirement of H.M. Inspectorate of Mines and Quarries for flameproof mining diesel engines specifically forbids the ducting of the emissions from the crankcase breather back into the air inlet of the engine. They specify a breather with a simple non-return valve or similar device.

The equation with which Mr Hall concludes is a well known one and is generally accepted if the hazard from the constituents is similar. In other words it would be appropriate if the various substances were producing the same sort of damage to the human lung. We doubt, however, if it would be appropriate if one of the constituents were damaging to the lung and the other affected the blood chemistry in a completely different way.

REFERENCE

(8) GODBERT, A. L. and LEACH, E. 'A preliminary survey of the pollution of mine air by nitrogen oxides from diesel exhaust gases', *S.M.R.E. Research Rept No. 265* 1970.

Mr J. P. S. Curran—As explained in the paper, it had always been our objective to design springs for an infinite fatigue life 'when operating between the two fixed stress levels corresponding to the needle on its seat and against its full lift stop'.

On this basis a spring which survived 10^7 cycles operation would survive the life of the engine providing no other adverse conditions were imposed, e.g. corrosion or increasing needle lift. In the majority of past applications the assumption of two fixed stress levels has been quite adequate and an infinite life is usually achieved.

The work of this paper, however, has shown that, for certain applications, a much wider stress range must be considered to cater for extra dynamic stress levels, if a limited fatigue life is to be avoided.

The particular R.N. application to which Mr Sampson has referred possibly approaches the operating conditions outlined in the paper where dynamic stresses are being superimposed on a spring originally designed on a fixed stress level basis. The 500 hours maintenance schedule Mr Sampson quotes would probably be equivalent to the spring being compressed more than 10^7 times. Hence the only solution—and one that allows safe prediction of (infinite) life—is to specify a spring with a slightly larger wire diameter so that stress levels are reduced.

Clearly a conservative view must be taken regarding service life to make allowance for the many unknown environmental and operating conditions that can occur.

List of Delegates

ADEY, A. J.	Simms Group Research & Development Ltd, London.
BANKS, J. S.	'Shipping World & Shipbuilders', London.
BASS, E. J.	Ford Motor Ltd, Basildon.
BELCHER, P. R.	Shell International Petroleum Ltd, London.
BENTLEY, G. H.	Metropolitan Water Board, London.
BERG, P. S.	AB Volvo, Gothenburg, Sweden.
BORYSOWSKI, K.	Mirrlees Blackstone Ltd, Stockport.
BOWEN, P. W.	Crossley Premier Engines Ltd, Manchester.
BOYLE, J.	Cummins Engine Ltd, Shotts.
BRESSER, K. J.	Rolls-Royce Ltd, Shrewsbury.
BROCKINGTON, P. A. C.	'Commercial Motor', London.
BROWN, C. D.	Cummins Engine Ltd, Darlington.
BRYAN, P.	Vauxhall Motors Ltd, Luton.
BULL, B.	Shell Research Ltd, Chester.
CALKIN, R. A.	Ford of Europe Inc., South Ockenden.
CASTLE, D.	Crossley-Premier Engines Ltd, Manchester.
CHAPMAN, T.	Cummins Engine Ltd, Darlington.
CLARKE, D. J.	Perkins Engines Ltd, Peterborough.
CLAYTON, J. C.	Birmid Qualcast (Foundries) Ltd, Birmingham.
CLEPHANE, J. J.	British Polar Engines Ltd, Glasgow.
CORKILL, D.	Perkins Engines Ltd, Peterborough.
CURRAN, J. P. S.	Bryce Berger Ltd, Gloucester.
DAVIDSON, W. R. S.	M.V.E.E., Christchurch.
DAVISON, C. H.	Vandervell Products Ltd, Maidenhead.
DAY, E.	Cummins Engine Ltd, Darlington.
DEACON, E. R.	Airesearch Turbocharger Division, Cheltenham.
DINGLE, W. R.	Ruston Paxman Diesels, Colchester.
DULLEY, T. A. C.	Development & Research Dept, Merz & McLellan, Esher.
EDWARDS, G.	Amalgamated Power Engineering Ltd, Manchester.
EDWARDS, J. V.	British Internal Combustion Engine Research Institute Ltd, Slough.
EKE, P. W. A.	North Thames Gas Board, Romford.
ELLIS, A. J.	Bryce Berger Ltd, Gloucester.
EVANS, E. N.	Mirrlees Blackstone Ltd, Dursley.
EVANS, R. W.	Ford Motor Ltd, Basildon.
FITZGERALD, W.	Esso Chemical Ltd, London.
FLEMING, C. R. J.	Ministry of Public Building and Works, London.
FOWLER, J. G.	Ford Tractor Division, Basildon.
FRASER-ANDREWS, C.	Rendel, Palmer & Tritton, London.
FRENCH, C. C. J.	Ricardo Engineers (1927) Ltd, Shoreham-by-Sea.
FRIEDRICH, H. G. F.	Gulf Research, Rotterdam, Holland.
FROY, R. K.	Shell-Mex & BP Ltd, London.
GALLOWAY, I. W.	Perkins Engines Ltd, Peterborough.
GARDINER, L. H.	Research and Technical Service Dept, Mobil Oil Ltd, Coryton.
GARNETT, P.	Hunslet (Holdings) Ltd, Leeds.
GARSIDE, J. V.	R. A. Lister Ltd, Dursley.
GLASSPOOLE, A. J.	English Electric Diesels Ltd, Newton-le-Willows.
GRIFFEY, M. F.	Ministry of Defence (Navy), Bath.
HACKER, M. F.	Mirrlees Blackstone Ltd, Dursley.
HÄGGH, B. J.	AB Volvo, Gothenburg, Sweden.
HALL, M. T.	British Rail Research, Swindon.
HARDENBERG, H. O.	Daimler-Benz AG, Stuttgart, W. Germany.
HARRIS, C.	Shell-Mex & BP Ltd, London.
HARRIS, T. D.	R. A. Lister Ltd, Dursley.
HEDGE, R. A.	British Rail, London.
HENEGHAN, S. P.	Coras Iompair Eireann, Dublin, Eire.
HENSHALL, S. H.	Mirrlees Blackstone Ltd, Stockport.
HODGE, D.	English Electric Diesels Ltd, Newton-le-Willows.
HODGSON, T. G.	British Leyland Motor Corp., Leyland.
HOLCOMBE, J. L.	Wellworthy Ltd, Lymington.
HOLLMAN, J. A.	Perkins Engines Ltd, Peterborough.
HOLMÉR, H. E. A.	AB Volvo, Gothenburg, Sweden.
HONNIBALL, J.	Ruston Paxman Diesels Ltd, Colchester.
HOPKINS, V. H. F.	Retired Consultant, Chichester.
HOWE, U.	Klöckner-Humboldt-Deutz AG, Cologne, W. Germany.
HOWELLS, H. E.	British Petroleum Ltd, Sunbury on Thames.
HOWES, P.	C.A.V. Ltd, London.
HUMPHREY, W. A.	Admiralty Engineering Laboratory, West Drayton.
JAMES, M. F.	R. A. Lister Ltd, Dursley.
JOHNSON, R. E. C.	North Thames Gas Board, Romford.
JONES, B. E.	Royal Military College of Science, Shrivenham.
JONES, R. A.	H. Leverton Ltd, Windsor.
KIMSTRA, K.	Stork Werkspoor Diesel, Amsterdam, Holland.
KRAUSHAR, C. A. A.	Perkins Engines Ltd, Peterborough.
LAKE, A. P.	Cummins Engine Ltd, Darlington.
LALOR, N.	I.S.V.R., University of Southampton.
LANGER.	Maschinenfabrik Augsburg-Nürnberg, Augsburg, W. Germany.
LANGTON, A. T.	Esso Research Centre, Abingdon.
LAW, D. A.	Wellworthy Ltd, Lymington.
LAWS, A. M.	Associated Engineering Developments Ltd, Rugby.
LAYFIELD, E. F.	British Rail, Derby.
LECKIE, G. G.	A.E. Developments Ltd, Rugby.
LEWIS, R. P.	Fodens Ltd, Sandbach.
LINDGREN, D. O. J.	Saab-Scania Automotive Group, Södertälje, Sweden.
LONGFOOT, G.	Wellworthy Ltd, Lymington.
LOWE, W.	Mirrlees Blackstone Ltd, Stockport.
LUETGE, H.	Cummins Diesel Deutschland, Essen, W. Germany.
LUNDSTRÖM, S. A. V.	AB Volvo, Gothenburg, Sweden.
LUNNON, C.	National Coal Board, Doncaster.
McATEER, P.	W. H. Allen Sons Ltd, Bedford.
MacKENZIE, A. K.	Ministry of Public Building and Works, Edinburgh.
MANNALL, J. R.	W. H. Allen Sons Ltd, Bedford.
MANNERS, D. S.	Esso Research Centre, Abingdon.
MANSFIELD, W. P.	B.I.C.E.R.I. Ltd, Slough.
MARSHALL, A.	Cummins Engine Ltd, Shotts.
MIDDLETON, I. C.	Ford Motor Ltd, Basildon.
MILLAR, D. H.	Ministry of Defence, M.V.E.E., Chertsey.
MOLYNEUX, P. H.	Lubrizol Great Britain Ltd, London.
MONK, R. A.	Perkins Engines Ltd, Peterborough.
MOORE, N. P. W.	Imperial College of Science and Technology, London.
MORGAN, F. C.	R. A. Lister, Ltd, Dursley.
MORRIS, J. J.	Perkins Engines Ltd, Peterborough.

LIST OF DELEGATES

MULHERN, G. J.	North of Scotland Hydro-Electricity Board, Edinburgh.
MUNRO, R.	Wellworthy Ltd, Lymington.
NASH, S. H.	Shell-Mex & BP Ltd, London.
NEVILLE, A. L.	Burmah Oil Trading Ltd, Bracknell.
NEWMAN, B. A.	Wellworthy Ltd, Lymington.
NORRIS, P. L.	British Railways Board, Swindon.
NOTTLEY, P. S.	M.V.E.E., Chertsey.
NUSTROM, C. G.	AB Nynas Petroleum, Nynashamn, Sweden.
O'NEILL, G.	Caterpillar Tractor Ltd, Glasgow.
PCHELAROV, V. V.	Bulgarian Embassy, London.
PEGLEY, A. C.	Ford of Europe Inc., Ockendon.
PETROOK, W.	British Railways Board, Derby.
PINKARD, G.	R. A. Lister & Co. Ltd, Dursley.
POWNEY, R. A.	Vauxhall Motors Ltd, Luton.
PRIEDE, T.	I.S.V.R., University of Southampton.
RATCLIFFE, F.	C.E.G.B., Hastings and Ashford Power Stations.
RAY, W. H.	A.O.L., Cobham.
READ, G.	Simms Motor Units Ltd, London.
RICHARDS, I.	M.V.E.E., Chertsey.
ROWE, L. F.	Ministry of Public Building and Works, London.
RUSSELL, M. F.	C.A.V. Ltd, London.
RUTISHAUSER, L. F.	Esso Research Centre, Abingdon.
SAMPSON, W. H.	Ministry of Defence (Navy), Bath.
SCOTT, G. W.	Caltex Ltd, London.
SCOTT, T. E.	British Railways Board, London.
SENIOR, N. R.	Perkins Engines Ltd, Peterborough.
SHAW, A. K.	Atomic Energy Laboratory, U.K.A.E.A., West Drayton.
SHORT, D. P.	Weslake Ltd, Rye.
SMALLEY, D.	British Leyland Motor Corp., Leyland.
SMITH, R.	General Motors Ltd, Wellingborough.
SMITHIES, R. R.	British Leyland Motor Corp., Leyland.
SOAR, A. A.	School of Mechanical Engineering, Bordon.
SWINSON, P. J.	Orobis Ltd, London.
TAYLER, A.	Sulzer Bros. Ltd, London.
TAYLOR, K. M.	Burmah Castrol, Bracknell.
THOLEN, P.	Klöckner-Humboldt-Deutz AG, Cologne, W. Germany.
THOMAS, M. O.	Texaco Ltd, London.
TIMONEY, S. G.	University College, Dublin, Eire.
TINDAL, M. J.	King's College, London.
TOVEY, T. J.	Kennedy & Donkin, Woking.
TRACY, H. G. H.	British Polar Engines Ltd, Glasgow.
TRYHORN, D. W.	Sir W. G. Armstrong Whitworth (Engineers) Ltd, Slough.
TUFFILL, H. C.	Sir Alexander Gibb & Partners, London.
TURNER, R.	British Petroleum Ltd, London.
TYRRELL, E.	Ministry of Technology, London.
VAUGHAN, R. A.	Kennedy & Donkin, Woking.
VANWAY, J. W.	Cummins Engine Ltd, Darlington.
VULLIAMY, N. M. F.	Perkins Engines Ltd, Peterborough.
WALDER, C. J.	Ricardo Ltd, Shoreham-by-Sea.
WALKER, J. H.	North Thames Gas Board, London.
WALKER, S. T.	British Petroleum Ltd, London.
WATERS, A. W.	Ministry of Defence, M.V.E.E., Chertsey.
WATERS, P. E.	I.S.V.R., University of Southampton.
WATKINS, L. H.	Road Research Laboratory, Crowthorne.
WATKINSON, P. R.	Cummins Engine Ltd, Darlington.
WESTON, J. W. R.	R.N.E.C., R.N., Manadon, Plymouth.
WILLIAMS, D.	Simms Motor Units Ltd, London.
WILLIAMS, M. A.	North Thames Gas Board, London.
WILSON, J. V. D.	Burmah Research & Development, Bracknell.
WING, R. D.	Imperial College of Science & Technology, London.
WOOD, D. J.	Cummins Diesel Sales & Service Ltd, London.

Index to Authors and Participants

Names of authors and numbers of pages on which papers begin are in bold type.

Bresser, K. J. 159

Castle, D. 44, 176
Clarke, D. J. 10, 175
Curran, J. P. S. 148, 183

Dingle, W. R. 159
Dulley, T. A. C. 160

Eke, P. W. A. 1

Fowler, J. G. 160
French, C. C. J. 161

Garside, J. V. 162
Glasspoole, A. J. 162

Häggh, B. 55, 176
Hall, M. T. 164
Henshall, S. H. 165
Holmér, E. 55, 176
Howells, H. E. 81, 179
Howes, P. 73, 179

Lalor, N. 63, 177
Law, D. A. 119
Lowe, W. 18, 175
Lunnon, C. 106, 183

Mackenzie, A. K. 90, 180
Mansfield, W. P. 165
Millar, D. H. 139

Petrook, W. 25, 175
Priede, T. 63, 177

Ratcliffe, F. 112
Russell, M. F. 98, 181

Sampson, W. H. 32, 166
Scott, T. E. 166
Senior N. R. 10, 175
Stewart, W. A. 25, 175

Tryhorn, D. W. 168

Vaughan, R. A. 169
Vulliamy, M. 170

Walder, C. J. 172
Walker, J. H. 1
Walker, S. T. 81, 179
Waters, P. E. 63, 177
Williams, M. A. 1
Wood, B. 160

Subject Index

Titles of papers are printed in capital letters.

Acceleration, fuel input, and turbo-blower speed, locomotives, 28
Acoustic cladding, fighting ships, 32, 159, 161, 162
Additives, fuel, effect on exhaust emissions, 86, 106, 161, 164, 179
Admiralty Standard Range of engines, and service problems, 32, 41, 161, 162
Air cleaners, 57, 115, 144, 159, 169, 177
Aircraft, use of liquid natural gas as a fuel, 8
Air pollution, *see* Exhaust emissions
Aldehydes, production, 84, 106, 164, 183
ALIGNMENT INVESTIGATION FOLLOWING A MEDIUM-SPEED MARINE ENGINE CRANKSHAFT FAILURE, 44
Alignment, marine engines; 36
    investigation following crankshaft failure, 44, 162, 165, 166, 176
Allen dual-fuel engines, performance, 6
Alternators, power station and desalination plant, 90, 160, 161, 170, 180
Anti-knock properties of natural gas, 2
APPLICATION OF DIESEL ENGINES IN THE ROYAL NAVY, 32
ASPECTS OF DIESEL EXHAUST EMISSIONS, ESPECIALLY IN CONFINED SPACES, 106
Automatic watchkeepers, naval engines, 40

Base engine design, 10, 159, 175
Bearings; failure, 44
    lubrication problems, 32, 41
Blow-by, 132, 160, 163
Bore and speed, effect on noise, 63, 171, 173, 177
British Rail locomotives, 25, 159, 161, 162, 166, 175
British Standards; derating formulae, 18, 19, 166, 175
    noise measurement, 98

Camshaft and valve train, industrial engine, 11
Cam-turning machine, piston ring production, 126
Canvey Island Methane Terminal, engines installed, 6
Carbon deposit control, 57, 92
Carbon monoxide formation; effect of fuel quality, 82, 161, 164, 179
    effect on the human body, 106, 164, 183
Cladding, acoustic, naval vessels, 32, 159, 161, 162
Collapse of piston ring materials, 123, 160, 163
Combustion system design, 70, 106, 171, 173, 177
Commissioning trials, power station, Ascension Island, 95
Communication between customer and supplier, 10, 44, 90, 119, 139
Compatibility properties, piston ring materials, 122, 160, 163
Component failures, locomotives, 30
Compression ratio, turbocharged road transport power unit, 58, 159, 177
Constant horsepower engine, 59, 171, 177
Constant Position Mounting System, naval vessels, 34
Control, diesel generators, 94, 160, 161, 170, 180
Cooling, charge-air, and air-to-air, turbo-charged engines, 18, 166, 175
Cooling fan noise, 66, 171, 178
Cooling, piston rings, 119, 160, 163
Cooling systems; locomotives, 25, 159, 161, 162, 166, 175
    military vehicles, 141, 161, 162, 163, 168
    naval applications, 32, 40, 44, 159, 161, 162
Cost; Ascension Island desalination plant, 170, 180
    choice of engine for road transport, 55, 159, 170, 172, 177
    electricity generating plant, running costs, 112, 113, 114, 115
    natural gas, dual-fuel engines, 1

Covers and panels, vibration and noise reduction, 100, 171, 173, 181
Crankcase panels, vibration reduction, 98, 171, 173, 181
Crankshaft; development, 32, 40
    failure, marine engine, 44, 162, 165, 166, 176
    industrial engine design, 11
    service problems, ASR1 engines, 41
CRITICAL FACTORS IN THE APPLICATION OF DIESEL ENGINES TO FIGHTING VEHICLES, 139
CRITICAL FACTORS IN THE APPLICATION OF DIESEL ENGINES TO RAIL TRACTION, 25
CRITICAL FACTORS IN THE APPLICATION OF DUAL-FUEL ENGINES, 1
Cyclic regularity, industrial engine, 10
Cylinder configuration and number, effect on vehicle engine noise, 69, 171, 173, 177
Cylinder head gaskets, road vehicles, service problems, 58, 171, 173, 177
Cylinder liners; naval vessels, vibration and wear, 39, 41
    surface-finish control, 130, 160, 163

Desalination plant and power station, design and commissioning, 90, 160, 161, 170, 180
Design; crankshafts, 44, 162, 165, 166, 176
    fuel-injection equipment, 73, 165, 174, 179
    piston rings, 119, 160, 163
Design of engines for; fuel limitations, 81, 161, 163, 164, 179
    generation of electricity, 90, 112, 160, 161, 169, 170, 180
    industry, 10, 159, 175
    military vehicles, 139, 161, 162, 163, 168
    natural gas, 1, 161
    naval application, 32, 159, 161, 162
    rail traction, 25, 159, 161, 162, 166, 175
    road transport, 55, 98, 159, 161, 170, 171, 172, 173, 176, 181
    site ambient conditions, 18, 166, 175
    underground operation, 106, 164, 183
DEVELOPING A DIESEL FOR AN INDUSTRIAL MARKET SECTOR, 10
DIESEL ENGINE PISTON RING DESIGN FACTORS AND APPLICATION, 119
DIESEL ENGINES AS A SOURCE OF COMMERCIAL VEHICLE NOISE, 63
DIESEL GENERATION FOR ASCENSION ISLAND, 90
DIESELS AND THE GENERATION OF ELECTRICITY, 112
Dual-fuel engine development, 1, 161

EFFECT OF AMBIENT AND ENVIRONMENTAL ATMOSPHERIC CONDITIONS, 18
Electricity generation; and associated desalination plant, 90, 160, 161, 170, 180
    obtaining economy and reliability, 112, 169
Enclosures, engine, noise reduction, 32, 159, 161, 162
Engine test tilt rig, 13
Engines; Admiralty Standard Range, 32, 161, 162
    Allen dual-fuel, 6
    base design, 10, 159, 175
    constant horsepower, 59, 171, 177
    Deltic, Foden, and Ventura, acoustic cladding, 37
    English Electric dual-fuel, 6
    installed at Canvey Island Methane Terminal, 6
    Mirrlees National dual-fuel, 6
    Pielstick PC 2, crankshaft failure, 44, 162, 165, 166, 176
    Volvo turbo-charged, 55, 159, 161, 170, 172, 176
English Electric dual-fuel engines, operating experience, 6
Environment; and de-rating of engines, 18, 166, 175
    contamination of air flow, power station, 114
    dust problem, electricity generating plant, 90, 160, 161, 170, 180

SUBJECT INDEX

effect on military vehicle operation, 139, 145, 161, 162, 163, 168
effect on the alignment of ships' machinery, 47, 162, 165, 166, 176
enclosed, treatment of exhaust and fire prevention, 106, 164, 183
temperature variation and its effect, 25, 73, 81, 87, 161, 164, 165, 174, 179
European requirements, power output of commercial vehicles, 172, 177
Exhaust brake, turbo-charger, 57
Exhaust conditioner and flame-trap, underground diesel, 108
Exhaust emissions; control, road vehicles, 56, 66, 159, 161, 171, 173, 177
effect of fuel quality, 81, 161, 164, 179
fuel-injection system regulation, 74, 174, 179
military vehicles, 144
physiological effects, 106
treatment for underground diesels, 107, 109, 164, 183
Exhaust silencing, 28
Explosion prevention underground, diesel exhaust treatment, 107, 164, 183

Failure; cylinder head gaskets, vehicle engine, 58, 171, 173, 177
locomotive components, 30
marine engine crankshaft, 44, 162, 165, 166, 176
Fatigue tests, piston rings, 120, 121, 160, 163
Fighting vehicles, installational and environmental conditions, 139, 161, 162, 163, 168
Filter; air, diesels for generating electricity, 115
lubricating oil, industrial engine, 13
Filtration, turbo-charger lubricant, 28, 159, 175
Finishing part components, industrial engine, 12, 159, 175
Fire prevention underground, diesel exhaust treatment, 107, 164, 183
Flame-proofing, industrial engine, 10
Flame-trap and exhaust conditioner for underground diesel, 108
Flywheel and housing, industrial engine, 12
Fork lift truck, design criteria, 14
Fouling, locomotive turbo-charger, 25
Foundations, 117, 170
Fuel; additives, effect on exhaust emissions, 86, 161, 164, 179
consumption, road transport vehicles, 56, 159, 171
economy, electricity generation, 113, 169
for military vehicles, 147, 161, 162, 163, 168
gaseous, dual-fuel engines, 1, 161
quality, effect of, 81, 161, 164, 179
residual, 88, 114, 163, 169, 179
underground engines, 106, 164, 183
Fuel/air ratio, 114, 169
Fuel injection equipment; design, 11, 73
naval engines, 42, 159, 162, 163
noise, 172, 181
rail traction, spring failure, 148, 165, 183
timing, 106
FUEL-INJECTION SYSTEM REQUIREMENTS FOR DIFFERENT ENGINE APPLICATIONS, 73
FUEL LIMITATIONS ON DIESEL ENGINE DEVELOPMENT AND APPLICATION, 81
FURTHER CONSIDERATIONS IN INJECTOR DESIGN FOR HIGH SPECIFIC OUTPUT DIESEL ENGINES, 148

Gas, *see* Natural gas
Gas Council natural gas transmission system, 3
Gasket failures, 55
Gearbox noise, 67, 171, 178
Gear train, fork lift truck, 14
Generator module, 37
Governing, 76, 165, 174, 179

Handling liquefied natural gas, 7
Heat balance, locomotive engine, 26, 159, 161, 175
Heat collapse, piston rings, 123, 160, 163
Heat forming, induction, piston ring production, 126
Heat recovery, power station, for desalination plant operation, 90, 160, 161, 170, 180
History, development of dual-fuel engines, and the use of liquefied natural gas, 1, 7
Hydrocarbon emission, fuel factors affecting, 84, 161, 164, 179

Idling, effect of, turbo-charged locomotives, 28, 159, 175
Ignition improvers, 87, 161, 164, 179

IMPROVEMENTS TO CONVENTIONAL DIESEL ENGINES TO REDUCE NOISE 98
Industrial applications; designing for, 10, 159, 175
effect of fuel quality and use of residual fuels, 81, 88, 161, 163, 164, 179
Installation; industrial diesels, 10, 159, 175
military vehicles, problem of housing air, exhaust, and cooling systems, 139, 141, 161, 162, 163, 168
naval vessels, shock and vibration mountings and acoustic cladding, 32, 159, 161, 162
rail traction, accommodation of cooling systems and exhaust silencers, 25, 159, 161, 162, 166, 175
Instrumentation, engine, 170
Inversion, industrial engine, 14

Jigs, need for accuracy, marine engine crankshaft, 49

Legal requirements; exhaust emissions, 55, 63, 73, 81, 106
noise, 55, 63, 73, 98
U.K., vehicle size, weight, and speed, 63
Liquefaction of natural gas, and its handling and storage, 7
Liquid natural gas, *see* Natural gas, liquid
Load cell method, piston ring pressure-pattern measurement, 124
Loading; fluctuations, electricity generation, 90, 112, 160, 161, 170, 180
rail traction, 29, 161, 175
variations, fighting vehicles, 140, 161, 162, 163, 168
Lubricant temperature, effect on crankshaft alignment, 44
Lubrication; diesels for generating electricity, 115, 170
fighting ships, oil checks during service, 32, 38
industrial engines, 13
locomotive turbo-chargers, 28, 159, 175
military vehicles, 147, 161, 162, 163, 168
oil consumption, 132, 160, 163
piston rings, 119
to eliminate scuffing, 55, 122, 127, 132, 160, 163

Machinery, auxiliary, power stations, 112, 169
Maintenance; locomotives, 30, 162, 175
military vehicles, 146
naval engines, 40, 162
power station diesels, 90, 116, 169
reduction, dual-fuel engines, 1
underground diesels, 106
Volvo turbo-charged engine, 57
Marine engines; crankshaft failure, 44, 162, 165, 166, 176
effect of fuel quality, and use of residual fuels, 81, 161, 164, 179
fuel-injection system design, 73, 165, 179
naval applications, 32, 161, 162
possible use of liquid natural gas as a fuel, 8
Materials, piston ring, properties, 119, 160, 163
Mechanical efficiency, and de-rating and correction for ambient conditions, 19, 166, 175
Mining, treatment of exhaust and fire prevention measures, 106, 164, 183
Mirrlees National dual-fuel engines, operating experience, 6
Monitoring; chemical, lubrication system, 115
naval engines, 40
Mounting, marine engines, 33, 162, 165, 166, 176
M.V. *Manchester Port* crankshaft failure, 44, 162, 165, 166, 176

Natural gas; and the development of dual-fuel engines, 1
liquid, storage and handling, 7
Noise, prediction, automotive engine, 68, 71, 171, 173, 177
Noise, reduction; by modification of engine structure, 98, 171, 173, 181
fighting ships, 32, 159, 161, 162
fuel-injection system, 74
locomotives, 28, 31
military vehicles, 139
road transport, 56, 63, 159, 171, 173, 177

Oil mist and vapour formation, 165, 183
Outage causes, diesel generators, 117
Oxides of nitrogen; effect of fuel quality on production, 81, 85, 161, 164, 179
physiological effect, 106, 164, 183
Oxides of sulphur emission, effect of fuel quality, 86, 161, 164, 179

SUBJECT INDEX

Panels and covers, vibration and noise reduction, 100, 171, 173, 181
Particles, diesel exhaust, physiological effects, 107
Pay-off potential, road vehicles, 56, 159, 170, 172, 177
Physiological effects of exhaust pollutants, 106, 164, 183
Piston and ring scuffing, 58, 159
Piston rings; design, production, and material properties, 119, 160, 163
    early 'bed-in', industrial engines, 11
Pistons; comparison, oil-cushioned and standard, 38, 161
    ultrasonic testing, 117, 169
Power output; European requirements for commercial vehicles, 172, 177
    fighting vehicles, 139, 161, 162, 163, 168
    fuel-injection system design, 74, 165, 174, 179
Power pack, military vehicles, 139
Power station; Ascension Island, diesel generators and associated desalination plant, 90, 160, 161, 170, 180
    designing engines for reliability and economical running, 113, 169
Power take-offs, fork lift trucks, 14
Pre-setting fuel pumps, 77, 174, 179
Pressure-pattern measuring rigs, piston rings, 124, 125
Problems in service, ASR1 engines, 41, 161, 162
Protective equipment/Automatic Watchkeepers, naval engines, 40

Quality control, piston ring manufacture, 132, 160, 163

Radiator size, turbocharged diesel, 55
Radio relay station, diesel engines for electricity supply, 90, 160, 161, 170, 180
Rail traction; 25, 159, 161, 162, 166, 168, 175, 176
    spring failures, fuel-injection equipment, 148, 165, 183
    underground operation, 164, 183
Rating and fuel delivery, 73, 165, 179
Rating for; military vehicles, 139, 161, 162, 163, 168
    naval applications, 32
    rail traction, 29, 31, 168, 176
    site ambient conditions, derating formulae, 18, 166, 175
    underground operation, 108
Reliability; diesels for generation of electricity, 90, 115, 116, 170
    military vehicles, 139
    naval engines, 41
    rail traction, 30, 31, 162, 175
    Volvo road transport engine, 57, 159, 171
Re-odorants in fuel, 87
Residual fuels for industrial and marine applications, 88, 114, 163, 169, 179
Road Traffic Act regulations, commercial vehicles, 63
Road transport; effect of fuel quality, 81, 161, 164, 179
    fuel-injection system requirements, 73, 165, 174, 179
    noise reduction, 63, 98, 171, 173, 177, 181
    piston rings, 119, 138, 160, 163
    Volvo turbo-charged diesels, 55, 159, 161, 170, 172, 177
Rocker cover, vibration reduction, 98, 171, 173, 181

Safety of personnel, treatment of diesel exhaust emissions in confined spaces, 106, 164, 183
Sankey diagram, generation/waste heat system, Ascension Island power station, 96
Scuffing, 55, 122, 127, 132, 160, 163
Sealed-for-life cooling system, locomotives, 28, 159, 161, 175
Service duty; fighting vehicles, 139, 161, 162, 163, 168
    fuel-injection equipment, 73, 165, 174, 179
    generation of electricity, 90, 112, 160, 161, 170, 180
    naval applications, 32, 159, 161, 162
    rail traction, 25, 159, 161, 162, 166, 175
    underground engines, 106, 164, 183
Sewage plants, dual-fuel engines, 1
Shock/vibration mountings, naval vessels; 33
    Constant Position Mounting System, 34

Smoke; effect of fuel quality, 81, 161, 164, 179
    fuel injection system adjustment, 74, 174, 179
    limit, non-turbo-charged engines, 20, 166, 175
Spark-ignition and dual fuel engines, comparison, 4
Speed and bore, effect on noise, 63, 171, 173, 177
Starting; diesel generators, 94, 112, 170, 180
    fuel-injection system requirements, 77
    low-temperature, 10, 58, 159, 173, 177
    turbo-charged diesel, 55
Storage, liquefied natural gas, 7
Strain gauge method, piston ring pressure-pattern measurement, 125
Stress, piston rings, 120, 160, 163
Sulphur dioxide, physiological effect, 106, 164, 183
SUMMING-UP, 155
Sump; industrial engine, 13
    vibration reduction, 98, 171, 173, 181
Surface coatings and surface finish control, piston rings, 127, 160, 163

Tappet covers, vibration reduction, 98, 171, 173, 181
Telescopic sightings, marine engine crankshaft failure investigation, 47, 49, 162, 165, 166, 176
Temperature, piston ring, assessment, 128, 129, 160, 163
Tests; alignment, marine engines, 44, 162, 165, 166, 176
    engine test tilt rig, 13
    fatigue strength of chrome platings, 40, 161
    performance, Admiralty Standard Range engines, 32
    production, fuel-injection systems, 78, 174, 179
    quality of lubricating oil in service, naval engines, 38
    rating of military vehicle engines, 140, 161, 163, 168
    standard method, maximum vehicle noise, 65, 171, 173, 177
    ultrasonic, pistons of power station diesels, 117, 169
Thermal cycling, rail traction, 25, 29, 161, 175
Thermal deformation, marine engine crankshaft failure, 44, 162, 165, 166, 176
Thermocouple installation, cylinder liner temperature measurement, 129
Timing and fuel-injection rate, 73, 106, 174, 179
Timing cover, vibration reduction, 98, 171, 173, 181
Tractors; fuel-injection system requirements, 73, 165, 174, 179
    noise reduction, 98, 171, 173, 181
Training, use of apprentices in power station annual overhaul, 116
TURBO-CHARGED DIESEL AS A ROAD TRANSPORT POWER UNIT, 55
Turbo-chargers; choice of for site ambient conditions, 20, 166, 175
    failure, power station, 117, 170
    fouling, locomotives, 28, 31, 159, 161, 166, 175, 176
    road vehicles, characteristics, and matching with engine, 55, 57, 59, 159, 161, 171, 172, 176

Underground diesels, exhaust control and treatment, 106, 164, 183

Valve wear; effect of residual fuels, 89, 163, 179
    turbocharged diesel, 58, 173, 177
Ventilation of tunnels used by diesels, 109, 164, 183
Vibration; power station diesels, 117, 169
    reduction, engine panels and covers, 100, 171, 173, 181
    tester, naval vessels, 42, 161
Vibration and shock mountings, naval vessels, 33
Viscous TV dampers, 42, 161
Volvo turbo-charged road vehicle engine development, 55, 159, 161, 170, 172, 176

Water jacket panels, vibration reduction, 98, 171, 173, 181
Water leaks, locomotives, 26
Water washing, locomotive turbochargers, 28, 159, 175
Wear; cylinder liners, 41
    properties, piston ring materials, 122, 160, 163
    valve, 58, 89, 163, 173, 177, 179